T0136961

Progress in IS

More information about this series at http://www.springer.com/series/10440

Volker Wohlgemuth · Frank Fuchs-Kittowski
Jochen Wittmann
Editors

Advances and New Trends in Environmental Informatics

Stability, Continuity, Innovation

Editors
Volker Wohlgemuth
Environmental Informatics
Hochschule für Technik und Wirtschaft Berlin (HTW Berlin), University of Applied Sciences
Berlin
Germany

Jochen Wittmann
Environmental Informatics
Hochschule für Technik und Wirtschaft Berlin (HTW Berlin), University of Applied Sciences
Berlin
Germany

Frank Fuchs-Kittowski
Environmental Informatics
Hochschule für Technik und Wirtschaft Berlin (HTW Berlin), University of Applied Sciences
Berlin
Germany

ISSN 2196-8705 ISSN 2196-8713 (electronic)
Progress in IS
ISBN 978-3-319-83117-6 ISBN 978-3-319-44711-7 (eBook)
DOI 10.1007/978-3-319-44711-7

Softcover reprint of the hardcover 1st edition 2016

Printed on acid-free paper

This Springer imprint is published by Springer Nature
The registered company is Springer International Publishing AG Switzerland
The registered company address is: Gewerbestrasse 11, 6330 Cham, Switzerland

Preface

This book covers the main research results of the 30th edition of the long-standing and established international and interdisciplinary conference series on leading environmental information and communication technologies (EnviroInfo 2016). The conference was held on 14–16 September 2016 at the Hochschule für Technik und Wirtschaft Berlin (HTW Berlin), University of Applied Sciences. The EnviroInfo conference series under the patronage of the Technical Committee on Environmental Informatics of the German Informatics Society looks back on a history of 30 conferences. Thus, one important thread in Berlin was a retrospective on the experiences made and the lessons learned in the field of environmental informatics during the last years. Basic topics that were covered under the central focus *Environmental Informatics—Current trends and future perspectives based on 30 years of history* were applications of geographical information systems, environmental modelling and simulation, risk management, material and energy flow management, climate change, tools and database applications and other aspects with regard to the main topic ICT and the environment.

Due to the interdisciplinary character of environmental informatics, one important goal of this conference was to bring experts from industry, research and education together to exchange ideas and proposals for solution of urgent problems and needs in the field on environmental protection and its IT support.

The editors would like to thank all contributors to the conference and these conference proceedings. Special thanks also go to the members of the programme and organizing committee. Especially, we want to thank all "helping hands" and students of the HTW Berlin. Last, but not least, a warm thank you to our sponsors and to the HTW Berlin for being the host of this year's conference.

Berlin, Germany
June 2016

Volker Wohlgemuth
Frank Fuchs-Kittowski
Jochen Wittmann

EnviroInfo 2016 Organizers

Chairs

Prof. Dr. Volker Wohlgemuth, Hochschule für Technik und Wirtschaft Berlin, Germany
Prof. Dr.-Ing. Frank Fuchs-Kittowski, Hochschule für Technik und Wirtschaft Berlin, Germany
Prof. Dr.-Ing. Jochen Wittmann, Hochschule für Technik und Wirtschaft Berlin, Germany

Programme Committee

Arndt, Hans-Knud
Bartoszczuk, Pawel
Brüggemann, Rainer
Düpmeier, Clemens
Fischer-Stabel, Peter
Fuchs-Kittowski, Frank
Funk, Burkhardt
Geiger, Werner
Göbel, Johannes
Greve, Klaus
Hilty, Lorenz M.
Hitzelberger, Patrik
Hönig, Timo
Jensen, Stefan
Karatzas, Kostas
Kern, Eva
Kleinhans, David

Knetsch, Gerlinde
Knol, Onno
Kremers, Horst
Lang, Corinna
Lorenz, Jörg
Mattern, Kati
MacDonell, Margaret
Marx Gómez, Jorge
Möller, Andreas
Müller, Berit
Müller, Ulf Philipp
Naumann, Stefan
Niemeyer, Peter
Niska, Harri
Ortleb, Heidrun
Page, Bernd
Pattinson, Colin
Pillmann, Werner
Rapp, Barbara
Riekert, Wolf-Fritz
Schade, Sven
Schreiber, Martin
Schweitzer, Christian
Simon, Karl-Heinz
Sonnenschein, Michael
Susini, Alberto
Thimm, Heiko
Vogel, Ute
Voigt, Kristina
Wagner vom Berg, Benjamin
Widok, Andi
Winter, Andreas
Wittmann, Jochen
Wohlgemuth, Volker

Contents

Part I
Design, Sustainability and ICT

Analysis of Product Lifecycle Data to Determine the Environmental Impact of the Apple iPhone

Hans-Knud Arndt and Chris Ewe

Abstract The increasing awareness of environmental protection, e.g. in the course of climate change, also affects products of the information and communication technology industry along their lifecycle. Companies have to consider how their processes and products can be designed correspondingly the expectations of their stakeholders as well as the public. Therefore they have to measure and analyze environmentally concerning data to improve their service provision. Using the Apple iPhone as an example this paper will execute such an analysis to evaluate its environmental impact. The data investigation is performed by reconditioning, analyzing and interpreting the data as well as giving potential causes and correlations with other datasets. Basing on the performed analysis a generalized model for the assessment of the environmental impact of ICT products can be enabled.

Keywords Environmental protection • Sustainability • Product lifecycle • Design • Data analysis • Data visualization • Reporting

1 Sustainability, Environment and Design in the ICT Industry

The Nature Conservancy states on their website that it's crucial for humanity to raise the awareness for environmental protection as one of the central parts of sustainability (The Nature Conservancy 2016). Major impacts of an inadequate further procedure would be the greenhouse gas caused consequences of the climate change. They state: "With rapid climate change, one-fourth of Earth's species could be headed for extinction by 2050" (The Nature Conservancy 2016). Direct impacts,

H.-K. Arndt (✉) · C. Ewe
Otto-Von-Guericke Universität Magdeburg, Magdeburg, Germany
e-mail: hans-knud.arndt@iti.cs.uni-magdeburg.de

C. Ewe
e-mail: chris.ewe@st.ovgu.de

V. Wohlgemuth et al. (eds.), *Advances and New Trends in Environmental Informatics*, Progress in IS, DOI 10.1007/978-3-319-44711-7_1

which threaten animal's as well as human's life's are: higher temperature, rising seas, stronger storms, increased risk of drought, fire and floods or the increase of head related diseases (The Nature Conservancy 2016).

Especially companies have a strong impact on effects like air pollution or the consumption of resources. Therefore, they have to design their products and processes to leave a preferably small footprint. This also applies to products of the information and communication technology (ICT) throughout their entire lifecycle. From procurement to production, usage and disposal, ICT products and their processes should be designed to have a minimal environmental impact in order to meet climate goals (Arndt 2013).

Smartphones, as a continuously growing part of the ICT, also have to meet these requirements (Statista 2015). Along their lifecycles they have to consider, which materials in which amounts should be used, which services should be offered or how the hardware can be disposed at the end of use. Manufacturers have to think about how the user interacts with the product to e.g. save material while using less hardware. Accurately analyzing this sustainability data will help companies to decide better how to influence their footprints. One company who already gathers data along the lifecycle of their smartphone is Apple. Collecting, preparing, analyzing and interpreting this data as well as examining possible causes is subject of the following chapters.

2 Apple, iPhone and Report Data

2.1 Apple Environmental Politics and the iPhone

According to Apple, a preferably low environmental product impact is one of the key factors in their lifecycle design (Apple Inc. 2015a). They state: "We don't want to debate climate change. We want to stop it" (Apple Inc. 2015a). To comprehend this, the company provides data in form of reports, which demonstrate the product impact on the environment (Apple Inc. 2015b). Apple was chosen as an example, because they provide lifecycle data for their products, which can be used for such an analysis. The iPhone as the product example was chosen because of its relevance for the company's success, its broad range of models with many data to analysis as well as its popularity as a smartphone (Börsen-Zeitung 2015). For the analysis all iPhone models with an environmental report have been chosen, which are: 3G (2008), 3GS (2009), 4 (2010), 4s (2011), 5 (2012), 5c (2013), 5s (2013), 6 (2014), 6+ (2014), 6S (2015), 6S+ (2015), SE (2016) (Apple Inc. 2015b; Spiegel 2013). The structure of the iPhone reports and the definition of Apple's understanding of the dataset are subject of the following section.

2.2 *Environmental Reports—Product Lifecycle Data*

The core of the reports is formed by criteria that enable the quantification of the environmental awareness. For our analysis we focus on the greenhouse gas emissions (GGE), which are one key factor to ensure environmental protection. "A greenhouse gas is any gaseous compound in the atmosphere that is capable of absorbing infrared radiation, thereby trapping and holding heat in the atmosphere. By increasing the heat in the atmosphere, greenhouse gases are responsible for the greenhouse effect, which ultimately leads to global warming." (Livescience.com 2015) All other given values are considered as influential factors, which could lead to possible explanations for the GGE development.

Apple's reports show the total amount of GGE as well as the share of each phase of the lifecycle in kg of carbon dioxide equivalents (CO_2e) (Apple Inc. 2015b). The company's understanding of the product lifecycle slightly differs between different models. In the following we show exemplary the definitions for the model 3G (Apple Inc. 2015b):

- **Production**: Extraction, production and transportation of raw materials and the manufacturing, transport and assembly of all parts and product packaging.
- **Transport**: Air and sea transportation of the finished product and its packaging from the manufacturer of the product to the distribution center. The transport of the product from the distribution center to the end customer is not included.
- **Customer use**: For the power consumption of the user, a three-year period is assumed. Use scenarios were modeled based on data that reflects intensive daily use of the product. Geographic differences in the power grid need to be considered on a continental scale.
- **Recycling**: Transport from the collection to the recycling centers as well as the energy that is used for mechanical sorting and for shredding of the components.

3 Environmental Impact Analysis of the Apple iPhone

3.1 *Course of Analysis*

1. **Collection and Cleaning of the Data**: The data have to be prepared to enable an access of analysis tools. Therefore we collect all data in Microsoft Excel.
2. **Analyzing the Data**: We use Microsoft Power BI to upload the Microsoft Excel sheets and analyze them by visualizing the datasets in different comprehensive ways (Microsoft Inc. 2016).
3. **Interpretation of Findings and possible Explanations**: We show the facts that have been provided by the visualizations and also give answers to explain the results. Possible explanations are given by comparing the emission data with other given data like the product weight.

3.2 Greenhouse Gas Emissions

The reports illustrate the total GGE as well as the single lifecycle stage GGE. While the total is given in kg CO_2e, the shares of the single phases are displayed as percentage from the total. To analyze the given data we implemented a Microsoft Excel sheet, which contains the total GGE as well as the converted values for each stage. The resulting is shown in Table 1.

In the next step the spreadsheet had been uploaded to the analysis tool. Therewith, it was possible to create a variety of visualizations, which show the development in the course of time. A visualization in form of a line chart can be seen in Fig. 1.

The ordinate shows the single models ordered by their release data from left to right. The abscissa shows the GGE in kg CO_2e. In the following the graphs will be interpreted:

(1) **Total Emissions**

Starting with 55 kg CO_2e of model 3G the emissions vary only by ±10 kg CO_2e until model 4S. Model 5 shows the first stronger increase by 20 %. From model 5s to 5c a decrease is shown, while the 5c has even less emissions than the first iPhone. With the release of iPhone 6 and 6+, these values strongly increased to 95 kg CO_2e and 110 kg CO_2e. It can be seen that iPhone 6+ causes twice as much emissions compared to 3G. The models 6S and 6S+ decrease these values in each case by 15 kg CO_2e. The latest model SE is with 75 kg CO_2e slightly above the median which lies at 71.25 kg CO_2e.

(2) **Lifecycle Phase Emissions**

It indicates e.g., that model 3G caused 24.75 kg CO_2e during the production process, which corresponds to 45 % of the total. The values illustrate that, except

Table 1 Total GGE and single lifecycle phase GGE of iPhone models

Mod.	Total	Production	Transport	Customer use	Recycling
3G	55	24.75	26.95	2.75	0.55
3GS	55	24.75	26.95	2.75	0.55
4	45	25.65	15.30	3.60	0.45
4S	55	33.00	17.05	3.85	1.10
5	75	57.00	13.50	3.00	1.50
5c	50	36.50	10.50	2.00	1.00
5s	65	52.00	9.10	3.25	0.65
6	95	80.75	10.45	2.85	0.95
6+	110	89.10	15.40	4.40	1.10
6S	80	67.20	8.00	4.00	0.80
6S+	95	79.80	10.45	3.80	0.95
SE	75	61.50	10.50	2.25	0.75

Source According to Apple Inc. (2015b)

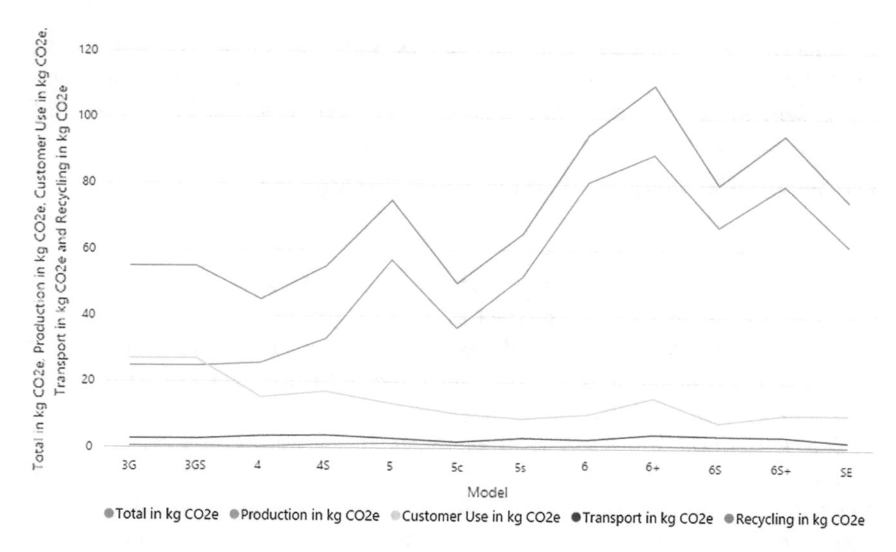

Fig. 1 Emissions in kg CO_2e—total and single lifecycle phases. *Source* According to Apple Inc. (2015b)

the first two models, the largest emission causer is the production phase followed by customer use, which is the largest for 3G and 3GS. However, the course demonstrates that the customer use emissions had been reduced in total over time with only slightly peaks in between. On the contrary, the production GGE show an increase over time. Transport and recycling show a continuously low impact and can therefore be considered as the areas with the lowest improvement potential.

(3) **Total Emissions compared to Production and Customer Use**

Figure 1 illustrates that the total GGE and the production GGE graph are running almost simultaneously. As an example the model 5s can be used. While total and production emissions increase from the predecessor model, the customer use emissions decrease. For that reason we identified the production phase as the most critical lifecycle stage to perform improvements, which will be conducive to a progressive environmental politic at Apple. At the first models the customer use had been in this role. Since we have no access to evaluate why the production of model 6 emits 80.75 kg CO_2e, more than 30 times as much as the production of model 4, future studies should access even better data for a deeper investigation.

3.3 *Possible Causes of the Shown Developments*

The illustrated development depends on different factors. For example, the internal processes could have changed or some new findings in production processes could effect to the emissions. Also the replacement of materials could contribute to a change. Since models 5 (except 5c), aluminum was used (Apple Inc. 2015b).

This material is highly recyclable, but could possibly cause more emissions during another phase (Pehnt 2010). Another example is the change of size and weight of the models, which is in turn also connected to the used materials. In the following, some of the presumptions shall be exemplary verified.

(1) Influence of Product Weight on Total Emissions

(a) Weight Development

Table 2 shows the weight of each model in gram. The weight is the sum of all materials, which are shown as product parts in the reports.

The individual changes from one model to the next are depicted in Table 3. Here, the gram numbers were taken, since they can be better compared to the emissions.

The table is read from left to right. For example, the weight from 3G to 3GS has increased by 2 g. If the weight has decreased the number is written in red. The upper right from the diagonal shows the timely development seen from early to later. Below the diagonal the opposite view is shown with inversion results.

(b) Greenhouse Gas Emission Development

In Table 4 the GGE changes are shown in the same way we worked out the weight changes in Table 3.

Table 3 e.g. shows that the GGE from 3G to 4 have changed by −10 kg CO_2e.

(c) Comparison of the Developments

The comparison reveals the relation between weight and emission changes over several generations or even from predecessor to successor model. Exemplary all four cases are shown for changes over several generations (colored in the tables):

Table 2 Weight of iPhone models

Mod.	3G	3GS	4	4S	5	5c	5s	6	6+	6S	6S+	SE
Weight (g)	133	135	135	140	112	132	112	129	172	143	192	113

Source Apple Inc. (2015b)

Table 3 iPhone weight changes in gram*Source* According to Apple Inc. (2015b)

Mod.	3G	3GS	4	4S	5	5c	5s	6	6+	6S	6S+	SE
3G	-	+2	+2	+7	-21	-1	-21	-4	+39	+10	+59	-20
3GS	-2	-	0	+5	-23	-3	-23	-6	+37	+7	+57	-22
4	-2	0	-	+5	-23	-3	-23	-6	+37	+8	+57	+26
4S	-7	-5	-5	-	-28	-8	-28	-11	+32	+3	+52	-27
5	+21	+23	+23	+28	-	+20	0	+17	+60	+31	+80	+1
5c	+1	+3	+3	+8	-20	-	-20	-3	+40	+11	+60	-19
5s	+21	+23	+23	+28	0	+20	-	+17	+60	+31	+80	+1
6	+4	+6	+6	+11	-17	+3	-17	-	+43	+14	+63	-16
6+	-39	-37	-37	-32	-60	-40	-60	-43	-	-29	+20	-59
6S	-10	-7	-8	-3	-31	-11	-31	-14	+29	-	+49	-30
6S+	-59	-57	-57	-52	-80	-60	-80	-63	-20	-49	-	-79
SE	+20	+22	-26	+27	-1	+19	-1	+16	+59	+30	+79	-

Table 4 iPhone GGE changes in kg CO_2e*Source* According to Apple Inc. (2015b)

Mod.	3G	3GS	4	4S	5	5c	5s	6	6+	6S	6S+	SE
3G	-	0	-10	0	+20	-5	+10	+40	+55	+35	+40	+25
3GS	0	-	-10	0	+20	-5	+10	+40	+55	+25	+40	+25
4	+10	+10	-	+10	+30	+5	+20	+50	+65	+35	+50	+30
4S	0	0	-10	-	+10	-5	+10	+40	+55	+25	+40	+20
5	-20	-20	-30	-10	-	-15	-10	+20	+35	+5	+20	0
5c	+5	+5	-5	+5	+15	-	+15	+45	+60	+30	+45	+25
5s	-10	-10	-20	-10	+10	-15	-	+30	+45	+15	+30	+10
6	-40	-40	-50	-40	-20	-45	-30	-	+15	-15	0	-20
6+	-55	-55	-65	-55	-35	-60	-45	-15	-	-30	-15	-35
6S	-35	-25	-35	-25	-5	-30	-15	+15	+30	-	+15	-5
6S+	-40	-40	-50	-40	-20	-45	-30	0	+15	-15	-	-20
SE	-25	-25	-30	-20	0	-25	-10	+20	+35	+5	+20	-

1. Weight increased and GGE decreased—e.g. 3G to 4 (marked blue)
2. Weight decreased and GGE decreased—e.g. 3G to 5c (market green)
3. Weight decreased and GGE increased—e.g. 4 to 5s (marked yellow)
4. Weight increased and GGE increased—e.g. 5s to 6+ (marked red)

The comparison of the values show that there is no explicit correlation between weight and GGE. This result clarifies that the consideration of individual factors can be insufficient for such a complex problem. On the one hand, a higher weight would also explain a higher material consumption and hence e.g. a higher demand for fossil fuels. On the other hand, the change of materials also requires considering their emissions and potentially adopted production processes, which could have a lower environmental impact than before. A heavier smartphone with improved processes can thus cause fewer emissions.

(2) **Influence of Used Materials on the Emissions**

Since the used materials affect the resulting emissions by the product itself as well as the emissions in the manufacturing process, it's important to research how different materials could have an impact on total and single lifecycle emissions. With the set of data we can try to answer questions, like:

1. Does the use of aluminum positively influence the GGE in the recycling stage?
2. Is there a material, which has a significantly high impact on the total GGE?
3. Does the use of polycarbonate significantly impair the environmental balance?
4. Does the 3G with 14 % plastic has higher GGE than the 4S which 2 % plastic?

Following we exemplary research the first two questions:

According to Apple, aluminum is used because it is almost indefinitely recyclable and has therefore a strong impact on this phase (Apple Inc. 2016; Hall et al. 2015). However, looking at the recycling phase the graph reveals that there is no clear relationship between the use of aluminium and a significantly decrease of recycling emissions. For some models the emissions are higher than looking at most models who hasn't used the material, like 3G or 3GS. Figure 2 shows these findings and marks the aluminium models with red cycles.

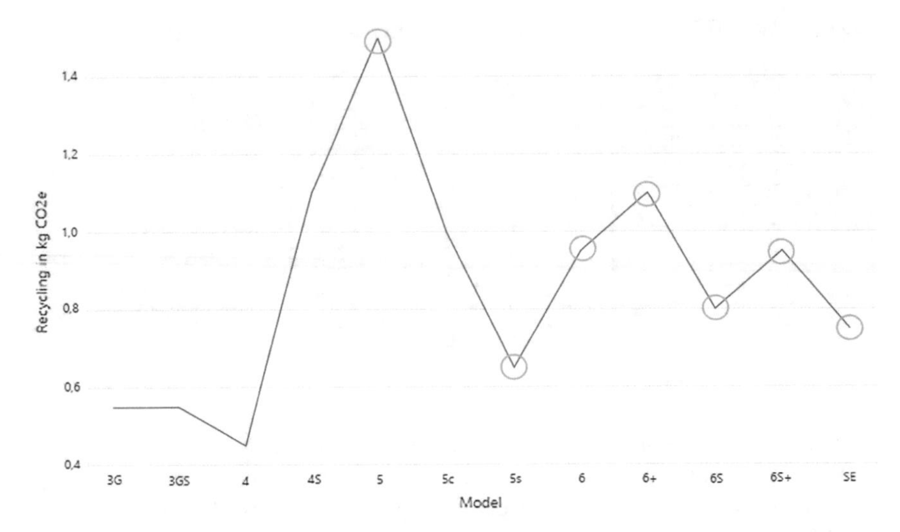

Fig. 2 Recycling GGE with aluminium containing models marked. *Source* According to Apple Inc. (2015b)

Again, the complexity of the problem is shown since a basically solid assumption that aluminium would be always conductive to emissions at least in the stage of recycling had to be refuted. One explanation approach could be, that the emission causing factors in the lifecycle phase of recycling are based on other indicators like machine emissions or similar.

The second question searches for materials, which have a significantly high impact on the total GGE. We would consider this as true for a material, which whenever weight increases the total emissions also increase. To answer this we have a look at the GGE developments as well as the development of the single components. What we are looking for are two graphs, one from the emission figure one from the material figure, which always go in the same direction. This finding would indicate a significant impression of the component on the emission graph and therefore reveal a possible connection of them both.

The comparison corresponds to a two dimensional table with changing values from one model to another. Doing so we found that the graphs of aluminium and emissions seem quite similar. Figure 6 illustrates both graphs as single and overlap-ping. The developments in the overlapping graph are given without an outline scale since the lines have different standards. The total emission graph starts at 40 kg CO_2e and the aluminum graph starts at 0 g since there are models without this material (Fig. 3).

Since aluminum was not part of the first four models we start to research the graphically development by having a look on the change from 4S to 5. Table 5 shows the changes of the consecutive models, so e.g. from 5s to 6 the change is +5 since model 5 has 21 g and model 6 has 26 g of aluminum.

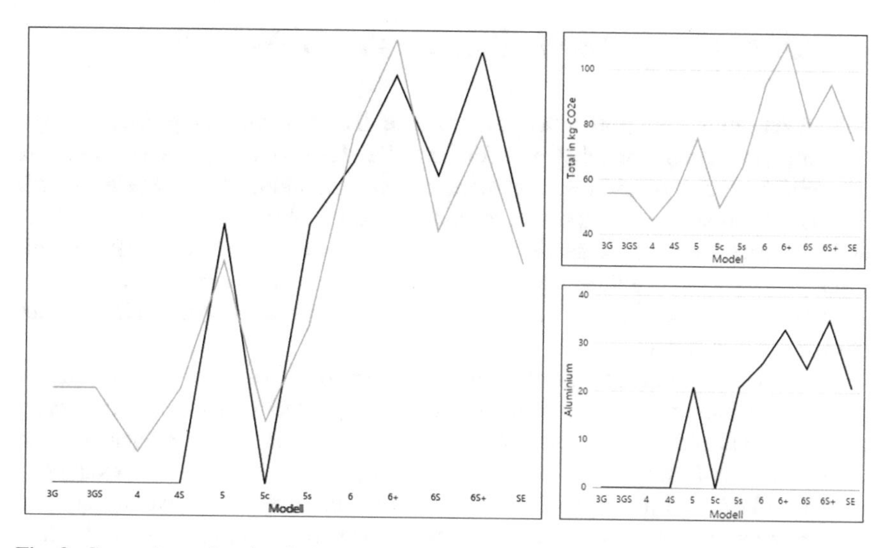

Fig. 3 Comparison of total emissions and the material aluminum. *Source* According to Apple Inc. (2015b)

Table 5 Aluminium and GGE changes of consecutive models

Mod.	Changes of aluminium in g	Changes of emissions in kg CO_2e
4S	0	+10
5	+21	+20
5c	-21	-25
5s	+21	+15
6	+5	+30
6+	+7	+15
6S	-8	-30
6S+	+10	+15
SE	-14	-20

Source According to Apple Inc. (2015b)

It can be seen that whenever the used aluminium increases or decreases the emission graph is going in the same direction. According to our findings it seems that aluminium causes a lot of emissions in total. Aluminium is a very good example to again show the complexity of this topic. For models 5 and 5s the same amount of aluminium is used, but the emissions do not increase by the same value. This shows that more factors must be taken into account than just one by one, because from this perspective it seems that using aluminium to produce a smartphone is not as useful as desired.

4 Conclusion and Implications for Future Research

Taking environmental protection actions like the reduction of greenhouse gas emissions, is an important challenge for the ICT industry. Companies have to figure out solutions to emit as little emissions as possible along the lifecycle of their products. Therefore, it's important to measure and analyze environmental data to make decisions on a reasonable base. On the example of the Apple iPhone this paper analyzed the sustainability related data given by Apple. For our analysis we concentrated on the GGE. Other data had been used as an explanation for the development of them.

As could be revealed, the production phase is the decisive factor with the highest potential for improvement. The curves of total and production emissions run nearly parallel. We also found that the product weight has no mandatory influence on the GGE. Afterwards we took the material share dataset to research two exemplary questions. First, we found that the models, which used aluminum caused in some cases more emissions in the lifecycle phase of recycling than most of the models, which haven't used this material. This finding was unexpected since aluminum promises to be highly recyclable and maybe still is because other, at this time, unknown factors caused this growth. The second question was, if at least one of the used materials has a significant impact on the total GGE. We found that the aluminum graph seemed very similar to the emission graph and analyzed the changes between the consecutive models. By contrasting this changes to the emission changes we proved that the graphs run always in the same direction. This proved that aluminum seems to have a profound impact on the total emissions. Another finding was that the impact of using an equal amount of aluminum did not result in an equal emission change. This showed that there has to be taken more into account than a single material to explain the total emission changes.

Due to the complexity of the topic as well as the relations between the data, further research should be conducted to answer even deeper questions. Also, already answered question could be modified so that e.g. the question of material influence could be extent to combinations of two or more materials. Further papers could also include more environmental data for explanations of the emission development like the products energy consumption. Another critical point of our analysis is the given dataset, which comes directly from Apple. Since Apple is the producer of the iPhone, independent measurements would be desirable but not feasible. Furthermore the product lifecycle defections have to be standardized since different models have slightly different views on some stages. Ultimately the performed analysis could lead to an environmental data dashboard which would enable the real time monitoring of environmental data. This dashboard could also be used for benchmarking purposes e.g. between the iPhone and another smartphone, which could be another implication for further research approaches.

However, the paper has shown that sustainability in the context of environmental protection is a crucial topic nowadays. Customers become more and more aware of these aspects. Their recognition of achievements in this field will also create competitive advantages, so that technology companies have to be even more aware of this challenge. To make progress companies have to measure, analyze and report those data. The performed analysis can be one way to investigate the datasets and give deeper insights. The development over time reveals facets one can't see by the first view so e.g. materials, which seem to be conductive must be questioned to get products and processes with a preferably low environmental impact.

References

Apple Inc. (2015a). Environmental responsibility. http://www.apple.com/environment/. Accessed 20 March 2016.

Apple Inc. (2015b). Product Reports, http://www.apple.com/environment/reports/. Accessed 04 April 2016.

Apple Inc. (2016). Environmental responsibility. http://www.apple.com/environment/finite-resources/. Accessed 24 April 2016.

Arndt, H.-K. (2013) Umweltinformatik und Design - Eine relevante Fragestellung? In: M. Horbach (Ed.) *INFORMATIK 2013: Informatik angepasst an Mensch, Organisation und Umwelt. 43. Jahrestagung der Gesellschaft für Informatik, Koblenz, September 2013. Gesellschaft für Informatik* (Vol. P-220, pp. 931–939), Bonner Köllen Verlag.

Börsen-Zeitung. (2015). Apple verdient 18 Mrd. Dollar - Höchster Quartalsgewinn aller Zeiten. https://www.boersen-zeitung.de/index.php?li=1&artid=2015019001. Accessed 2 April 2016.

Hall, S., Grossman, S., Johnson, S. C. (2015). Recycling von Aluminium - Kein anderes Material ist ähnlich vielseitig und bietet vergleichbare ökologische Vorteile wie Aluminium. http://www.novelis.com/de/seiten/raw-materials-recycling.aspx. Accessed 24 April 2016.

Livescience.com. (2015). Greenhouse gas emissions: Causes & sources. www.livescience.com/37821-greenhouse-gases.html. Accessed 13 April 2016.

Microsoft Inc. (2016). Microsoft Power BI. https://powerbi.microsoft.com. Accessed 13 April 2016.

Pehnt, M. (2010). Energieeffizienz – Definitionen, Indikatoren, Wirkungen. In: M. Pehnt (Ed.) *Energieeffizienz* (pp. 1–34). Berlin/Heidelberg/New York: Springer.

Spiegel. (2013). Apple-Handy: Alle iPhone-Modelle im Überblick. http://www.spiegel.de/fotostrecke/apple-smartphones-alle-iphone-modelle-im-ueberblick-fotostrecke-101297-8.html. Accessed 04 April 2016.

Statista. (2015). Innovations- und Design-Centrum (IDC), Absatz von Smartphones weltweit vom 1. Quartal 2009 bis zum 4. Quartal 2014 (in Millionen Stück). http://de.statista.com/statistik/daten/studie/Z246300/umfrage/weltweiter-absatz-von-smartphone-nach-quartalen/. Accessed 28 March 2016.

The Nature Conservancy. (2016). Climate change: Threats and impacts. http://www.nature.org/ourinitiatives/urgentissues/global-warming-climate-change/threats-impacts/. Accessed 11 April 2016.

Sustainable Software Design for Very Small Organizations

Stefanie Lehmann and Hans-Knud Arndt

Abstract Very small organizations rarely use software to organize their business processes, because there are comparatively few known solutions for their particular problems. On the one hand, these software solutions are often associated with high costs and on the other hand, particularly very small organizations do not have enough resources to find suitable solutions. These organizations miss a lot of opportunities: Through the use of the Information and Communication Technique (ICT) business processes can be processed not only resource-efficient due to reduced paper consumption. The ICT also helps to make business processes more efficient. To find such a sustainable software solution for very small organizations, requirements are identified. To determine a structure for such software to document business processes, it is important to apply design aspects for the implementation of the software. A definition of such design aspects can be achieved by the properties of a sheet of paper, which help to organize business processes in very small organizations. By analogy of the dominant metaphor of paper we specify requirements of sustainable software design for very small organizations.

Keywords Sustainability · Software design · Very small organization · Business process

1 Motivation

Very small organizations such as start-ups, associations or even societies are characterized by few employees and limited financial and time resources. Typically, the information transfer in these organizations is focused on handwritten notes on a sheet of paper. However, it is obvious that very small organizations should be able

S. Lehmann · H.-K. Arndt (✉)
Otto-von-Guericke Universität Magdeburg, Magdeburg, Germany
e-mail: hans-knud.arndt@iti.cs.uni-magdeburg.de

S. Lehmann
e-mail: stefanie.lehmann@ovgu.de

V. Wohlgemuth et al. (eds.), *Advances and New Trends in Environmental Informatics*, Progress in IS, DOI 10.1007/978-3-319-44711-7_2

to easily transfer information from handwritten notes into an Information and Communication Technique (ICT)-based system. To do so, on the one hand reduces risks and losses of knowledge of business processes, such as caused by non-readable handwritings. On the other hand, based on ICT the exchange of knowledge can be done more efficiently. The limited resources of small organizations must be considered for such changes. In this paper, we discuss how sustainability of software design can support very small organizations during their day-to-day operations. This helps to improve the business processes especially in organizations, e.g., against the background of frequent change of personnel. One of our goals is that staff members can quickly see the important facts of the business processes. Today, most very small organizations fulfill their documentation of business processes in their own way, which may change over time. The consequences are uncertainties due to the documentation of business processes, which might hinder the sustainable operation when business processes need to be rethought or errors occur due to incorrect documentation. Therefore, staff members of very small organizations should be encouraged to handle information in a more uniform style by the use of a structured form and only one system.

Thinking of such an ICT-based solution, one idea might be to create something completely new (e.g., programmed by students during an internship). Instead, characteristics of a sustainable software design should be determined to identify existing sustainable software solutions (e.g., open source) especially for very small organizations.

2 Paper as the Most Important Information Carrier

Even in today's information age, especially very small organizations benefit from paper as the most important information carrier. Only occasionally ICT has prevailed over paper to transfer information in small organizations.

An established example to provide information with the help of technology instead of paper is shown by the step from a letter to an email (Hoffschulte 2011). There are similarities and differences between letter and email as a medium for providing information.

Similarities or parallels between letter and email are that, e.g., both present a structure by which the information for transmission is being performed. A significant difference between them lies in the technological implementation of the email: using technology information can be faster or even immediately distributed. In addition, digitalized information like emails can easily be reproduced. Furthermore, information can be transmitted more environmentally friendly by reducing paper consumption through emails.

Like in letters and emails, information is exchanged in business processes. Consequently, the general rule is that for business processes an ICT support is suitable for more efficiency and effectiveness in contrast to the use of paper, even in very small organizations. The use of ICT enables an improvement of business processes. But in particular an ICT solution for very small organizations must be as paper-like as possible. On the other hand, many current pen-based methods of ICT (e.g., like the Apple Pencil for iPad Pro Apple Inc. 2016) serve the dominant metaphor of paper very well. Especially very small organizations have a need for long-lasting, sustainable software and in consequence also hardware solutions. Therefore, pen-based methods might be too modern and show little progress in the context to improve business processes of very small organizations. So this argument might guide them to a more conservative strategy.

Against this background the question about the best paper-orientated software design for very small organizations arises. In order to find an approach, design aspects can help here. In the past, it has been shown that the Ten Principles for Good Design stated by Dieter Rams (Vitsoe Ltd. 2016) can be used to evaluate the design of software (Arndt 2013).

3 Requirements of Sustainable Software Design for Very Small Organizations

To represent knowledge-intensive business processes by a technology, the Ten Principles for Good Design by Dieter Rams could be helpful. In this step, the advantageous properties of paper-based information should be transferred into digitalized ICT-based information. Sustainable design requirements on ICT are described for the technological representation of typical paper-based information of very small organizations by using the various aspects of the Ten Principles of Good Design.

These principles provide a framework of requirements for good design of products, but are also suitable for the design software (Arndt 2014). In the following, the description of the principles is used to show analogies to the properties of paper-based information for ICT-based information in the context of very small organizations. Each principle is considered in detail. Consequently, there will be a derivation of requirements for ICT to represent information and, therefore, also business processes which can be described by such information.

3.1 Good Design Is Innovative

This principle implies that a technological development could optimize the utility value of a product. This innovative design is always produced in association with innovative technology. From the transmission of these statements to the support of business processes follows that with a modern ICT-based method instead of a

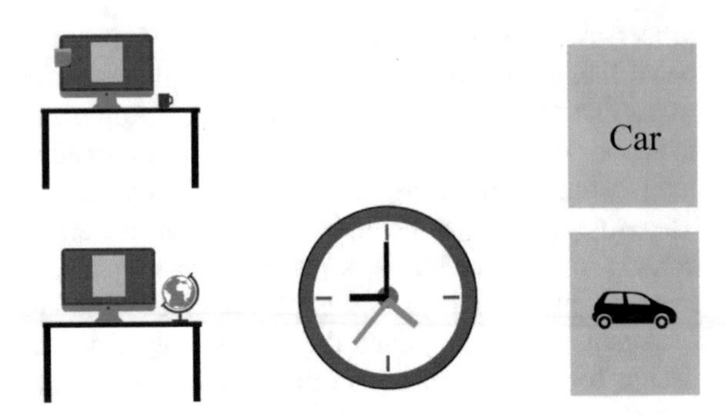

Fig. 1 Visualization of requirements based on the principle "good design is innovative". *Source* According to Vector Open Stock (2015a, b)

paper-based method the business processes can be improved: The working times and places of work can be made more flexible, because the staff members are not tied to a single sheet of paper on which the information is stored. Moreover, thanks to modern technology, a flexible design of business processes is possible: Like on a sheet of paper there should no precise work instructions needed how knowledge has to be left. In addition, information cannot only be saved in the form of texts, but also, e.g., with the help of graphics.

Figure 1 outlines the requirements on an ICT to be innovative to support the business processes: Like a sheet of paper the information could be used at different workstations, at different times and the business processes could be designed according to the users' needs, if the information is made available by ICT.

3.2 Good Design Makes a Product Useful

In addition to the primary functions, also psychological and aesthetic features of a product must be observed. Such a product is also ICT. Compared to paper-based information, all information can be detained for a business process to document it and to protect it against information overload. The data must always be up to date, which helps in distributing information. Furthermore, the data must be manageable: A sheet of paper has no restrictions in this sense, because its contents can be monitored at a glance. When a page is full, another sheet for additional information could be added. Transferring to ICT, restrictions have to be avoided for mapping information: either by a character limit or by the dimensions of the screen.

Figure 2 depicts the requirements of ICT to be useful for supporting the business processes: Like with a sheet of paper there is no limitation, data can be sorted and searched, and information can not only be taken from each user, but the user can also add or delete data when the data are updated.

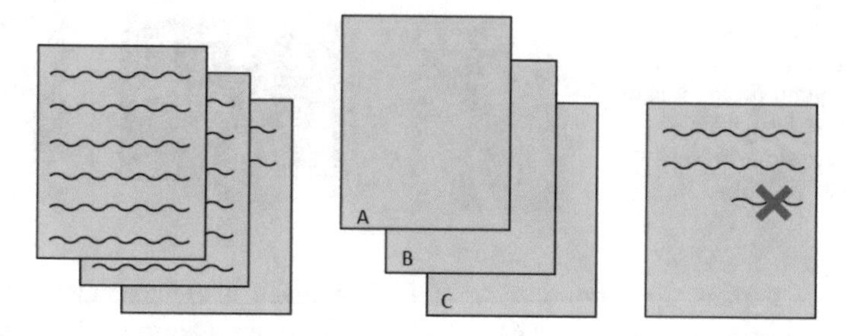

Fig. 2 Visualization of requirements based on the principle "good design makes a product useful"

3.3 *Good Design Is Aesthetic*

Devices that are used every day characterize the environment and affect the well-being. Consequently, in managing business processes with ICT, the operation should be as effective as or even better than by using a piece of paper. Just as the use of paper has prevailed in the daily office, also the use of a technological solution for information should be accepted and change the business processes positively. Thus, the business processes can be improved and thereby working time could be abridged.

Figure 3 visualizes the requirements of ICT to realize the aesthetic aspect for supporting the business processes: The procedures and processes can be improved, so that working hours can be reduced, and the user feels comfortable at his workplace, e.g., because there is less paper chaos.

3.4 *Good Design Makes a Product Understandable*

Following this principle, the structure of ICT has to be illustrated in a reasonable manner and without explanations. To organize information, it is established that the use of the data structure of lists is generally understandable. By analogy with a sheet or stack of paper the structure of the data is recognizable: A stack of paper can be sorted without many explanations. Expanding on the complete, global ICT, also its other functions should be as self-explanatory as possible. Limited on the natural language the use of the mother tongue within the system provides to understand this.

Fig. 3 Visualization of requirements based on the principle "good design is aesthetic". *Source* According to Freepik (2011), Vector Open Stock (2015b)

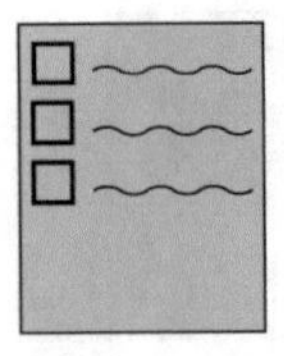

Fig. 4 Visualization of requirements based on the principle “good design makes a product understandable”. *Source* According to Vector Open Stock (2015b)

Figure 4 depicts the requirements of ICT to make it understandable for supporting the business processes: The use of lists illustrates a sorting of the data. The execution of the functions should be understandable to the user; the functions of ICT should be understood through the use of the mother tongue.

3.5 *Good Design Is Unobtrusive*

Products, which serve a purpose, have an instrumental character and should therefore have a neutral style. If the purpose is to manage business processes in a group, the focus is on the information contained in these business processes and the system itself moves into the background. If the design is as flat as possible, this corresponds to the design of a simple white paper where the notes are left on. Even if a system defines a certain structure, the user should be allowed a self-realization. So he/she can leave the notes in their own way. E.g., similar to a sheet of paper, the background color of information could be changed for communicating and sending various signals.

If the system acts as a tool, it should be apparent what kind of information is put in and is put out, is being processed. For this representation of business processes a simple Kanban Board can be useful, as outlined in Fig. 5.

3.6 *Good Design Is Honest*

A product has an honest design, if it does not break with the expectations of the users. Hence, also a knowledge management system should not promise more than that it collects information and distributes it for reuse. All other guaranteed

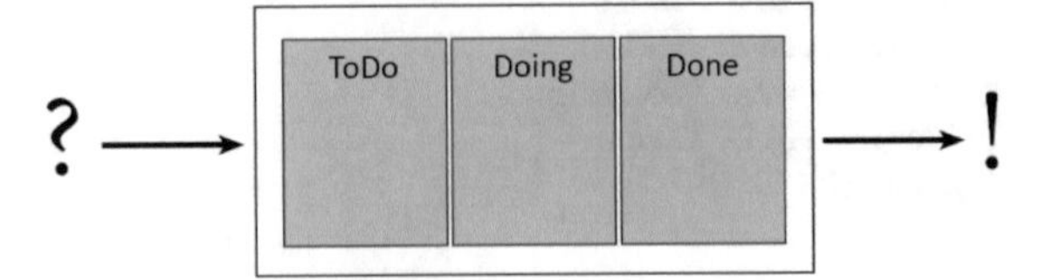

Fig. 5 Visualization of requirements based on the principle “good design is unobtrusive”

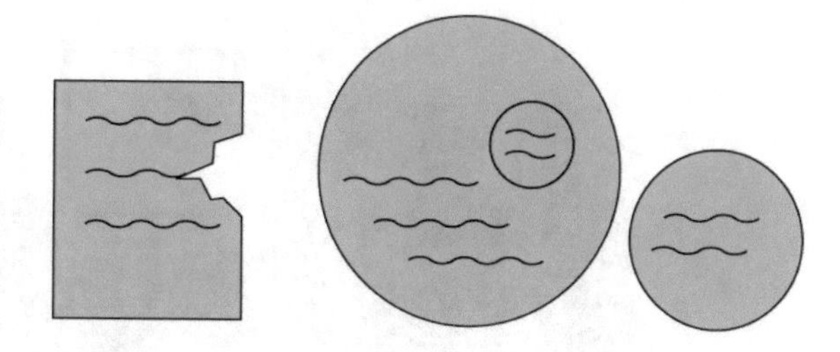

Fig. 6 Visualization of requirements based on the principle “good design is honest”

properties should be fulfilled: from the permanent availability of the data down to the smallest unit of a function. So each non-working unit, up to a system crash, counteracts the promise of the system. It may therefore be no loss of information. By analogy to a stack of paper also information should be saved in a data structure of lists. This data structure should be recognizable and no false ideas should be formed by unnecessary design. Figure 6 visualizes cases that contradict the thesis: data is lost or the structure of the data is not recognizable.

3.7 *Good Design Is Long-Lasting*

A long-lasting ICT should neither be modern or stylish, nor should it appear antiquated. In analogy to a white sheet of paper also its use and design has been proven over a long period. It will dispense with the use of design trends. For a long-term use and to counteract the throwaway society, analogous to paper an ICT should not only be able to archive information, but its design should also be durable. Figure 7 outlines the cases that are to be improved by the technology.

3.8 *Good Design Is Thorough Down to the Last Detail*

The design must not be left at discretion or be randomly. Thoroughness and accuracy of the design consider the use by the user. A uniform design can reduce

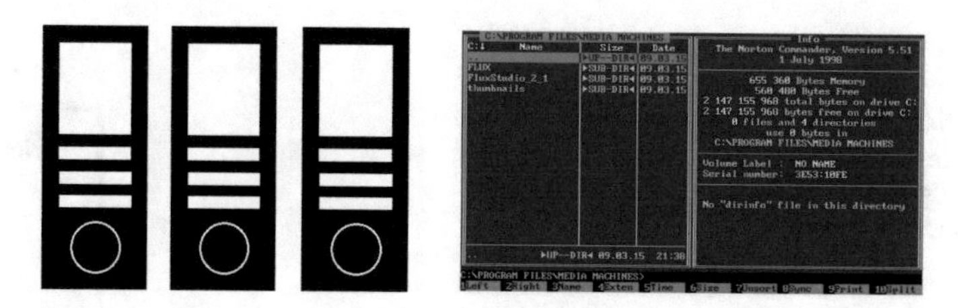

Fig. 7 Visualization of requirements based on the principle “good design is long-lasting”. *Source* According to Engineering Student Resource (2016)

Fig. 8 Visualization of requirements based on the principle "good design is thorough down to the last detail". *Source* According to Vector Open Stock (2010)

the cognitive load of the user. Accordingly, the design is consistently implemented. In analogy to a sheet of paper, there are uniform formats. Transferred to ICT, therefore, each user should have the same functions and views, the various sub-pages should not differ from the layout and all elements should be equivalent. Figure 8 describes how this thesis is shown in an ICT: On the one hand, like in a conference all users have the same views and possible functions of the data and on the other hand, the design of any information is equal.

3.9 Good Design Is Environmentally Friendly

It is important to conserve resources to preserve the environment. Where information is organized ICT-based, a significant amount of paper can be economized. Apart from the physical conservation, also resource-saving design aspects have to be observed during the transition from paper to ICT-based information: Designing has to be done on a minimization of physical and visual pollution. It should contain no unnecessary elements in the layout. Figure 9 shows the goal of this principle: For slightest visual stress the extreme values of a white sheet of paper and a black font for the strongest contrast would be the use case for an ICT to strive.

Fig. 9 Visualization of requirements based on the principle "good design is environmentally-friendly"

ToDo

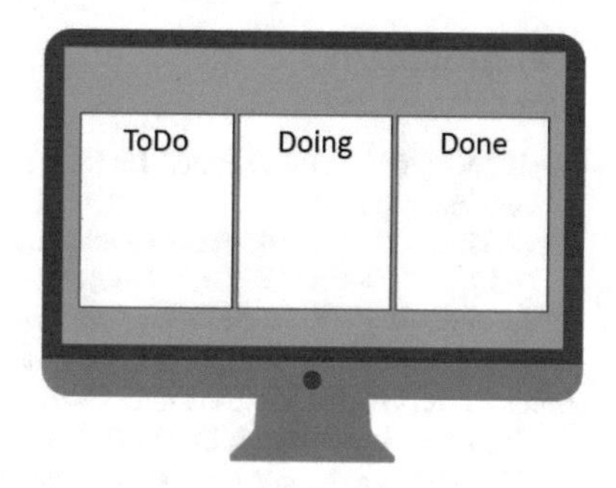

Fig. 10 Visualization of requirements based on the principle "good design is as little design as possible". *Source* According to Vector Open Stock (2015b)

3.10 Good Design Is as Little Design as Possible

The most representative statement about sustainability is the statement of Bauhaus "Less is more". Dieter Rams transferred this statement into "Less but better". In this case, the design is to focus on the essentials and go back to the pure, to the simple. Based on this description, the design of a paper is as simple as possible, if it is a white sheet. Here, all previously mentioned properties are kept, which were described with reference to the other theses: e.g., the user can design the sheet to meet his requirements or if the sheet is full, another one will be added. Once again, the analogy is transferred from the paper in the ICT for organizing the busi-ness processes. If this last thesis is transferred into the ICT, also the previous theses are considered: For the organization of business processes applies that if the design requirements of ICT fulfill the previous theses, the design is kept as simple as possible. Figure 10 shows a possible implementation of the design requirements according to the thesis.

4 Conclusion and Outlook

We presented an approach to provide requirements for existing software solutions. As a base, the properties of paper have been identified for shaping the business processes. By transferring these requirements to ICT, some design requirements on ICT for the organization of business processes result. Consequently, in order to improve the business processes in very small organizations and to conserve resources by using an ICT, the design is used as a key. So, on the one hand, the limited resources of very small organizations can be saved, and on the other hand, there is less paper consumption by using an ICT. In the future, specific technical possibilities should be considered in more detail for the representation of business processes or for an analog implementation of writing a note on a paper with a pen. In addition, a more accurate solution for very small organizations should be found based on the determination of the available resources.

References

Apple Inc. (2016). Apple Pencil for iPad Pro. Retrieved March 14, 2016, from http://www.apple.com/apple-pencil/.

Arndt, H.-K. (2013). Umweltinformatik und Design—Eine relevante Fragestellung? In M. Horbach (Ed.), *INFORMATIK 2013: Informatik angepasst an Mensch, Organisation und Umwelt. 43. Jahrestagung der Gesellschaft für Informatik, Koblenz, September 2013. Gesellschaft für Informatik* (Vol P-220, pp. 931–939). Bonner Köllen Verlag.

Arndt, H.-K. (2014). Big Data oder Grand Management Information Design? In E. Plödereder, L. Grunske, E. Schneider, D. Ull (Eds.), *INFORMATIK 2014: Big Data—Komplexität meistern. 44. Jahrestagung der Gesellschaft für Informatik, Stuttgart, September 2014. Gesellschaft für Informatik* (Vol P-232, pp. 1947–1956).

Freepik. (2011). Paper Crumpled Ball in Yellow. Retrieved March 14, 2016, from https://image.freepik.com/free-photo/paper-crumpled-ball-in-yellow_2493755.jpg.

Hoffschulte, G. (2011). Papierkrieg, Muthesius Kunsthochschule zu Kiel. Retrieved March 14, 2016, from http://www.gramufon.de/gramufon/papierkrieg/.

The Engineering Student Resource. (2016). Norton Commander DOS. Retrieved March 14, 2016, from http://www.eos.ncsu.edu/e115/images/NortonCommanderDOS.png.

Vector Open Stock. (2010). Business Presentation. Retrieved March 14, 2016, from https://www.vectoropenstock.com/vectors/pre-view/397/business-presentation.

Vector Open Stock. (2015a). Vehicles Icon Set. Retrieved March 14, 2016, from https://www.vectoropenstock.com/vectors/preview/74497/vehicles-icon-set.

Vector Open Stock. (2015b). Work Desk Home Cartoon with PC. Retrieved March 14, 2016, from www.vectoropenstock.com/vec-tors/preview/73664/work-desk-home-cartoon-with-pc.

Vitsoe Ltd. (2016). Dieter Rams: Ten principles for good design. Retrieved March 14, 2016, from https://www.vitsoe.com/gb/about/good-design.

Software Development Guidelines for Performance and Energy: Initial Case Studies

Christian Bunse and Andre Rohdé

Abstract Energy efficiency and -awareness are of growing importance in the field of information and communication technology. On the one hand, computing center aim for reducing their energy consumption in order to save money and improve their carbon footprint. On the other hand, mobile devices that are typically battery powered, have to be aware of their energy consumption in order to prolong their uptime while at the same time keeping a specific level of quality-of-service. Research has shown that especially software, running on a (mobile) device has a large impact onto the energy consumption of that device. Software developers should be aware of a software systems' energy related cost and about means for optimizing them. Best practices, community-believed guidelines for improving the quality (e.g., performance, energy efficiency, usability, etc.) of software, are one way to do so. This paper evaluates the effects of selected guidelines by a small experiment series and within a larger, commercial product. Results show that by systematically applying selected programming guidelines, energy consumption can be reduced by 9 % and performance by 1 % in average.

Keywords Energy efficiency · Software development · Evaluation

1 Introduction

Energy is a limiting factor of information and communication technology. Especially mobile and embedded devices do not have a permanent power supply, but depend on rechargeable batteries. Due to increases in hardware performance and the incorporation of additional hardware components (e.g., Near Field Communication) the energy needs of devices are still growing. Software (i.e., operating system and application) utilizes hardware and, thus, is key in improving energy consumption.

C. Bunse (✉) · A. Rohdé
University of Applied Sciences Stralsund, 18435 Stralsund, Germany
e-mail: Christian.Bunse@fh-stralsund.de

A. Rohdé
e-mail: Andre.Rohde@fh-stralsund.de

V. Wohlgemuth et al. (eds.), *Advances and New Trends in Environmental Informatics*, Progress in IS, DOI 10.1007/978-3-319-44711-7_3

Energy-aware software development, energy-aware algorithms and energy-aware resource substitution are only three examples of recently initiated research areas that try to reduce energy needs by optimizing the software rather than the hardware (Höpfner and Bunse 2011). A common problem of developers is the lack of (evaluated) guidelines, best-practices, or methods for the development of energy-aware software systems. Existing guidelines often have a focus on performance and user perceived quality-of-service. Improving energy consumption is, if addressed at all, expected to be a side-effect of improving performance. However, guidelines should not solely be based on common knowledge, but on measurement and experimentation. Otherwise, they cannot be systematically applied to achieve repeatable results.

In this paper, we report about the results of an empirical study that evaluates selected development guidelines for mobile and embedded systems. The goal is to equip developers with tools that do have a visible effect. Our evaluation approach is two staged (e.g., in-vitro and in-vivo). We first evaluate a set of guidelines in isolation by means of small, specifically designed systems. In a second step, combined guidelines and their effects are evaluated in the context of a commercial system.

The results of the study support our hypothesis. Applying carefully selected guidelines will significantly improve the energy consumption of a software system while keeping the effects to other quality factors low. These results complement our previous findings regarding the energy aspects of resource substitution mechanisms, and algorithms (Höpfner et al. 2012), or design patterns (Schirmer 2012). In summary, we state that by providing developers with efficient and effective guidelines they are enabled to systematically engineer energy-efficient software systems.

The remainder of this paper is structured as follows: Sect. 2 discusses related work. Section 3 presents background information regarding the examined quality factors as well as on measuring them. Section 4 introduces the underlying organization, principles and hypothesis of the study with a focus on the selected guidelines. Section 5 presents evaluation results and discusses threats to validity. Section 6 concludes the paper and provides a short outlook onto future work.

2 Related Work

Several research projects have been conducted regarding the energy consumption or -efficiency of (mobile) software systems. On the one hand, the focus is on the lower abstraction levels of communication. Feeney (2001) proposes new energy-aware routing techniques. (Gur) define energy-aware protocols for transmitting data in wireless networks. On the other hand, users and their behavior are subject to energy related research. Wilke et al. (2013) define micro-benchmarks for emailing and web browsing applications and Bunse (2014) evaluated the impact of keeping users informed about their energy-related behavior.

When it comes to software development, text books and websites provide a huge amount of "*Dos and Don'ts*", ranging from rather abstract tips such as "*Go for loose coupling and high cohesion*" to quite specific ones such as "*Avoid floating point numbers*". Unfortunately, such guidelines are of limited value unless they are (empirically) validated. Briand et al. (2001) examined the effects of applying design guidelines and conclude that object-oriented design documents are more sensitive to poor design practices, in part because their cognitive complexity becomes increasingly unmanageable. Bandi et al. (2003) and Hnatkowska and Jaszczak (2014) examined the impact of using idioms on source code maintainability and concluded that using idioms is a beneficial practice, especially in corrective maintenance.

Only recently the impact of development guidelines on quality factors such as energy or performance has been started. Seo et al. (2008) investigated the impact of java statements on the energy consumption of a system. Gottschalk et al. 2012 analyzed source code to identify energy-related code smells. Höpfner and Bunse (2011) examined the relation between algorithms and energy consumption (i.e., sorting algorithms). Khan et al. (2011) examined and confirmed that software optimization at the application level can help achieve higher energy efficiency and better thermal behavior. Hönig et al. (2013) claim that new concepts for energy-aware programming are needed and exploits symbolic execution techniques as well as energy estimates for program code. Bunse and Stiemer (2013) and Noureddine and Rajan (2015) explored the impact of design patterns to a system's global energy consumption. For example, observed that transformations for the Observer and Decorator patterns reduced energy consumption in the range of 4.32–25.47 %. This paper continues this research at the method level by evaluating abstract development guidelines.

3 Background

Performance, user-perceived quality, and energy are key quality factors of (mobile) software systems. According to Sneed (1987) such factors are closely interrelated and often reside in an antagonistic relationship. The goal therefore is to optimize one quality factor (e.g., energy consumption) while keeping the effects to other factors as low as possible. In this regard, measurement is a prerequisite for optimization and essential for baselining, as well as for the evaluation of effect sizes.

In computing, performance is defined by the speed a computer operates at. However, what is the speed (performance) of a software application? Due to the interactive nature of software systems, software performance can be defined as "*the effectiveness of the system, including throughput, individual response time, and availability*" (Van Hoorn et al. 2012). Response time (also called execution time) measures the time for running a specific code fragment while throughput measures the amount of work done in a specific time. We will use execution time since development guidelines manifest in small pieces of code and execution time is less dependent on other components.

In physics, energy is a property of objects which can be transferred to other objects. Energy comes in many forms which makes it difficult to define, measure, and extrapolate (Nigg et al. 2000). The energy consumption of software systems is basically the sum of the energy consumption of all utilized hardware elements. To simplify measurement, Johann et al. (2012) defines metrics that are based on measuring consumption at the battery level while executing the application.

4 Preparation

4.1 Development Guidelines

Performance and energy-efficiency are important factors for the commercial success of a product (aur). Therefore organizations such as Apple (2014), Google (2014), or Microsoft (2014) provide corresponding development guidelines. Interestingly, such guidelines are either GUI style-guides or related towards performance, while energy is viewed as a by-product. In the context of this study we selected six common guidelines (Google 2014) that claim to have an impact onto energy-consumption:

- **R1—Avoid unnecessary objects**. The creation of objects generates CPU load and allocates memory. It thus, slows a system while at the same time increasing energy needs (Google 2014).
- **R2—Static methods should be preferred**. According to Google (2014), the use of static methods will speed up invocations by 15–20 % and reduce energy needs.
- **R3—Use static final for constants**. Declaring constants to be "static final" hinders the generation of a class-initializer method. This avoids "expensive" field-lookups and thus, will save energy.
- **R4 Avoid internal properties**. Directly accessing the attributes of an object will improve performance and energy needs by 3–7 times (Google 2014).
- **R5—Use enhanced *for-loop* syntax**. Loop-mechanisms should be selected according to the underlying data structures. The traversal of a *ArrayList* using a *for-loop* is 3–4 times faster than by using a *for-each* loop (Google 2014).
- **R6—Avoid float**. In general, float operations are 2–3 slower than integer operations (Google 2014). Regarding performance there is no difference between float and double operations. However, doubles needs twice the memory of floats. Thus, doubles should be avoided as well.

4.2 Platforms and Devices

The measurement of properties such as performance or energy consumption is affected by external or environmental factors. One means to mitigate such effects is replication. While replication extenuates external impacts, variation helps in generalizing results. We replicated each measurement more than 100 times and performed measurements on different platforms using different operating systems.

In detail, we designed the study to test selected development guidelines (R1–R6) for two operating systems: iOS and Android. For each operating system we selected (in 2014) one or two typical devices following device distribution numbers as published by "*MacTechNews.de*" and *statista.com.*

- iOS test-systems were developed using ObjectiveC and were executed on devices that run "iOS 6.1.3". Regarding hardware we decided to use an Apple iPhone 4S (two cores, 512 MB RAM), and (in the second stage of the study) we additionally used an iPad 3 (two cores, 1 GB RAM).
- Android test-systems were developed using Java and were executed on devices that run "Android 4.3" or "CyanogenMod 10.2 Stable". Regarding hardware we used a Samsung Galaxy S3 (Quadcore with 1 GB RAM) and a Google Nexus 4 (Quadcore, 2 GB RAM).

4.3 Study Design and Systems

The goal of this study is to evaluate the effects of selected development guidelines onto the performance and energy consumption of software systems. First we examine the effects of guideline application in isolation. To do so, we developed a couple of prototype applications. These address only one of the six selected programming guidelines, are small, and do not involve GUI interaction.

Android programs were developed in Java while the systems for the iOS platform were developed in ObjectiveC. The systems are small (<100 LOC) and only address the guideline under test. Each system comes in two variants. One that implements a specific guideline, and one that does not. Measurement data is collected while executing each program several times. To make sure that the guideline and not a specific implementation is responsible for observed effects, we developed up to six variant implementations. But, evaluation results have not shown significant differences between them.

The first stage of this study evaluated the guidelines in isolation. The results initially confirm our assumption that by applying R1–R6, performance and energy consumption can be improved. However, these results are of limited value for the development of actual software systems. Such systems require the application of multiple, and possibly combined guidelines. But, combination makes the interpretation of measurement results difficult. We therefore decided (in the 2nd stage) to "optimize" a larger system and examine measurable effects. This will give an initial understanding and help to design follow-up studies.

In detail, we analyzed the source-code of a commercial product and applied, where possible, guidelines R1–R6. *Spraycan* is a drawing application primarily targeting iOS devices. Modifications of an existing system are limited (i.e., not all possible guideline combinations can be tested). For example, it appeared that the majority of float operations in *Spraycan* are used in the context of drawing operations. Drawing is based on the OpenGL framework which cannot be changed. Thus, guideline

R6 could not be generally applied throughout the app. We decided to examine effects based on three usage scenarios: T1 covers the "Home Feed" (i.e., starting, editing settings, and discussion). T2 covers loading and displaying single image previews and T3 loading and displaying the art-gallery of *Spraycan*.

4.4 Measurement

As discussed in Sect. 3 we use 'execution time' as basic performance metric. Following D. Stewart (InHand Electronics, Inc.) "many different methods exist to measure execution time, but there is no single best technique." We use time-stamps based on global system time. To mitigate external impacts, measurements were replicated and performance was calculated as the arithmetic mean of all single measurements.

Following Schirmer (2012), energy consumption was measured by software measurement-tools. On iOs we used *PowerGremlin* (PalominoLabs 2014). Powergremlin reads and stores battery values. On Android we used *PowerTutor* to measure energy consumption at varying levels of abstraction (Michigan 2014).

Additionally, we checked that the devices were in airplane mode, wireless communication and GPS were switched off, the battery was fully charged, the display was dimmed, and that the temperature of CPU and battery were "normal". During measurement, devices were not touched to avoid accelerometer reactions. Furthermore, we ensured that none of the in-built energy saving mechanisms were activated.

5 Evaluation

5.1 Research Hypothesis

Our study had two goals: (1) to explore the impact of selected, single development guidelines onto the performance and energy consumption of a software system. (2) to examine the effects of combining and applying several such guidelines to a larger system. The central question underlying this research therefore is: *Are there significant differences between software systems that have/have not been developed by following specific development guidelines (i.e., R1–R6)?*

Of particular interest are the quality aspects of performance and energy consumption in relation to computing tasks. Therefore, we defined a couple of assumptions prior to running the actual experiments:

1. Guideline-conform systems are more efficient regarding **comparative** and **total demand** performance and energy measures.
2. There is no difference between guideline-conform and non-conform systems regarding **comparative** and **total demand** performance and energy measures.
3. Guideline-conform systems require longer **execution times** and consume more energy than non-conform systems.

5.2 *Results*

According to the study design, results fall into two categories: First, we compare execution time and energy consumption of small systems that implement a single development guideline. Second, we examine the impact of applying combined guidelines onto the performance and energy consumption of a larger system.

Table 1 shows the results of evaluating guideline R1-R6 in isolation. Results regarding performance (in s) and energy (in J) are averaged across multiple runs. Negative values indicate that the non-guideline system "performed better" than the adapted/improved system.

One obvious result is that applying a single guideline not necessarily results in the desired effect. This is especially true for the iOS platform. It appears that R2 should only be used on the Android platform (average savings of 37 %) while on iOS performance was degraded by 25 % and energy consumption by 10 %. Interestingly, there are even some cases where the performance of a guideline-optimized system was worse than that of the original system, but energy consumption was slightly better (e.g., R3). In contrast, systems for the Android platform operate as expected, but reveal large differences regarding performance and energy consumption.

In the second stage, guidelines where applied (in combination) to the *Spraycan* application. Performance and energy consumption where then measured for three different usage scenarios. The results are shown by Table 2.

Table 1 Evaluation results—R1–R6

Guideline	Average distance			
	iOS		Android	
	Time (s)	Energy (J)	Time (s)	Energy (J)
R1	4.489	920.17	3.021	569.18
R2	−1.194	−83.43	1.212	160.2
R3	−0.003	2.66	1.621	325.06
R4	2.411	447.02	3.864	587.68
R5	3.004	704.57	1.814	366.8
R6	1.817	172.07	0.773	217.56

Table 2 Evaluation results: T1–T3

Scenario	Average distance			
	iOS		Android	
	Time (s)	Energy (J)	Time (s)	Energy (J)
T1	0.38	61.86	0.43	83.56
T2	−0.13	−31.40	0.06	230.07
T3	0.41	42.66	0.46	64.35

Measurement results of the second stage do not reveal large effect sizes compared to those of the first. However, the results support our hypothesis. Carefully selecting and applying guidelines R1–R6 will result in moderate improvements. Merged into one single value for the complete study, it can be stated that energy consumption can be improved by 16.14 %. However, such a value is misleading since there are significant differences between devices and platforms. Selecting the "right" guidelines is key. If wrongly chosen, optimization sometimes led to unwanted effects (T2).

5.3 Discussion

In general, the results support our hypotheses. Regarding hypothesis one we can confirm that in average guideline-conform software systems are more energy-efficient and have an improved performance. When calculating the gross average effect it appears that energy consumption can be reduced by 9 % and performance by 1 %. In turn, we can decline hypotheses two and three.

In the majority of cases the application of a specific guideline results in (small) performance gains and larger energy savings. The largest effects have been occurred regarding guideline R1. R1 recommends to avoid many, short-lived objects. Energy-consumption on Android could be improved by factor 19 and on iOS by factor 73. The smallest effect was observed for guideline R2. While, the guideline "saved" 37 % of energy on Android, energy consumption was increased on iOS by 10 %. The latter was probably caused by the ObjectiveC compiler. Generalized across all guidelines and operating systems the difference between guideline-compliant and non-compliant systems is 2.1 s and 380 J. Thus, in general by applying the guidelines systems can be improved regarding performance and energy-consumption.

However, the effects are not always as expected. In particular, optimization on the iOS platform produced inconsistent results (i.e., R2). One reason might be that the selected guidelines (R1–R6) are based on performance tips for Android systems. Although the guidelines do not specifically address OS properties, we observed differences between operating systems. One reason for these differences, might stem from the fact that ObjectiveC is a compiled language while Java is interpreted. According to Apple the ObjectiveC compiler applies optimization at the Compiler-

Table 3 Guideline recommendations

Guideline	iOS	Android
R1	+	+
R2	–	+
R3	–	+
R4	+	+
R5	+	–
R6	+	+

Level (Apple 2014). Although we set the XCode optimization level to none, the compiler still performs some optimization steps that might have led to the observed effect.

One interesting observation was made concerning the deviation of measurement results when replicating a test-case. While iOS measurements where quite stable (mean deviation: 2.78 J), deviations on Android are larger (mean deviation: 49.57 J). We assume that reasons are hidden, unidentified processes of the OS.

The second stage showed that there are positive effects of applying multiple guidelines in combination. However, improvements are not as significant as one might have expected following the results of stage one. In general, we observed a difference of 0.31 s and 83.71 J across all variants and platforms. This supports our assumption that guidelines should be carefully selected.

Regarding iOS there are large differences between devices. In detail, there are large differences regarding scenario T2 when executed on an iPad or an iPhone. Since T2 addresses the loading and displaying of preview images, we assume that the difference might stem from iOS internal memory handling. However, we did not find a provable source for this assumption. Another notable result in this direction is that in sum savings on the iPad were larger than those on the iPhone.

Table 3 provides recommendations regarding guideline to use ("+") and guidelines to use with care ("–"). Please note: These recommendations may not be generally valid without knowing the results of a (meta-) analysis.

5.4 *Threats to Validity*

The authors view the results of this study as exploratory due to threats that limit generalization. In detail, the following threats have been identified:

Regarding performance, we decided to use 'execution time' as metric. However, performance has more than one facet and results might be different when using other metrics. Another problem might be that measurement creates overhead that may falsify data. However, our results indicate that the overhead is negligible.

One problem might have occurred regarding data validity. Typically, measurement problems as well as the impact of environmental aspects are addressed by replication. Unfortunately, running the measurements 100 times takes approximately 20 min. Therefore, we had to limit the number of replications. In addition, we cannot exclude that other factors (e.g., battery level) too might have influenced the measurements.

One threat might have occurred due to a biased selection/implementation of prototypes. This might have resulted in a selection that is not representative of the population. By covering different guidelines and combinations, we believe that the results show a trend that has to be further studied. Also the selection of devices might have had an impact. The differences between Android and iOS as well as those between iPhone and iPad are indicators for this need. In addition, specific hardware properties (e.g., cryptography) might have an impact too. Interestingly, there are guidelines that seem to work in general and others that are of limited use.

Another possible threat that hinders generalization of results is that not all possible guideline combinations (e.g., R1 + R2, R1 + R5 + R6, etc.) can be observed in the *SprayCan* system. In addition, scenarios T1–T3 do not allow to examine specific combinations in isolation. This threat is currently addressed by a follow-up experiment series that aims at evaluating such combinations in detail.

6 Conclusion

In this paper we examined the effects of code-level software optimization onto performance and energy consumption. We believe that our results fill an important gap, and have the potential for building the basis of an effective developer-toolbox.

We have set up an evaluation infrastructure and tested it in two experimental runs. The infrastructure includes software-based energy measurement tools, and a utility for measuring performance. The infrastructure has been used to evaluate the effect of development guidelines. Such guidelines are available across all level of abstraction: From architectural patterns down to code idioms. We selected six common guidelines at the code level and evaluated them in a two staged approach: (1) Isolation and (2) Combination.

The results support the hypotheses: By systematically applying the guidelines, energy consumption can be reduced by 9 % and performance by 1 % in average. However, there are several open issues that warrant further research. First, we have to extend the evaluation of guidelines to all level of abstractions as well as to combination effects. Second, the study has to be replicated across devices and operating systems to allow for meta-analysis.

Interestingly, Google (2014) clearly states that "Before you start optimizing, make sure you have a problem that you need to solve." Thus, evaluation needs baselining and measurement. A first step, might be the creation of energy benchmarks to systematically assess and compare the effects of energy optimization.

References

Apple: Performance tips. (2014). http://developer.apple.com/-PerformanceTips.html.

Bandi, R., Vaishavi, V., & Turk, D. (2003). Predicting maintenance performance using object-oriented design complexity metrics. *IEEE TSE*, *29*(1), 77–87.

Briand, L., Bunse, C., & Daly, J. (2001). A controlled experiment for evaluating quality guidelines on the maintainability of oo-designs. *IEEE TSE*, *27*(6).

Bunse, C. (2014). On the impact of user feedback on energy consumption. In *28th International Conference on Informatics for Environmental Protection*, Oldenburg, Germany.

Bunse, C., & Stiemer, S. (2013). On the energy consumption of design patterns. *Softwaretechnik-Trends*, *33*(2).

Feeney, L. (2001). An energy consumption model for performance analysis of routing protocols for mobile ad hoc networks. *Mobile Networks and Applications*, *6*(3).

Google: Performance tips. (2014). http://developer.android.com/perf-tips.html.

Gottschalk, M., Josefiok, M., Jelschen, J., & Winter, A. (2012). Removing energy code smells with reengineering services. In *42. GI Jahrestagung*, Braunschweig, Germany.

Hnatkowska, B., & Jaszczak, A. (2014). Impact of selected java idioms on source code maintainability – empirical study. In *Ninth International Conference on Dependability and Complex Systems*, Brunów, Poland.

Hönig, T., Eibel, C., Schröder-Preikschat, W., Cassens, B., & Kapitza, R. (2013). Proactive energy-aware software design with SEEP. *Softwaretechnik-Trends*, *33*(2).

Höpfner, H., & Bunse, C. (2011). Energy awareness needs a rethinking in software development. In *6th International Conference on Software and Data Technologies*, Seville, Spain.

Höpfner, H., & Schirmer, M. (2012). Software-based energy requirement measurement for smartphones. In *42nd GI Jahrestagung*.

Höpfner, H., Schirmer, M., & Bunse, C. (2012). On measuring smartphones' software energy requirements. In *7th International Conference on Software Paradigm Trends*, Rome.

Johann, T., Dick, M., Naumann, S., & Kern, E. (2012). How to measure energy-efficiency of software: Metrics and measurement results. In *First International Workshop on Green and Sustainable Software*, Zurich, Switzerland.

Khan, M., Hankendi, C., Coskun, A., & Herbordt, M. (2011). Software optimization for performance, energy, and thermal distribution: Initial case studies. In *International Green Computing Conference and Workshops*, Orlando, USA.

Michigan, U. (2014). A power monitor for android based mobile platforms. http://ziyang.eecs.umich.edu/projects/powertutor/index.html.

Microsoft: Performance guidelines. (2014). https://technet.microsoft.com/cc835002.

Nigg, B., MacIntosh, B., & Mester, J. (2000). Biomechanics and biology of movement. *Human Kinetics*.

Noureddine, A., & Rajan, A. (2015). Optimising energy consumption of design patterns. In *International Conference on Software Engineering*, Florence, Italy.

PalominoLabs. Powergremlin. (2014). https://www.openhub.net/p/powergremlin.

Seo, C., Malek, S., & Medvidovic, N. (2008). Component-level energy consumption estimation for distributed java-based software systems. In *Component-Based Software Engineering, 11th International Symposium*, Karlsruhe, Germany.

Sneed, H. (1987). *Software management*. Müller GmbH.

Van Hoorn, A., Waller, J., & Hasselbring, W. (2012). Kieker: A framework for application performance monitoring and dynamic software analysis. In *3rd International Conference on Performance Engineering*.

Wilke, C., Richly, S., Götz, S., & Assmann, U. (2013). Energy profiling as a service. In *43rd GI Jahrestagung 2013*, Koblenz, Germany.

Green ICT Research and Challenges

Roberto Verdecchia, Fabio Ricchiuti, Albert Hankel, Patricia Lago and Giuseppe Procaccianti

Abstract Green ICT is a young and pioneering field. Therefore, as often pointed out in the literature, studies evaluating the main research activities and the general direction of this new and continuously evolving research field are scarce and often incomplete. This study presents a quantitative analysis, through a systematic literature review, of the main activities, trends and issues that can be found in the Green ICT literature. The research reports the analysis of various characteristics of the studies gathered for this review, such as addressed type of effect and year of publication. It also led to the identification of the most recurrent issues of the research and development of Green ICT strategies. Finally, this study proposes a new category of effect (*people awareness*) that, even if often addressed by the field, is not included in current Green ICT frameworks.

Keywords Green ICT · Systematic literature review · Research · Challenges

1 Introduction

In academic research, the environmental impact of ICT is an important topic, spanning across multiple disciplines. ICT is seen as both a relevant contributor to CO2 emissions due to its increasing carbon footprint (Murugesan 2008), and as an enabler for reducing the footprint of other sectors through "smart" systems (e.g. smart build-

R. Verdecchia · F. Ricchiuti · A. Hankel · P. Lago (✉) · G. Procaccianti
Vrije Universiteit Amsterdam, Amsterdam, The Netherlands
e-mail: p.lago@vu.nl

R. Verdecchia
e-mail: r.verdecchia@student.vu.nl

F. Ricchiuti
e-mail: f.ricchiuti@student.vu.nl

A. Hankel
e-mail: a.c.hankel@vu.nl

G. Procaccianti
e-mail: g.procaccianti@vu.nl

V. Wohlgemuth et al. (eds.), *Advances and New Trends in Environmental Informatics*, Progress in IS, DOI 10.1007/978-3-319-44711-7_4

ings, smart grids). We define Green ICT as a combination of activities that minimise the negative impact of ICT on the environment and optimise the positive impact ICT can have. Or, in other words, as any activity that considers the direct, indirect and systemic impact of ICT on the environment (Berkhout and Hertin 2001). We especially want to see how research activities relate to the possible effects ICT can have. For this we use the framework proposed by Hilty (2008). Accordingly, the effects of ICT on the environment can be classified as follows:

- **First order** or *primary* effects: effects of the physical existence of ICT (environmental impacts of the production, use, recycling and disposal of ICT hardware).
- **Second order** or *secondary* effects: indirect environmental effects of ICT due to its power to change processes (such as production or transport processes), resulting in a modification (decrease or increase) of their environmental impacts.
- **Third order** or *tertiary* effects: environmental effects of the medium- or long-term adaptation of behavior (e.g. consumption patterns) or economic structures due to the stable availability of ICT and the services it provides

What follows is a systematic literature review focused on the effects of ICT on the environment. We look into how research activities relate to these effects, what common issues are in the papers selected for our study and—given those analyses—the direction of the Green ICT research field and emerging research gaps.

This paper is structured as follows: after this introduction, Sect. 2 presents our research method and its protocol. Section 3 presents our results and in Sect. 4 we discuss them providing useful insights. Finally, Sect. 5 concludes the paper.

2 Methodology

Our chosen research method for this study is the Systematic Literature Review (SLR) (Keele 2007), a secondary study aimed at collecting and analyzing evidence from the existing state-of-the-art. Our goal is to identify current trends and challenges in the Green ICT field, and possibly gaps for further research. In this section we describe the protocol adopted for our SLR, i.e. how we searched for publications, selected our primary studies and classified those studies.

Research Question

Green ICT is a relatively young research field that has been explored only in the last two decades. Therefore, as mentioned by Jenkin et al. (2011), an overview of the literature of this thematic has not been fully developed. This research is intended to give a comprehensive overview of the research activities that have been carried out during the short timespan of the topic. The general research question of this study has been therefore defined as follows:

How do Green ICT research activities relate to the effects of ICT on the environment?

Literature Search

In order to select the research activities related to the effects of ICT on the environment two distinct groups of keywords were created. At least one of the keywords of each group had to be in the title of a study in order to be included in this literature review. The resulting query, written in pseudocode, follows:

TITLE: (*ICT or "Information technology" or "Information system"*) *AND TITLE*: (*green or sustainable or sustainability*) *AND* (*effect or impact or influence or dematerialization or e-waste or rebound or "societal change"*)

Expanding the research by taking into account also studies that included the keywords in the abstract could have led to interesting outcomes. Nevertheless it resulted infeasible due to the query syntax of some digital libraries. In fact, some of the online libraries, as *Google Scholar*, limit the query to two search options: the title of the paper or a full text search.

Regarding the digital libraries adopted for this research, as customary *Google Scholar, ACM, IEEE, Web of Science* and *Aisel* were selected to carry out the selection of primary studies. Since the research field of Green ICT can be considered as relatively new, we decided not to insert any time constraints on the publication date of the studies. Zotero[1] reference management tool was utilized to keep track of the papers gathered for this research.

Selection of Primary Studies

Executing the research query in different digital libraries resulted in a preliminary set of 1178 studies. An overview of the distribution of these studies among the different repositories is shown in Table 1. The second column of the table reports the total number of studies that were collected in the various digital libraries, while in the third column the papers that were selected for this study are stated.

As detailed in Table 1, the selection of the literature by means of queries executed on different digital libraries led to a large set of studies. In order to further refine the research by identifying only the papers relevant to answer the research question presented in Sect. 2, inclusion and exclusion criteria were generated. The papers that were gathered in the first phase of the literature selection were then evaluated against these criteria. A comprehensive list of the utilized inclusion and exclusion criteria can be found as on-line material.[2] The whole process resulted in the selection of 122 primary studies.

---

[1]https://www.zotero.org/.

[2]https://goo.gl/JmwSfp.

Table 1 Distribution of studies among digital libraries

Digital library	Total	Selected
Google scholar	971	67
IEEE xplore	62	19
Web of science	64	13
AISeL	40	14
ACM	41	9

Classification of Primary Studies

In order to identify the effects addressed in the primary studies, we adopted a manual evaluation process. This process consisted in analyzing the selected resources and identifying which category of effects they primarily address in their research. Subsequently the studies were mapped on the effects they primarily focus on. Papers that only marginally address an effect, e.g. by only mentioning it in a few sentences, were not mapped to such effect. The focus of this process was to map the papers exclusively to the effects actively researched by the study. Therefore effects that were merely reported as introductory text, related works etc. were not taken into account during the mapping.

The definitions of the various categories of effects reported in Fig. 1 were taken from the description of the framework provided by Hilty (2008). It is important to notice that, in various cases, multiple effects were addressed in the same research. Therefore studies could not be mapped exclusively to a single category of effect. For this research these latter type of researches were associated to all the effects they primarily address.

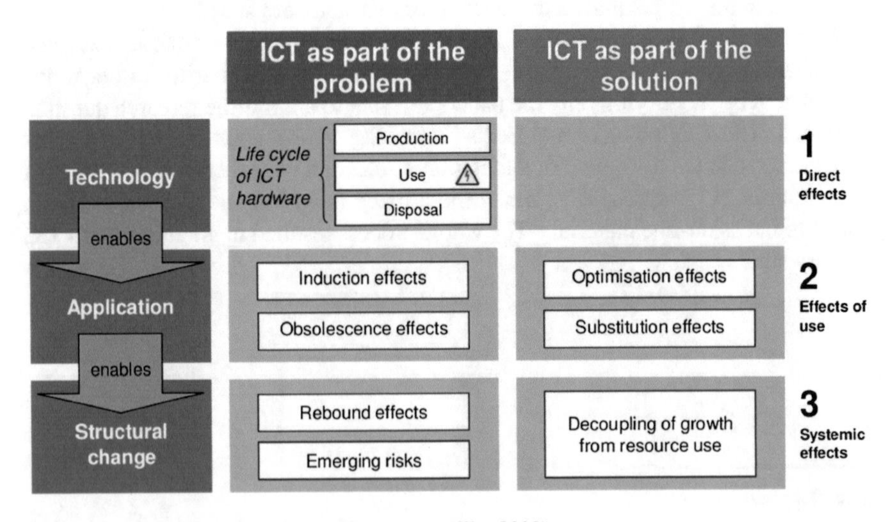

Fig. 1 Conceptual Framework of ICT impacts (Hilty 2008)

The process of the identification, selection and classification of the issues throughout the literature review was done as follows. In a first phase, all the issues regarding the application of Green ICT practices that were found in the studies were annotated. Secondly the listed problems that were strictly domain dependent, i.e. specific to the particular case studies, were discarded. After this process the most recurrent issues were merged into macro-classes that uncovered similar problems. Finally, the issues that resulted to be mapped exclusively to a single study were discarded, as this issues were considered as irrelevant to answer the research question.

3 Results

Studies per Year

Figure 2 shows on the x-axis the years in which the studies were published and on the y-axis the relative number of studies. As depicted in the figure, all the studies selected for this research were published from 2003 onwards. Since Green ICT is a relative young research field, this result is not that surprising. However, there is a clear increase of publications from 2009, which is not that surprising either, especially if we consider the increasing attention to sustainability and energy efficiency in global concerns and international agendas, and the recognition of the potential transformational power of ICT in achieving sustainability objectives.

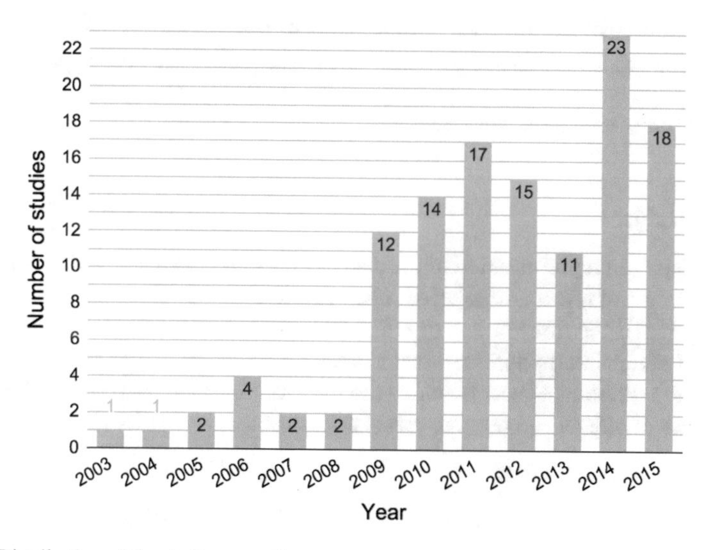

Fig. 2 Distribution of the studies over time

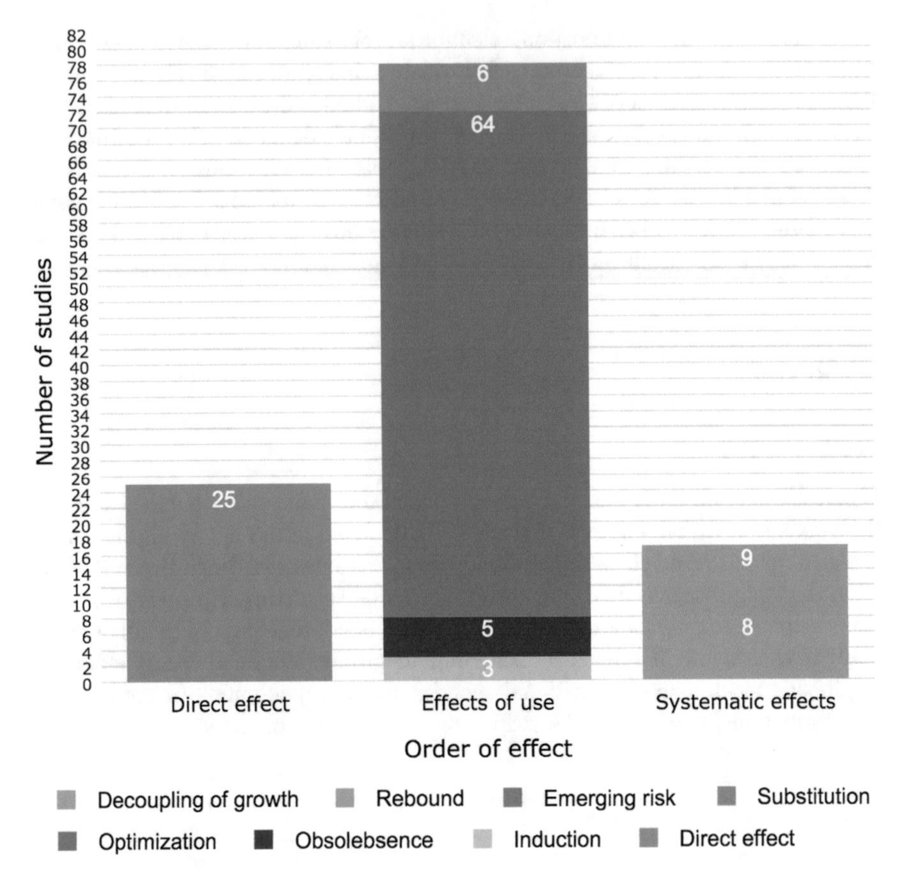

Fig. 3 Number of studies categorized by effect

Studies per Effect

Figure 3 depicts a representation of the distribution of the primary studies among the different orders of effect and relative subcategories.

On the x-axis the three orders of effect are reported, while the y-axis denotes the number of studies that addressed these orders. The colors of the stacked bar chart represent the different effects in which the three orders are divided according to the three orders of effects and relative sub-classes described by Hilty (2008).

Researched Effects per Year

A detailed representation of the years in which the distinct effects of the framework were researched is given by Fig. 4.

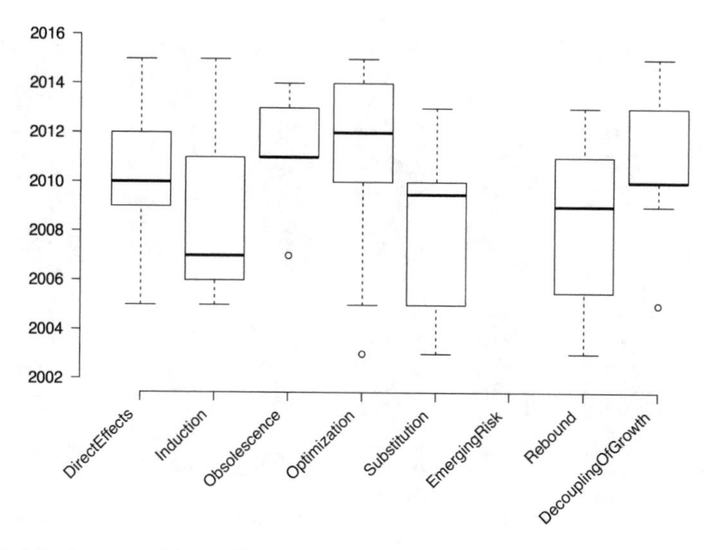

Fig. 4 Publication year of the studies categorized by effect

Recurrent Issues

Seven distinct general classes of issues were identified: *Green ICT is a complex subject in rapid evolution, Lack of research and documentation, Lack of metrics and standards, Lack of incentives and regulations, Lack of people awareness, High cost or unclear return on investment, Need of complex data.*

The recurring issues were mapped to the effects described by Hilty (2008) according to the study in which the problems appeared and the effect described in the research. This was done in order to identify if a specific category of issues could be traced back to a particular category of effects.

Figure 5 depicts the appearance of the different issues among the effects. Effects that did not report any problem classifiable in one of the above mentioned classes of issues were omitted from the figure.

As shown in Fig. 5, the *lack in research and documentation* is the most common problem among the different levels and was found in a total of 22 papers among the different effects. *Lack of people awareness* was the second most common issue and was reported by 13 distinct papers. *High cost or unclear return on investment* was mentioned in several papers, but focused exclusively on two types of effect. On the contrary *lack of incentives and regulations* was reported in the same amount but by studies focusing on different effects. Other problems, such as *lack of metrics/standards* and *Complex subject in rapid evolution* were mentioned only by a minority of the studies.

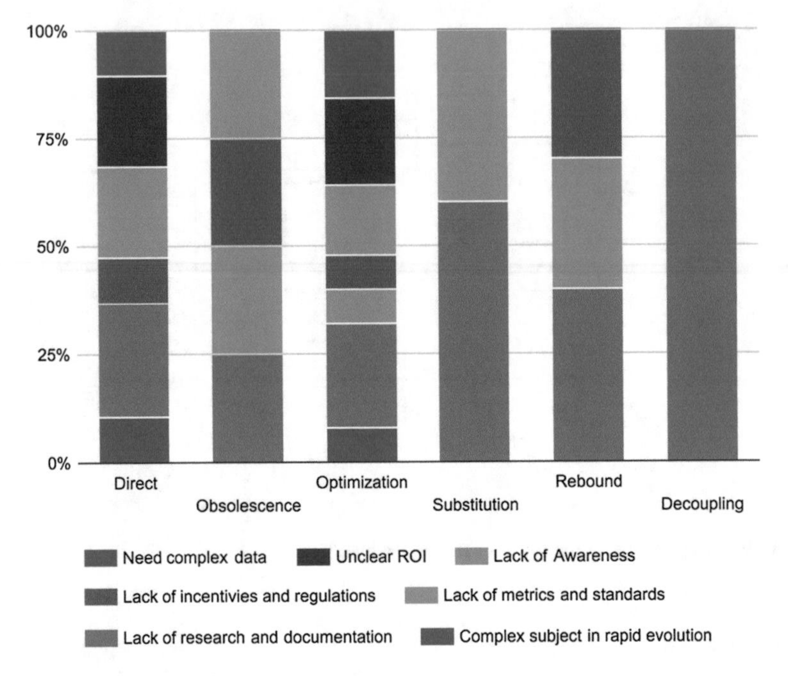

Fig. 5 Distribution of the issues among effect

4 Discussion

The analysis presented in Sect. 3 gives a comprehensive view of the most important characteristics of the data set used as inputs to answer our research question, i.e. "*How do Green ICT research activities relate to the effects of ICT on the environment?*".

Conflicting Trends

Figure 2 shows a general increasing trend in the number of researches carried out per year. The exceptions are for years 2012, 2013 and 2015. Regarding 2015, the lower number of researches found can be attributed to the fact that, when the research was conducted during 2015 and therefore, if the research would be conducted again in the future, this number is likely to be higher. A more extensive evaluation is needed to understand the reasons behind the drop in 2012 and 2013 since we could not find an obvious explanation from our results.

People Awareness Strategy Effects

During the mapping of the ICT environmental effects reported in the primary studies to the ones described by Hilty (2008), we discovered an interesting gap. A class of papers could not be classified as they did not report any of the effects described in the framework. These papers focus on something referred to by Lago and Jansen as *peo-*

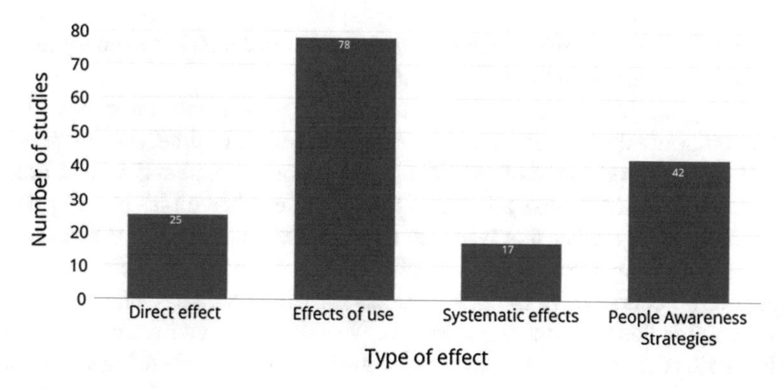

Fig. 6 Number of studies categorized by type of effect

ple awareness strategies (Lago and Jansen 2010), which can be defined as strategies to use ICT to give people insight in their energy consumption and suggest alternative consumption models (paraphrased from Lago and Jansen 2010). This latter type of researches could be discarded from this study by including an additional exclusion criteria. Nevertheless, the high number of papers tackling this topic clearly suggests that this new type of effect addresses a need. Therefore, we have decided to include them.

As shown in Fig. 6, the research carried out in this field of Green ICT is not negligible, constituting about 34.4 % (42 studies) of the total number of primary studies. Notice that, as explained in Sect. 2, single studies may cover multiple types of effects.

The broad definition of the effects as described by Hilty (2008) allows the categorization of a wide range of ICT effects on the environment. Unfortunately, effects resulting from *people awareness strategies* do not fall under any of the specific categories of effects in the framework. Of course, this requires a revision of Hilty's framework.

In particular, we suggest to expand the framework by including *People Awareness effects* as new category. This causes the inclusion of a new sub-class of effects that falls under the second order of effects, constituting an additional class of positive impacts of ICT on the environment.

Research Direction and Research Gaps

As depicted in Fig. 3, *optimization* was the most researched effect in the primary studies, by contributing more than half of the total number of studies. This was followed by studies on direct effects. These two effects combined made up 74 % of the total primary studies. In contrast, none of the studies focused on *emerging risk* effects. Other complex effects more strictly related to the Green ICT theory, such as *Induction* and third order effects, have been only marginally addressed. As noted in Sect. 2, the third group of keywords of the research query contained terms specifically focused on including studies on effects of the third order. The scarcity of papers addressing this order of effects can therefore not be attributed to the query formulation, which suggests an important research gap.

Regarding the years of publication, surprisingly maybe, the studies focusing on the third order of effects were spread over a longer period of time, starting relatively early when compared to the others. While studies on the first and second order of effects appeared later in time, all remain actively researched topics (see Fig. 4).

Considering the studies that take into account *direct* effects, this subject seemed to be considered only partially and was often used to introduce specific optimization techniques. Of course there are many studies focusing on energy efficiency in ICT that were not included in our literature review as they were addressing only a small part in detail, such as processors in servers. In-depth analyses focusing solely on the direct effects of ICT and relating them to the bigger picture, such as e-waste, were scarce in our primary studies and tend to give an overview rather than analyzing them in depth.

Studies on *optimization* effects often addressed very specific topics, such as ad-hoc motorway wireless networks (Feng and Elmirghani 2009) or RAM optimization techniques (Kawahara 2011). As detailed in Sect. 3 a recurrent issue of the Green ICT research field is the lack of knowledge. Nevertheless in this case an overview of actual optimization solutions seems to be missing (there are overviews of possible solutions): a comprehensive overview of the optimization techniques of the various optimization sub-fields, such as data center optimization, smart building approaches etc.

While the development of standards, metrics and tools is crucial, having an overview of the state-of-the-art research activities of the various optimization fields has to be considered paramount. In this direction, in previous work (Lundfall et al. 2015) proposes a tool to estimate the economic impact of Green ICT practices from the state-of-the-art. In the field of software architecture, we are building a catalogue of reusable tactics for energy-aware software solutions (Lewis and Lago 2015; Procaccianti et al. 2014).

Regarding the *systemic* effects of ICT on the environment, as reported in Sect. 3, very little related research has been conducted. Only few of the selected papers focused exclusively on these effects, and these often did not report any theoretically or empirically proven data. Therefore it is difficult to identify specific research gaps of the third order of effects, as it seems that the implication of these *systemic* effects requires much more exploration to be understood.

As reported in Sect. 3, during the literature review process a set of common issues shared among the selected studies was determined. The problems belonging to this set recurred often in the literature and were explicitly reported. Nevertheless none of the selected studies focused primarily on documenting and analyzing these easily standardizable issues. This might be an indicator of a general lack of research and understanding of these problems. A more in-depth study on this topic could potentially lead to a comprehensive documentation of the common issues of Green ICT strategies deployment and research, enabling further analyses of the identified problems.

5 Conclusion

This systematic literature review provides an overview of the trends, problems and research gaps (hence open challenges) that are uncovered from the Green ICT literature. To do so, we have used a framework created by Hilty (2008) that categorizes positive and negative effects of ICT on the environment. As a co-product, we could provide feedback on how the existing literature relates to this framework.

Green ICT is a relatively new research field and the interest in related topics is constantly growing. Section 3 provides evidence of this growing interest. We could also show that the corpus of literature on this topic is heterogeneous, presenting several research gaps. A major part of the literature is mostly focused on *direct effects* and *optimization effects*, leaving a large number of other effects not well explored, if researched at all, as is the case of *emerging risks* effects. Furthermore, we found a set of papers did not fit Hilty's framework and classified these as *people awareness effects* (Lago and Jansen 2010).

While carrying out this SLR, it became clear that much work is still needed to establish a precise and complete framework aimed to evaluate effectively the wide range of possible impacts of ICT on the environment. The next step is to further investigate and refine our key findings so that we can contribute to attaining a more comprehensive framework. In addition an in-depth description of the studies found in the SLR to carry out a more qualitative review could be of value as well.

References

Berkhout, F., & Hertin, J. (2001). Impacts of information and communication technologies on environmental sustainability: Speculations and evidence. *Report to the OECD, Brighton, 21*.

Feng, W., & Elmirghani, J. M. H. (2009). Green ICT: Energy efficiency in a motorway model. In *2009 Third International Conference on Next Generation Mobile Applications, Services and Technologies* (pp. 389–394).

Hilty, L. M. (2008). *Information technology and sustainability*. Norderstedt: Books on Demand GmbH.

Jenkin, T. A., Webster, J., & McShane, L. (2011). An agenda for green information technology and systems research. *Information and Organization*, *21*(1), 17–40.

Kawahara, T. (2011). Challenges toward gigabit-scale spin-transfer torque random access memory and beyond for normally off, green information technology infrastructure. *Journal of Applied Physics*, *109*(7).

Keele, S. (2007). Guidelines for performing systematic literature reviews in software engineering. In *Technical report, Ver. 2.3 EBSE Technical Report*. EBSE.

Lago, P., & Jansen, T. (2010). Creating environmental awareness in service oriented software engineering. In *Service-Oriented Computing*, number 6568 in Lecture Notes in Computer Science (pp. 181–186). Springer.

Lewis, G., & Lago, P. (2015). Architectural tactics for cyber-foraging: Results of a systematic literature review. *Journal of Systems and Software*, *107*, 158–186. Sept.

Lundfall, K., Grosso, P., Lago, P., & Procaccianti, G. (2015). The green practitioner: A Decision-Making tool for green ICT. In *Proceedings of 29th International Conference on Informatics for Environmental Protection (EnviroInfo 2015) and Third International Conference on ICT for Sustainability (ICT4S 2015)*. Atlantis Press.

Murugesan, S. (2008). Harnessing green IT: Principles and practices. *IT Professional*, *10*(1), 24–33. Jan.

Procaccianti, G., Lago, P., Lewis, G. A. (2014). A catalogue of green architectural tactics for the cloud. In *IEEE 8th International Symposium on the Maintenance and Evolution of Service-Oriented Systems and Cloud-Based Environments (MESOCA)* (pp. 29–36).

Some Aspects of Using Universal Design as a Redesign Strategy for Sustainability

Moyen M. Mustaquim and Tobias Nyström

Abstract Sustainability is something that unites humankind and the important 2015 UN Climate Change Conference manifested this and was described by many as our last chance. A shifting towards sustainability through design is a challenge for managers and policymakers of organizations since the existing system or product could be complex and may have difficulty to adopt such a shift. This paper explores how organizations and their designers and developers could benefit from having a predictable process to follow for conducting such a shift, since numerous challenges are associated with costs and revenues. While universal design (UD) is a design philosophy closely associated with the sustainable design, an advanced perspective of UD could be implied as a redesign strategy for existing design and may be used as a radical design and innovation strategy for sustainability. In this paper, some of the aspects of UD as a redesign strategy for sustainability are addressed. Based on the previous theoretical frameworks, a UD approach for redesigning towards sustainability was formulated and discussed.

Keywords Sustainability · Information system design · Universal design · Universal design for redesign

1 Introduction

Organizations face continual challenges and are in fierce competition to keep up to date with constantly changing market needs, due to rapid technological developments. Following or being ahead of any current trends is also seen as a major reason for investment to make a company unique and differentiate them from others through innovation. Sustainability is well known for its power to differentiate products and

M.M. Mustaquim · T. Nyström (✉)
Uppsala University, Uppsala, Sweden
e-mail: tobias.nystrom@im.uu.se
URL: http://orcid.org/0000-0002-4326-2882

M.M. Mustaquim
URL: http://orcid.org/0000-0003-0598-7257

V. Wohlgemuth et al. (eds.), *Advances and New Trends in Environmental Informatics*, Progress in IS, DOI 10.1007/978-3-319-44711-7_5

their design since it add values that can be appreciated by the customers. Moreover, society as a whole naturally feels different social and environmental pressure and responsibilities (Sherwin 2004) and organizations running business are no exceptions. With different strategies of adopting current trends, redesign could be seen as central for organizations' strategy while dealing with sustainability. This is because redesign could be an ideal incremental innovation strategy for organizations that take into consideration complex parameters of the business process, which are important for an organizations commercial and economic success. Although, at one time design could be sufficient for an easy situation, Weick (1993) emphasized the importance of redesign in a turbulent environment by referring to Kilmann et al. (1976). When designing a product with sustainability in mind, redesign strategy is equally important since in a redesign process different factors like customer satisfaction and user experience could be easily accessible, which are important for the success in terms of sustainability. The risks associated with redesign are also lower compared to a completely new design in a radical innovation process. While design could have a lot of potential in relation to sustainability, the present lack of different designer engagement is seen as a missed opportunity for business from the sustainability perspective. Therefore, new radical thinking and solutions are required for the sustainability achievement through design, by demanding more engagement of the designers in the design for sustainability issues (Sherwin 2004). In this respect, UD has been well known to be interrelated with the notion of sustainability and in recent times it has been often strictly argued that sustainable design cannot be achieved without being universally designed, and vice versa. UD as a design strategy has notably shown its potential in design for sustainability issues. For instance, Mustaquim and Nyström (2013) revealed how UD principles as a base could generate sustainable IT system design principles by thinking outside associated accessibility issues. Although UD is being used as a design philosophy, it has failed to show any evidence of being used as a redesign strategy with respect to sustainability. This absence of using UD as a redesign approach initiated the underlying research question of this paper—What could be the aspects of UD as a redesign strategy for sustainability?

A brief background presented in Sect. 2 analyzes the status and relationships of UD with sustainability to explore the scope of UD as a redesign strategy. These analyses lead towards exploration of UD as a redesign process for sustainability, presented in Sect. 3 in which previous theoretical frameworks are used. Some design principles are then derived (Sect. 4) that could be useful for designers to practice redesign for sustainability. Finally, discussions with future work aspects and conclusions are drawn in Sect. 5.

2 Background

2.1 *Sustainability and Universal Design*

Redesigning strategy and UD has not been found to be discussed in the previous literature (searching 05-20-2016 Scopus and Web of Science with keywords: 'UD, universal design, design for all, and inclusive design' combined with 'redesign and re-design') and therefore should be considered as a very new and interesting concept to explore. However, considering UD previously proved to be useful in research conducted by Mustaquim and Nyström (2013, 2014) when UD was viewed as a useful resource for sustainability achievement. Historically, UD as a concept was introduced as a solution to accessibility problems detected in the environment and products and it resulted in seven design principles (Story et al. 1998) that should permeate the design process. The accessibility problems became noticeable when the demography changed and more people got older or had other disabilities.

There is ambiguity surrounding the definition of sustainability and the expression is dependent on the research field, context, and personal cognition (Glavic and Lukman 2007). We recognize that the current processes of design, consumption, and production could be unsustainable and may thus have a negative impact on the multidimensionality attribute of sustainability. It should therefore be of highest priority for all levels of analysis (global, regional, closest society, and individuals) to try to strive towards equilibrium or to reverse human impact. The impact is sometimes measured as our global footprint; the footprint measures the human demand on biological productivity by assessing how much biologically productive sea and land area is necessary to maintain the given consumption of the human population at a specific point in time (Wackernagel and Rees 1996). The calculation of the aggregated global footprint is very complex and yields only a rough estimation, (Turner et al. 2007) but it is clear that our present consumption is far from sustainable. The number of planets demanded by all humans has an increasing trend line, which showed 1.47 planets in 2012 (Global Footprint 2013). In previous research, Nyström and Mustaquim (2014) thus perceived and stressed the importance of not neglecting the complexity, dynamics, and multidimensionality that demands a holistic view on sustainability issues. This is especially important when setting goals and strategy for design or redesign. In the previous research, UD showed the possibility of analyzing sustainability, but was not considered as a redesigning concept since usually UD is practiced in a way to be a quick solution when addressing accessibility problems in the environment or products. The value of UD could be larger if it is applied as something with the possibility of changing the context of a design and thus having a larger impact on the outcome of the sustainability issues. The latter use of UD towards a radical design outcome is thus the focus of this paper.

2.2 *Universal Design as a Redesign Strategy*

UD as a redesign strategy for a focused scope of interest like sustainability is the attention of this paper and therefore will be discussed here. Although the term "design for sustainability" is well known, it often aims to redesign existing products or services towards any sustainability-focused view. This, however, still triggers incremental innovation and is quite useful for many underdeveloped or developing economics because the associated risk would be less for this kind of redesign. But for a long-term commercial success, radical innovation is important and as soon as we bring the knowledge from design science into this process, the needs of user-centered design, etc. come into consideration. Using UD from a design science perspective is a novel idea for sustainability with which the associated risk could still be low since a total new radical shift for designing towards sustainability would not be initiated. Thus, the large-scale radical innovation could be achieved by the conjunction of several smaller incremental innovation outcomes achieved previously through UD, and that is one way to view the redesign using UD.

It is also important to realize that redesign for sustainability could again mean going back to the roots of a specific problem in design and then finding an improved solution, i.e. practicing radical design strategy. In addition, redesign was recommended to be started by practicing radical design (Jarvenpaa and Stoddard 1998). The UD philosophy is not a revolutionary process and therefore it is not possible to come up with a result suddenly reflecting UD in a design phase. Instead it should be incorporated from the beginning of a design. Hence, UD could be seen as a radical design process, which should be ideal for redesign.

Norman and Verganti (2014) addressed the hill-climbing paradigm to describe how incremental and radical innovation can take place and also explained that there is a need to change the frames of different solutions for a radical design movement. Both UD and sustainability are well known for creating several solution frames within their respective domains. What is needed now is to do something with both UD and sustainability that has not been done before and thereby triggering and unfolding the radical design movement. Norman and Verganti (2014) also claimed that most radical innovations were done without the help of design research and therefore are a weakness of any radical design thought. Understanding and incorporating an advanced concept of UD as a design philosophy in the form of redesign strategies could therefore work as a trigger in a radical design movement for sustainability. By UD as a redesign strategy, it would be possible to generate a long-term breakthrough radical design map for bringing changes in the sustainability research. That is to say, while several incremental designs and innovations through human-centered design and design-driven research have contributed different solutions by traditionally looking into UD, it is now time to take a radical look into this concept as a redesign strategy and by doing this some novel interpretation could be initiated within the interest of sustainability and its research domain.

3 Universal Design as a Redesign Process for Sustainability

Previous discussions clarified that in discussing the potential of UD as a redesign strategy for sustainability it would be justifiable to discourse how UD could be used as a design philosophy for radical design and innovation for sustainability. In this section, the arguments for this will be presented and discussed. Tidd and Bessant (2013) illustrated in their framework, based on the Henderson-Clark model, (Henderson and Clark 1990) the dimensions of innovation when they showed how different ranges of choices could occur at diverse levels(component level and system level) and the degree of innovation (incremental and radical). Utterback and Abernathy (1975) presented in their dynamic model an innovation life cycle that described the innovation process from the appearance to the disappearance, compared to Tidd and Bessant who presented a static model of innovation theory. In this section, UD as a radical design and innovation strategy for sustainability will be discussed within the context of this framework and life cycle. Figure 1 shows the different dimensions of innovation (adapted from Tidd and Bessant 2013) discussed in Sect. 3.1, while Fig. 2 shows the revised life cycle of innovation (adapted from Utterback and Abernathy 1975) discussed in Sect. 3.2.

3.1 *Dimensions of Universal Design as a Radical Innovation Strategy*

Figure 1 illustrates how UD could be used as a component to system level. The kinds of changes in terms of innovation for sustainability or sustainable actions that an organization could experience are also highlighted. The figure could be seen as a matrix displaying six dimensions of UD, dependent on the radicalness versus incremental degrees of innovation and from the component to system abstraction level.

An existing design of a product or service could be improved using UD as a design philosophy with which the new design could be more sustainable than the previous one. In the component level of a system this could be seen as an incremental innova-

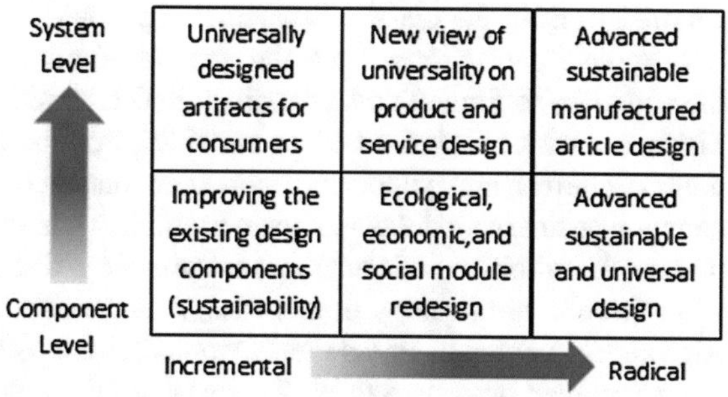

Fig. 1 Dimensions of UD as innovation strategy of sustainability (adapted from Tidd and Bessant 2013)

	Incremental		Radical
System Level	Universally designed artifacts for consumers	New view of universality on product and service design	Advanced sustainable manufactured article design
Component Level	Improving the existing design components (sustainability)	Ecological, economic,and social module redesign	Advanced sustainable and universal design

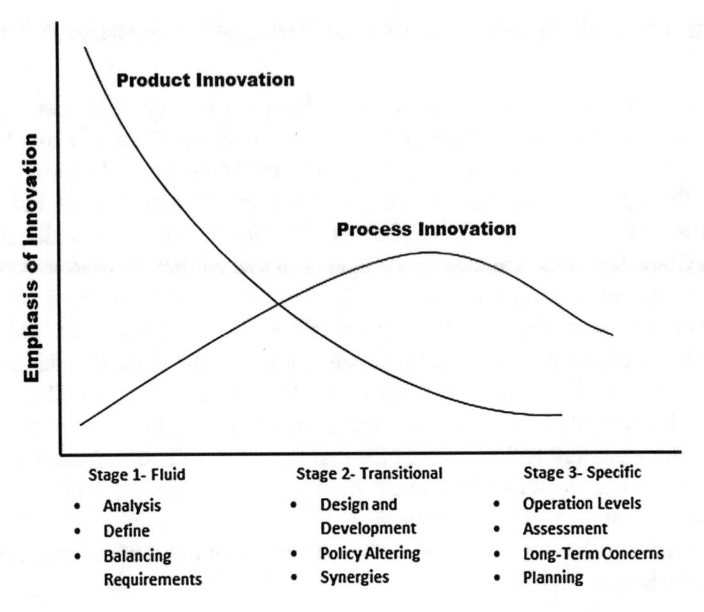

Fig. 2 Innovation life cycle for UD as a redesign strategy for sustainability (adapted from Utterback and Abernathy 1975)

tion since the output action after using UD would actually be improving (sustaining) what an organization was previously doing. On the other hand, at the system level this incremental innovation could result in sustainable artifact designs for the consumers. If UD was used, this artifact could be called 'universally designed' where the boundary of the universality could be predefined within the context of sustainability. For example, during the improvement of a home for the elderly in which UD was already practiced, the design could be more sustainable if UD was used for further addition of different functions instead of a total redesign of the home. This way it would be used as a redesign tool at the component level.

A step from the incremental innovation stage towards radical innovation would be possible when UD could be thought to be used in a more complex way, introducing a new concept to the organizations. At component level, the addition of different parameters from the classical sustainability could be seen as a new way of using UD for the organizations. Thus, the concept of universality could be extended from the boundary of limited understanding of sustainability from the system-level view. The end product design could be benefited from an attempt to build an advanced notion of universality supporting a broader concept of sustainability. For example, a complex business model could be an example when the addition of a component for making the whole models outcome sustainable is a challenging task. However, with UD as a redesign strategy in this case, the organizations could obtain new knowledge about how the business model would behave. Adding or removing elements at the component or system levels for sustainability could therefore be easier, as new knowledge about their business model could be realized by the organization in this example case.

Finally, radical innovation could be achieved when organizations would consider an advanced concept of UD in their design at a component level. Of course knowledge of these concepts should be built by doing things in an improved way that the organizations are aware of, that is to say, by successive performances of incremental innovation. Since radical innovation aims for unique knowledge, the component level could be represented by the new way of looking at sustainability. Similarly, at the system level the sustainable artifacts designed by organizations for the end users could be seen as the radical innovation. For example, a complex integrated business system could be thought of as an example in which the presentation of a complete new knowledge of the world could be attempted by the organization. In this case, the different component levels of the complex system could be redesigned by using UD as a redesign strategy. By doing this, a resulting system could reflect a radical innovation result, while achieving and maintaining the goals of sustainability too.

It is important to realize here that the matrix explained above should not be confused with the use of a design philosophy, like UD as a new addition to different ranges of services. Instead, the existing component or system levels of the system could be augmented by using UD as a redesign concept, thereby triggering radical innovation towards the objective of sustainability.

3.2 *Innovation Life Cycle for Universal Design as Redesign Strategy*

Innovation often occurs from unknown circumstances and it is thus important to realize different innovation opportunities over time. Figure 2 was adapted from Utterback and Abernathy (1975) showing three phases of the innovation cycle with the corresponding stages associated with them, while UD could be considered as a redesign strategy and is described in this section.

The fluid state reflects the initial uncertainty stage in which UD could be seen as a redesign strategy to realize the new configuration and required features of the system to be designed. In this phase, it is also important to understand how any technical knowledge will be used to design a configuration to be delivered in the form of an artifact to users. In the context of UD for sustainability, analyzing the existing product, design, and users and thereby defining the required configuration by keeping the requirement balanced within the technological ability dimensions are some of the stages that could be associated in the fluid phase. Realizing the proper configuration is tricky in this case and therefore the fluid phase would require a trial and error strategy for finding the right setup for an organization. A sustainable product innovation using UD would be highly benefited from the fluid phase of the innovation life cycle. An example of this could be realizing the setting up of the proper sustainable actions needed to be designed by an organization using UD. Thus, understanding the correct parameter of sustainability that needs to be focused on or achieved using redesign should be accomplished in the fluid phase.

Once the fluid phase is initiated properly, the next step will be a transitional phase called 'the dominant design' phase in which design and development, alternation of policy, and handling different synergies happen. The rules for defining sustainability for a process or product design using UD should be considered in this phase, and therefore it is very important for an organization since escaping from this boundary defined in this phase would be very difficult once the process in this phase had been started. The dominant design phase could be ideal for focusing process innovation for sustainability using UD. An example could be the setting up and designing of different selected sustainability actions through UD and thus focusing the specific scope of an organization.

Finally, the specific state of the innovation cycle would focus on fine-tuning the end artifacts. Operation-level analysis, assessment of the designed product or service, long-term benefits or problems, and thereby planning for new product design are some of the associated stages here. Both product and process could be considered here to be designed. Using UD could be beneficial as a redesign strategy for sustainability in this phase, since it would reduce cost, which is one of the primary focuses of this phase. Cost reduction of the designed product or service and analyzing for further redesign could be examples of this specific phase, e.g. products could be efficiently built by using more sustainable products and using less raw materials and other resources. Here the complexity of sustainability is exposed, namely to balance between cost and sustainability and finding a good solution (Pareto optimality) since multi-objective optimization has no single optimal solution (Hwang and Masud 1979), and thus the decision-maker picks the trade-offs and chooses one solution (Coello 1999). Thus redesigning using UD for a sustainable outcome of a product or process could be seen as a scope of narrowing down possibilities in a successive way, whereby different possibilities and scopes of innovation should be emerged for an improved sustainable outcome through design.

4 Design Principles

The different UD dimensions and associated innovation phases presented in Sects. 3.1 and 3.2 were used as a base to structure six design principles that are presented in this section. The addressed design principles could be used to redesign an existing product or services for sustainability by triggering radical innovation and design in the long run. These design principles and their corresponding actions with the three phases of the innovation life cycle were shown in Table 1 in the form of a characteristic matrix described below. For each of the three stages (states) two design principles and their properties in bold text are allocated and are exclusively dedicated for each stage. None of the bold text are design principle properties belonging to the respective stage.

Table 1 Characteristic matrix for design principles and stages of innovation life cycle

Design principles properties	Fluid state	Transitional state	Specific state
Component-level definition	**Requirement analysis**	Redesign module	Module cost reduction
Systems level balance	**Balance user needs**	Artifact design	System productivity
Redesign in component level	Define sustainability	**Custom design**	Cost analysis
Transitional redesign of artifact	Product assessment	**Idea maturation**	Alter business strategy
Specific component level	Redefine requirements	Define boundary	**Cost reduction**
Low-cost sustainable artifact	Reanalysis of needs	Alter policy	**Advance productivity**

Principle One: Analyze and define sustainability problems with component level. Sustainability achievement is often a complex problem and analyzing the requirement from the root level is therefore important. Breaking a system into smaller components might bring novel solutions by new thinking patterns. Realizing and defining the key sustainability problem that an organization would like to see through redesign should therefore be the first stage of a redesign process for sustainability.

Principle Two: Balance sustainability requirements in system level. Analysis from the component level would move more towards system-level analysis, and a balance between the sustainability requirements for different components of the system is needed to avoid any unachievable needs at the system level. This will manifest the comprehensive sustainability goal and wrongful optimization caused by myopia at the component level will thus be avoided.

Principle Three: Redesign and develop sustainability modules at component level. Component-level redesign based on the boundary of sustainability is the next thing organizations should follow. This is a transitional phase and is often ambitious, and a dominant design strategy would be necessary to facilitate the right innovation. Here different earlier initiated design concepts associated with sustainability could grow with the help of UD in the form of running incremental innovation.

Principle Four: Alter policy and handle synergies for transitional redesign of the artifacts. In the next step, the focus should be on the artifacts, since redesigning different components or modules of a system would eventually lead to the end design. Organizations need to alter their policy and accept or reject different synergies as this comes from the proper understanding of sustainable artifacts by using the contextual meaning of UD. Here the alignment between business strategy and artifact is expressed and might need adjustments if mismatched. This could then give incentives to modify the strategy and gain or strengthen the competitive advantages of the organization.

Principle Five: Operation-level assessment of specific component redesign. Operation-level assessment and analysis of the design from the component level could help to reduce costs for future design and development. Organizations should therefore use redesign strategy at a specific level of the design for a specific component or set of components of the system. Vision for future design and requirements for the new or reanalysis can be realized here and thus the perceived and real meaning of sustainability could be realized in an improved way for the organizations by the use of UD.

Principle Six: Specific advanced low-cost, sustainable artifact design. More mature incremental innovation trying to take a radical shift in design should consider complex concepts like cost or improved productivity in the specific phase. Component-level analysis, although specific, with focused parameters of sustainability would then lead towards the artifact level. How to keep the artifacts' cost lower for their users and still provide sustainable results through the use of UD will therefore be the ultimate challenge.

5 Conclusions

It is often more cost effective to redesign something existing compared to initiate and construct something new. Also the associated risks and uncertainty is higher, and calculating return on investment is more difficult. The novelty in this research is reached by the use of UD as a redesign process towards a set sustainability goal, i.e. combining UD and sustainability. In their research, Norman and Verganti (2014) concluded that it is very much possible to reach a meaningful radical innovation result through design-driven research, and for doing this, research directions should be changed towards new meaningful interpretation for users. Following this, we state that the traditional way of interpreting UD seems to be in a loophole and by looking into UD as a redesign strategy for promoting sustainability through design is a novel concept and can produce meaningful results for the end users by triggering radical innovation, which was argued in this paper. Furthermore, the addressed aspects of UD in this paper could be interesting and important for business when identifying critical targets and accomplishing them with limited resources. Organizations can improve different associated items in their business model with the help of UD as a redesign tool for sustainability.

Improved understanding of sustainability could be achieved by not looking at only the environmental aspects of sustainability. Geels (2010) therefore identified from Stirling (2007) that sustainability is "a deliberate social learning process" and not merely a "technocratic challenge." Now this deliberate process cannot be designed at once, but searching outside technology in radical design and redesign could be supported by the idea of UD. The concept of sustainable human-computer interaction (HCI) can therefore extend beyond the persuasive and use traditional HCI design concepts to contribute to sustainability.

One weakness that should be further studied is the problem to implement a strategy for radical innovation, since most of them are unplanned. Many regard a strategy that includes radical change as futile, but the use of sustainability could bring novel ideas and thus enable radical innovations in the form of novel business models. Another problem is measuring the radicalness of an innovation, and at what stage the incremental innovation becomes radical. Future research based on findings from this paper could be in the form of investigating the factors found in the three stages of the innovation life cycle for UD and distinguishing if they are equally important for success in the pursuit of a redesigning strategy towards sustainability.

In this paper, we explored the possibility to using UD as a redesigning instrument for sustainability for the task of redesigning exiting artifacts, socio-technical systems, products, and services. Sustainability is presently a major business driver, and organizations are trying to keep it up. Proper tools for achieving sustainability are still in the form of an expanding growth, and technological development enables social, environmental, and economic improvement. A change is needed since limitless economic growth is not possible, as our planets capacity is limited following our present global economic footprint status. To compete in this challenging market redesign towards sustainability can be a critical movement that the organizations should embrace for adding value to their brands for achieving customer satisfaction. If an organization does not embrace sustainability, there is a risk of losing competitive advantages to competitors and missing the opportunity of increasing present human capital value and gain ahead of others. UD and its expanded approach as a redesign philosophy as presented in this paper could improve the chance of reducing the risk associated with redesigning and thereby may reduce investment costs with improved user experience. Further analysis with empirical data of redesign for sustainability would therefore give new insights about the presented process and design principles in this paper.

References

Coello, C. A. C. (1999). A comprehensive survey of evolutionary-based multiobjective optimization techniques. *Knowledge and Information Systems*, *1*(3), 269–308.

Geels, F. W. (2010). Ontologies, socio-technical transitions (to sustainability), and the multi-level perspective. *Research Policy*, *39*(4), 495–510.

Glavic, P., & Lukman, R. (2007). Review of sustainability terms and their definitions. *Journal of Cleaner Production*, *15*(18), 1875–1885.

Global Footprint Network. (2013). *The national footprint accounts* (2012th ed.). Oakland: Global Footprint Network.

Henderson, R. M., & Clark, K. B. (1990). Architectural innovation: the reconfiguration of existing product technologies and the failure of established firms. *Administrative Science Quarterly*, *35*(1), 9–30.

Hwang, C.-L., & Masud, A. S. M. (1979). Multiple objective decision making-methods and applications: a state-of-the-art survey (Vol. 164), Lecture Notes in Economics and Mathematical Systems. Heidelberg: Springer.

Jarvenpaa, S. L., & Stoddard, D. B. (1998). Business process redesign: Radical and evolutionary change. *Journal of Business Research, 41*(1), 15–27.

Kilmann, R. H., Pondy, L. R., & Slevin, D. P. (1976). *The management of organization design: Strategies and implementation*. New York: North-Holland.

Mustaquim, M., & Nyström, T. (2013). Designing sustainable IT system from the perspective of universal design principles. In C. Stephanidis & M. Antona (Eds.), *UAHCI: Design methods, tools, and interaction techniques for eInclusion, LNCS* (Vol. 8009, pp. 77–86). Heidelberg: Springer.

Mustaquim, M., & Nyström, T. (2014). Open sustainability innovation pragmatic standpoint of sustainable HCI. In B. Johansson, B. Andersson, & N. Holmberg (Eds.), *Perspectives in business informatics research, LNBIP* (Vol. 194, pp. 101–112). Cham: Springer International.

Norman, D. A., & Verganti, R. (2014). Incremental and radical innovation: Design research vs. technology and meaning change. *Design Issues, 30*(1), 78–96.

Nyström, T., & Mustaquim, M. (2014). Sustainable information system design and the role of sustainable HCI. In A. Lugmayr (Ed.), *Proceedings of the 18th International Academic MindTrek Conference (MindTrek '14)*. New York: ACM.

Sherwin, C. (2004). Design and sustainability. *The Journal of Sustainable Product Design, 4*(1), 21–31.

Stirling, A. (2007). Deliberate futures: Precaution and progress in social choice of sustainable technology. *Sustainable Development, 15*(5), 286–295.

Story, M.F., Mueller, J.L, Mace, R.L.: The universal design file: Designing for people of all ages and abilities, (revised ed.). NC State University—The Center for Universal Design (1998)

Tidd, J., & Bessant. J. (2013). *Managing innovation: Integrating technological, market and organizational change* (5th ed.). Wiley

Turner, K., Lenzen, M., Wiedmann, T., & Barret, J. (2007). Examining the global environmental impact of consumption activities Part 1: A technical note on combining input-output and ecological footprint analysis. *Ecological Economics, 62*(1), 37–44.

Utterback, J. M., & Abernathy, W. J. (1975). A dynamic model of process and product innovation. *Omega, 3*(6), 639–656.

Wackernagel, M., & Rees, W. (1996). *Our ecological footprint: Reducing human impact on the earth*. Gabriola Island: New Society Publishers.

Weick, K. E. (1993). Organizational redesign as improvisation. In: G.P. Huber & W.H. Glick (Eds.) *Organizational change and redesign: Ideas and insights for improving performance* (pp. 346–379). Oxford: Oxford University Press.

Part II
Disaster Management for Resilience and Public Safety

Development of Web Application for Disaster-Information Collection and Its Demonstration Experiment

Toshihiro Osaragi, Ikki Niwa and Noriaki Hirokawa

Abstract In the event of a devastating earthquake, a large number of streets will be damaged and/or blocked by collapsed buildings, and the use of emergency vehicles is expected to be paralyzed and unavailable. It is, therefore, important to quickly collect and utilize disaster-information for mitigation. We develop a Web application for collecting and sharing disaster-information collected by users in real time. Furthermore, we demonstrate that the system can provide effective information in real time for reducing the damage of disaster, by performing a demonstration experiment and a simulation carried out by assuming a devastating earthquake in densely built-up wooden residential area in Tokyo.

Keywords Disaster information · Web application · Information posting · Information sharing · Real time

1 Introduction

In the aftermath of a devastating earthquake, it is assumed that fires will break out at the same time in different places, and it will be difficult to fight fires since many street-blockages are presumed to be occurred, and rescue and fire-fighting activities

T. Osaragi (✉) · I. Niwa · N. Hirokawa
Department of Architecture and Building Engineering,
Graduate School of Environment and Society, Tokyo Institute of Technology,
Tokyo, Japan
e-mail: osaragi.t.aa@m.titech.ac.jp

I. Niwa
e-mail: niwa.i.ac@m.titech.ac.jp

N. Hirokawa
e-mail: hirokawa.n.aa@m.titech.ac.jp

V. Wohlgemuth et al. (eds.), *Advances and New Trends in Environmental Informatics*, Progress in IS, DOI 10.1007/978-3-319-44711-7_6

are expected to be obstructed. Quickly collecting and utilizing disaster-information might be effective for reducing the damages under such serious condition (Osaragi et al. 2014; Osaragi 2015).

In this paper, we develop a Web application for collecting and sharing disaster-information by users in real time (Niwa et al. 2015). Next, we evaluate the system by performing a demonstration experiment in which local residents collect disaster-information by using the system. Furthermore, we investigate the effects of disaster-information collection for supporting emergency vehicles by performing a simulation of fire-fighting.

2 Previous Research

In order to clarify the characteristics of the Web application to be developed in this study, first, we carry out a review for the relevant systems proposed in previous researches.

Based on recent information devices such as smartphones or tablets, various kinds of systems for collecting and sharing disaster-information have been proposed. Some of them are listed in Table 1. Kubota et al. (2013) developed a system for sharing disaster-information with the Open-source GIS, in which users sent disaster-information and pictures by email, and they used location information added to Exit information of pictures. Also, Stuart et al. (2014) proposed a real-time crisis mapping system of natural disaster by using the data of social media effectively. The disaster-information-sharing-system developed by Hiruta et al. (2012) is unique, since they did not require any servers for sharing the disaster-information through smartphones under the condition of interrupted communication. Iizuka et al. (2011) developed a mapping system of disaster-information, in which the information can be updated in real time in the university campus. Also, Matsuno (2011) proposed an autonomous wireless network system for sharing disaster-information in an individual evacuation area. Matthew et al. (2010) demonstrated an application of a disaster-information-sharing-system by applying it to volunteer-based geographic information and crowdsourcing disaster relief.

It is not always necessary to satisfy all aspects of requirements (A to F) shown in Table 1. In the emergency situation, however, disaster-information should be updated and shared in real time, since the phase of disaster is changing dynamically. Moreover, it is desirable that disaster-information is utilized not only for grasping the degree of damages, but also for forecasting near future for disaster mitigation. Namely, we need a system which considers secondary use of disaster-information.

Table 1 Examples of disaster-information sharing system proposed in the previous research

Researches	A Real time property	B Robustness in disaster	C Capability of information sharing	D Independence from devices	E Easiness of utilization	F Secondary use of data
(1) Kubota et al. (2013)	×	○	×	○	◎	×
(2) Stuart et al. (2014)	◎	○	○	×	○	×
(3) Hiruta et al. (2012)	×	◎	×	○	×	×
(4) Iizuka et al. (2011)	◎	○	×	○	○	×
(5) Matsuno (2011)	○	○	◎	×	×	×
(6) Matthew et al. (2010)	×	○	○	×	◎	×

◎ = sufficiently satisfied, ○ = partly satisfied, × = not considered

3 Real Time Synchronous Web Application

3.1 Overview of System

In this Chapter, we overview the developed system and describe its structure and detailed usage of the Web application.

Figure 1 shows the overview of our proposed system. Users access to the system using their own information-devices through a Web browser, and they post disaster-information (building-collapse, street-blockage, fire-outbreak), and pictures (black arrows in Fig. 1). The information and pictures posted by users are displayed on the map screen of all users in real time. They can understand the locations, kinds and levels of disasters immediately. Additionally, using the disaster-information stored on the database, we can execute a fire-spreading simulation and forecast how the fire will spread within a few hours (red arrows in Fig. 1).

Figure 2 shows an example of fire-spread in densely built-up wooden residential area in Tokyo. Thus, this system is expected to contribute for variety of purposes, such as the assistance for safe evacuation and decision making of firefighters. The fire-spreading simulation is just an example of damage forecast. We are planning to incorporate various functions for more effective secondary use of the disaster-information.

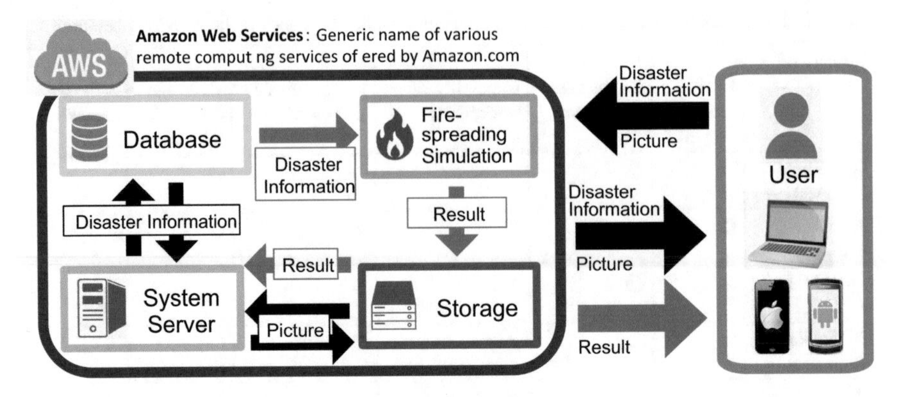

Fig. 1 Overview of web application developed in this research

Fig. 2 Example of fire-spreading simulation on the system

3.2 *Structure of System*

The feature of this system can be summarized as follows;

(1) AWS Cloud Server (EC2 by AWS) is allocated at multiple points and can temporarily change the processing capability.
(2) System Server (Node.js) can efficiently process a large amount of requests with Non-blocking I/O Model.
(3) Communication Standard (WebSocket) has the advantages in the real-time interactive communication with a dedicated protocol, low server load, and low delay.
(4) WebGIS (Google Map) can draw and map with API, and many people can use it with ease.
(5) Database (mongoDB) has capability for high-speed data processing with NoSQL and has high convenience by schema-less.
(6) Storage (S3 by AWS) can preserve the data by automatic backup and expand capacity without limitation substantially.

(7) Fire-spreading simulation (Tomcat7) works fast by multithread response in the environment of Java Servlet. Users can use the system irrespective of the kind of an information-device, since it is implemented as a Web application.

The system is operated on the cloud servers of Amazon Web Services (AWS; Amazon.com Inc. 2016), which can be used in all over the world. Hence, even if a large-scale disaster occurs in Japan, the system never stops. For information synchronization between users and the system in real time, Node.js and WebSocket are employed. In case that disaster-information is updated, it is distributed by push-service from server-side to client-side. Therefore, users do not have to repeatedly send the requests to the server, and synchronize disaster-information can be stably provided without time loss under the condition of weak communication network after the event. Furthermore, by using NoSQL for the database, much more data can be processed faster than the databases such as RDB.

3.3 *Usage of Web Application System*

Figure 3 explains the usage of the system, and Fig. 4 shows the screen-view of user's smartphone or tablet. Disaster-information is posted by specifying a location on the map and selecting a kind of damage. In case of dangerous situation, the details of damage can be added after posting. Selecting a marker on the map, the information window is expanded and we can check the disaster-information which is posted by all users. Posted pictures can also help us understand the concrete situation of disaster. Additionally, the system has a function of exporting the disaster-information as a CSV file format for secondary use of disaster-information.

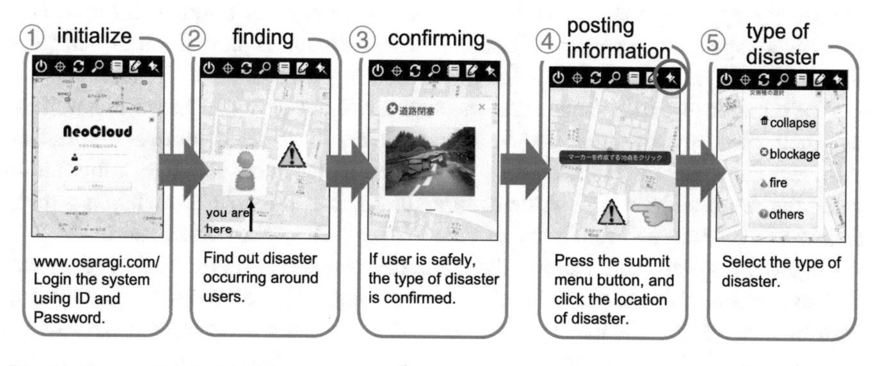

Fig. 3 Usage of web application system

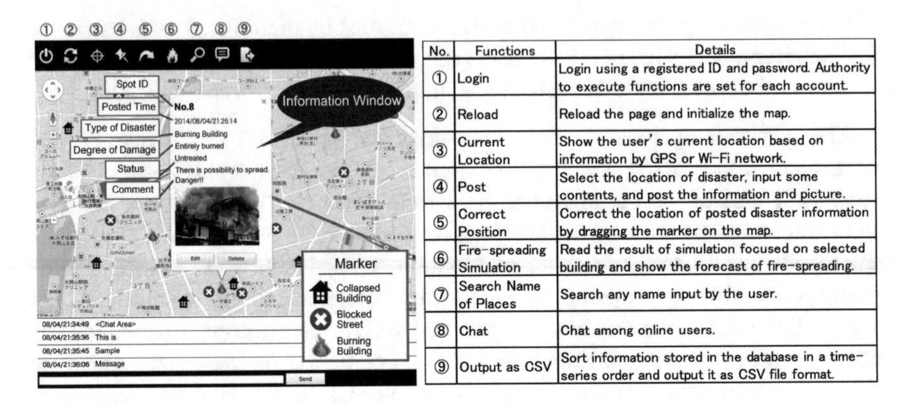

No.	Functions	Details
①	Login	Login using a registered ID and password. Authority to execute functions are set for each account.
②	Reload	Reload the page and initialize the map.
③	Current Location	Show the user' s current location based on information by GPS or Wi-Fi network.
④	Post	Select the location of disaster, input some contents, and post the information and picture.
⑤	Correct Position	Correct the location of posted disaster information by dragging the marker on the map.
⑥	Fire-spreading Simulation	Read the result of simulation focused on selected building and show the forecast of fire-spreading.
⑦	Search Name of Places	Search any name input by the user.
⑧	Chat	Chat among online users.
⑨	Output as CSV	Sort information stored in the database in a time-series order and output it as CSV file format.

Fig. 4 Screen view of web application

4 Performance of the System in Terms of Information Collection

In this Chapter, we evaluate the proposed system by performing a demonstration experiment in which local residents collect disaster-information by using the system.

We evaluate the system by performing a demonstration experiment in which we assume that a large earthquake occurs and local residents collect the disaster-information of virtual disasters (building-collapse, street-blockage, and fire-outbreak) by using the proposed system. For assuming locations of virtual disasters, we adopt some authorized property damage models that were built on the basis of the damage-survey on the major earthquakes that occurred in the past in Japan. The details of models applied here are described in Hirokawa and Osaragi (2016). By applying these models, the property damage of building-collapse and street-blockage are estimated. Demonstration experiment is carried out in Setagaya Ward, Tokyo, while we monitor the movement of participants in the laboratory.

The number of participants is 21, the staff members of Setagaya ward, and they use their own smartphones. As describe in later, the limited number of staff is sufficient for collecting information, since the efficient and effective information gathering is possible by this system. In addition, it is possible to maintain a high degree of accuracy and reliability of data, if they collect the information.

Figure 5 (top-left panel) shows that participants are moving for exploring virtual disasters around their locations, and others are watching their monitors to evaluate performance of the demonstration experiment. Figure 5 (top-right panel) is an enlarged view of the movement of one of the participants. The yellow circle indicates the area in which he/she can identify the virtual disaster in the monitor. Virtual disaster outside the yellow circle are invisible to him/her. We can see that

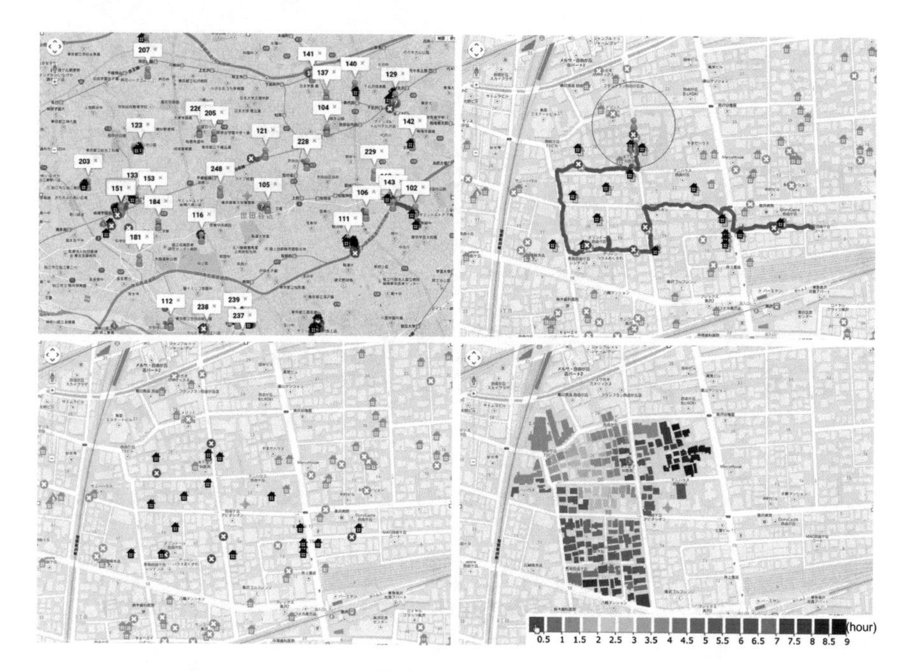

Fig. 5 Monitoring of users' disaster-information collection in the demonstration experiment

he/she is posting disaster-information in the visible range. Figure 5 (bottom-left panel) shows the correct information that he/she has collected within 15 min. Figure 5 (bottom-right panel) shows fire simulation, which is performed on the basis of the fire information he/she posted, to predict the fire spread of fire.

Figure 6 shows the relationships between the elapsed time and the number of posting of disaster-information. Although the number of participants is about 1/40,000 of local residents in this area, the amount of information collected is 1–2 % of the total disasters. We can confirm that disaster-information is effectively collected and posted during the short time period of 15 min.

It is important to confirm whether this system is available or not under the condition of disaster. Hence, we measure the time required for synchronization of posting disaster-information among all users (reciprocating time between a server and a client) in 100 times while virtually increasing the number of people to 3,000 step by step. Assuming the network band limitation after a disaster occurs, two kinds of transmission speed (1 and 15 Mbps) are examined. As the result of this experiment, we confirm that the elapsed time increases to less than 50 ms even if the number of users increases to a certain degree. That is, as long as a cloud server (AWS) used equivalent to the one we use in this experiment, it is possible to support the activities of emergency vehicles without losing real time property.

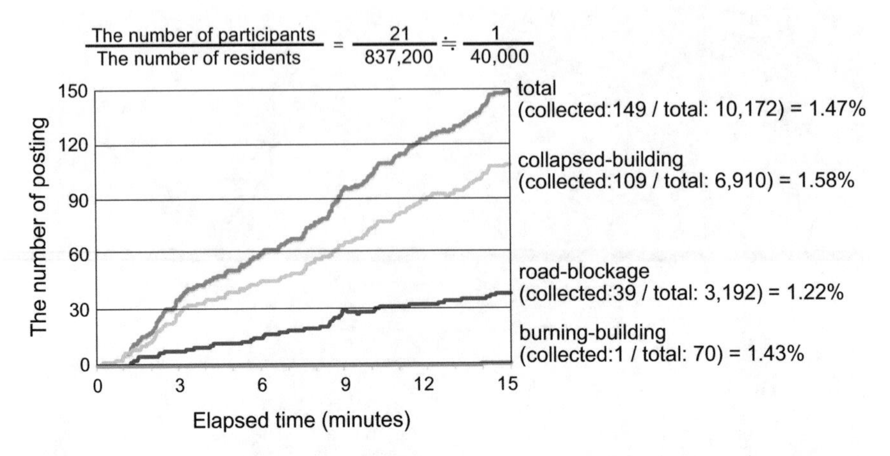

Fig. 6 Performance of users' activities for collecting disaster-information in the demonstration experiment

5 Evaluation in Terms of Supporting Emergency Vehicles

We evaluate the disaster-information collection in terms of activity support of emergency vehicles (Osaragi et al. 2015). Namely, by performing simulation of street-blockage and fire-brigade movement, we demonstrate the effects of collecting street-blockage information for reducing the access time of fire brigades. The assumptions in simulation are shown in Table 2.

Table 2 Assumptions in simulation

Scenario earthquake	North Tokyo Bay Earthquake (M 7.3)
Study area	Setagaya Ward in Tokyo Metropolitan
Number of people collecting information	(1) 0.3 % of local residents (2,698 people) (2) 0.5 % of local residents (4,496 people)
Number of simulations	100 times for both case (1) and (2)
Street-blockage	Consider only blockage caused by rubble of collapsed buildings
Local residents	Initial locations are uniformly distributed. They collect the information of street-blockage while walking to random* directions
Emergency vehicles	(Destination) Location of fire which will result in the most serious damages (Route) The fastest route to the location of fire

*It is difficult to model the movement of residents in the event of a disaster. Here, we assume a random walk for describing people's walk. However, more efficient information collection will be possible, if they check some high priority streets, which are more likely to be blocked by collapsed buildings

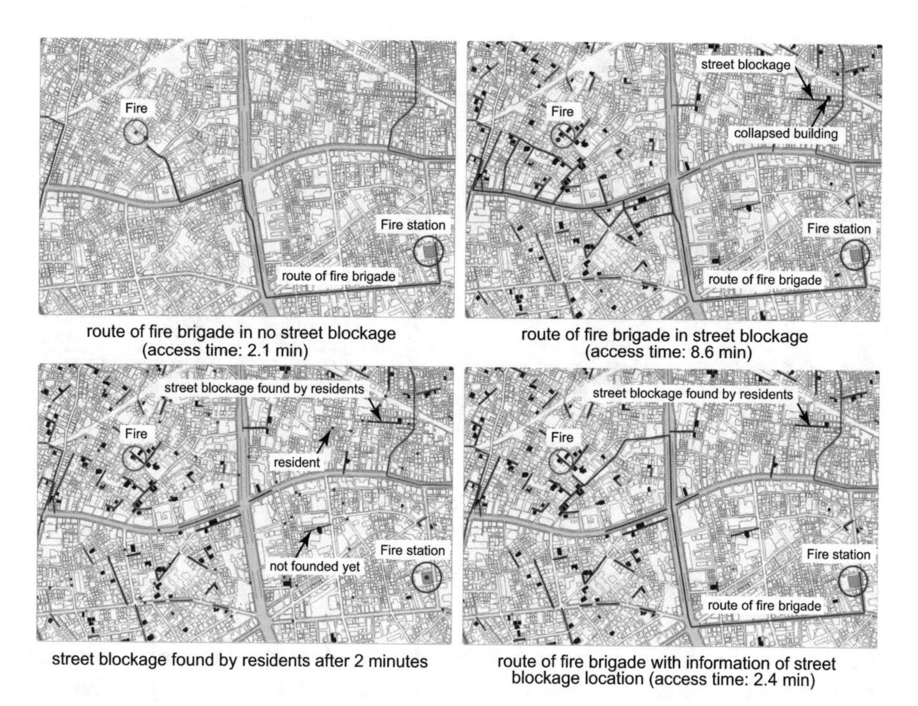

Fig. 7 Simulation of street-blockage-information collection by local residents, and its effects for reducing access time of emergency vehicles

Figure 7 shows an example of simulation. The access time of a fire brigade is 2.1 min in case of no street blockage (Fig. 7, top-left panel). However, it takes 8.6 min in case of street blockage caused by collapsed buildings (Fig. 7, top-right panel). Figure 7 (bottom-left panel) shows the information collection by residents. The number of people who collect street-blockage information is 0.5 % of local residents in the study area for 5 min. The access time of fire brigade is reduced to 2.4 min, if the information of street-blockage is used for choosing the route accessing to the location of fire (Fig. 7, bottom-right panel).

We perform this simulation for 100 times by changing the location of fire and street-blockage estimated by the property damage models. The result is shown in Fig. 8. Even if the number of people collecting information is only 0.3 % of residents of the study area, more than 80 % of information on street-blockage is collected after 10 min (Fig. 8, left panel). In this case, the mean access time of the fire brigade from the fire stations to the locations of fires is significantly shortened to 3.5 min (Fig. 8, right panel). Namely, it is almost equivalent to the time when all the disaster-information is known.

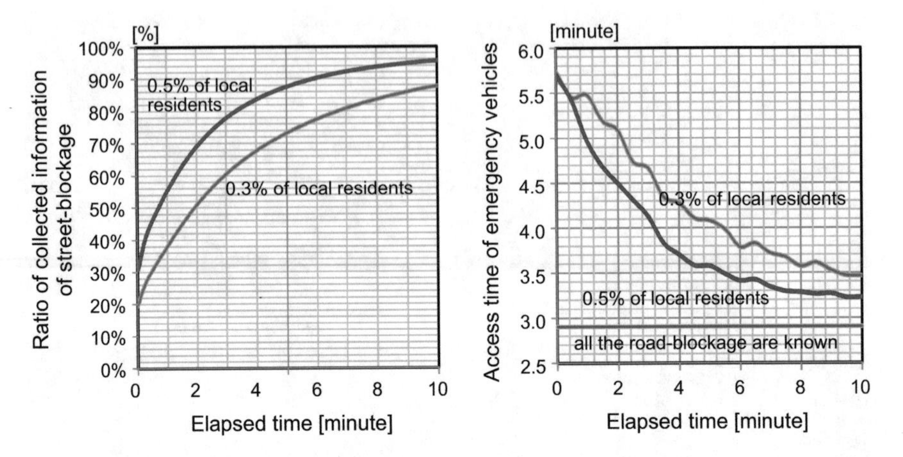

Fig. 8 Effects of reducing access time of emergency vehicles

6 Summary and Conclusions

We developed a Web application system, which enables users to post, share, and utilize the disaster-information in real time through the cloud server. By performing the demonstration experiment by local residents under the assumption of a devastating earthquake, we quantitatively demonstrated that we can collect and post disaster-information using the system. The high performance of disaster-information collection for short time period was shown. Also we demonstrated the effectiveness of secondary use of collected disaster-information, by performing simulation of movement of fire brigade under the condition of street-blockage after a large earthquake. The functions of posting and sharing disaster-information in the system can be easily arranged or customized for other disasters (such as flood, typhoon, etc.) by slightly modifying its User Interface.

However, many systems, which have been developed by investing large costs, are often not used in actual disaster. In the future, we are planning to develop a switch function in normal times and emergency, in order to use the application routinely.

Acknowledgments A portion of this work is supported by Council for Science, Technology and Innovation (CSTI), Cross-ministerial Strategic Innovation Promotion Program (SIP), "Enhancement of societal resiliency against natural disasters" (Funding agency: JST).

References

Amazon.com Inc. Amazon EC2 Pricing. Retrieved March 25, 2016, from http://aws.amazon.com/ec2/pricing/.

Hirokawa, H., & Osaragi, T. (2016). Earthquake disaster simulation system: Integration of models for building collapse, road blockage, and fire spread. *Journal of Disaster Research, Fuji Technology Press Ltd., 11*(2), 175–187.

Hiruta, M., Tsuruoka, Y., & Tada, Y. (2012). A proposal of a disaster-information sharing system. *IEICE Technical Report MoMuC2012–2, Information Processing Society of Japan, 112*(44), 5–8.

Iizuka, K., Suzuki, S., Ishikawa. M., Iizuka, Y., & Yoshida, K. (2011). Real time disaster situation mapping system with functionality of estimating current location. *IPSJ SIG Technical Report, Information Processing Society of Japan, 2011-IS-117*(2), 1–8.

Kubota, S., Matsumura, K., Yano, S., Kitadani, T., Kitagawa, I., & Ichiuji, A. (2013). A proposal of disaster-information sharing system using open source GIS. *Papers and Proceedings of the Geographic Information Systems Association, 22*, F-5–2 (CD-ROM).

Matsuno, H. (2011). Construction of autonomous wireless network system for sharing pre and post disaster information (in Japanese). *IEICE Technical Report SIS2011–43, Information Processing Society of Japan, 111*(342), 19–24.

Matthew, Z., Mark, G., Taylor, S., & Sean, G. (2010). Volunteered geographic information and crowdsourcing disaster relief: A case study of the haitian earthquake. *World Medical and Health Policy, 2*(2), 7–33.

Niwa, I., Osaragi, T., Oki, T., & Hirokawa, H. (2015). Development of real time synchronous web application for posting and utilizing disaster information. In *12th International Conference on Information Systems for Crisis Response and Management (ISCRAM 2015), Proceedings of the ISCRAM 2015 Conference*, Kristiansand, Norway.

Osaragi, T. (2015). Spatiotemporal distribution of automobile users: Estimation method and applications to disaster mitigation planning. In *12th International Conference on Information Systems for Crisis Response and Management (ISCRAM 2015), Proceedings of the ISCRAM 2015 Conference*, Kristiansand, Norway.

Osaragi, T., Morisawa, T., & Oki, T. (2014). Simulation model of evacuation behavior following a large-scale earthquake that takes into account various attributes of residents and transient occupants. In *6th International Conference on Pedestrian and Evacuation Dynamics (PED 2012), Pedestrian and Evacuation Dynamics 2012* (Vol. 1, pp. 469–484). Heidelberg: Springer.

Osaragi, T., Hirokawa, N., & Oki, T. (2015). Information collection of street blockage after a large earthquake for reducing access time of fire fighters (in Japanese). *Journal of Architecture and Planning, Architectural Institute of Japan, 80*(709), 465–473.

Stuart, M., Lee, M., & Stefano, M. (2014). Real-time crisis mapping of natural disasters using social media. *IEEE Intelligent Systems, 29*(2), 9–17.

Social Media Resilience During Infrastructure Breakdowns Using Mobile Ad-Hoc Networks

Christian Reuter, Thomas Ludwig, Marc-André Kaufhold and Julian Hupertz

Abstract Social media and instant messaging services are nowadays considered as important communication infrastructures on which people rely on. However, the exchange of content during breakdowns of the underlying technical infrastructures, which sometimes happens based on environmental occurrences, is challenging. Hence, with this paper, we examine the resilience of social media during breakdowns. We discuss communication options and examine ad-hoc functionality for the exchange of social media data between different actors in such cases. To address this, we have developed a concept, which makes use of mobile ad-hoc networks (MANETs) for the spontaneous exchange of information with smartphones. We implemented our concept as the mobile application Social Offline Map (SOMAP) and evaluated it within two iterations (1.0 and 2.0). Finally, we discuss our contribution within the context of related work and the limitations of our approach.

Keywords Infrastructure · Resilience · MANET · Social media · Environmental informatics

---

C. Reuter (✉) · T. Ludwig · M.-A. Kaufhold · J. Hupertz
Institute for Information Systems, University of Siegen, Siegen, Germany
e-mail: christian.reuter@uni-siegen.de

T. Ludwig
e-mail: thomas.ludwig@uni-siegen.de

M.-A. Kaufhold
e-mail: marc.kaufhold@uni-siegen.de

J. Hupertz
e-mail: julian.hupertz@uni-siegen.de

V. Wohlgemuth et al. (eds.), *Advances and New Trends in Environmental Informatics*, Progress in IS, DOI 10.1007/978-3-319-44711-7_7

1 Introduction

Breakdowns of critical infrastructures—for instance, after an interruption in power supplies based on environmental occurrences—constitute heavy problems. Especially the mental stress within the affected population caused by long-lasting breakdowns can be widely observed (Volgger et al. 2006). Furthermore, the coordination of response actions and the associated communication demand high efforts of the responsible authorities. During a power breakdown, interaction via ICT is only available for a certain period of time. Once the backup power supplies of network providers fail, information cannot be extensively distributed anymore.

Particularly social media have become important communication channels within diverse emergencies (Reuter et al. 2015b): For instance, social media was used during the 2009 Oklahoma Wildfires (Vieweg et al. 2010), 2011 Japanese Earthquake (Wilensky 2014), 2012 Hurricane Sandy (Hughes et al. 2014) or 2013 European floods (Kaufhold and Reuter 2016). These settings have to potential to impair the availability of infrastructure, e.g. during the European floods the local power plant of Magdeburg was threatened by water, or demand other kinds of effective local coordination to fill the gap between the start of an emergency and professional response or to support volunteers in overcoming an emergency. However, when networks fail, ordinary people usually cannot make use of this high amount of information shared in social media anymore. But even during long-lasting power or network breakdowns, mobile devices would still be able to communicate in a decentralized way due to its functionalities, such as Bluetooth or Wi-Fi-Direct in order to support both stability and continuity.

This *design case study* (Wulf et al. 2011)—composed of an empirical pre-study, the development of ICT and its evaluation in two cycles—aims to examine whether social media can remain "social" during breakdowns with the aid of mobile ad-hoc networks (MANETs) to assist direct ways of communication between mobile devices. With the term 'social' we refer to the ability of people to access, post and respond to social media data preferable in real-time, reflecting their use of social media sites in non-impacted times. This is in accordance with the definition of social media as a "group of Internet-based applications that build on the ideological and technological foundations of Web 2.0, and that allow the creation and exchange of user-generated content" (Kaplan and Haenlein 2010).

Based on theoretical concepts, existing applications in this field (Sect. 2) and the evaluation of our offline-map application SOMAP 1.0, which visualizes data from social media (Sect. 3), we present SOMAP 2.0; the enhanced version of SOMAP allows establishing ad-hoc networks by enhancing peer-to-peer (P2P) functions and provides new ways of direct communication (Sect. 4). The integrated network functionality was evaluated by a functional test supplemented by semi-structured interviews. The results of the evaluation are discussed and compared with regard to the current state of the art. The final conclusion comprises limitations and implications for further research in the field of MANETS and infrastructure breakdowns (Sect. 5).

2 Literature Review

When asked about London's power outage in 2003, many people were surprised afterwards that such breakdowns could actually happen (Brayley et al. 2005). According to Lorenz (Lorenz 2010), people have little awareness about power outages until they happen. Thus, ways to distribute relevant information to the people in such cases have to be found. Below, we discuss possibilities for direct communication and present related approaches together with existing applications in that field. Afterwards we outline and discuss the research gap being tackled in this paper.

Communication During Power Outages: The cellular network is a technology that enables people to communicate during power outages, because central stations are usually protected by backup power supplies (Hiete et al. 2010). However, in the course of a power outage this possibility collapses step-by-step or is overloaded after a while (Reuter 2014a). A study on a power outage in the Netherlands in 2007 emphasizes that the direct contact among neighbors is the most common source of information and communication (Helsloot and Beerens 2009). In opposite to the common expectation that people are passive victims in crisis situations (Hunt 2003), citizens often are the first responders on-site (Stallings and Quarantelli 1985) and require certain information (Ludwig et al. 2015). ICT to enable cooperative resilience (Reuter et al. 2016) as the ability to overcome crises of cooperation with the help of adaptability to modified realities by means of cooperation technology.

Depending on the situation, the visualization of information can be essential as Kaufhold and Reuter (Kaufhold and Reuter 2016) outlined with the example of Google Maps in the floods in Germany in 2013. According to Toriumi et al. (2013), social media are used to a greater extent during crisis situations especially via mobile devices. In regions affected by communication failures, information processing via these platforms is temporarily not possible. Even the activities of *digital volunteers* (Starbird and Palen 2011), who process data and information on online social media services, cannot be utilized in such situations. Here, mobile devices can be beneficial due to the availability of the cellular network in power outages. But the challenge of how they can be harnessed after central communication infrastructures collapse still remains. Promising technologies such as Bluetooth or Wi-Fi-Direct enable direct communication among different devices, so that the establishment of ad-hoc networks is conceivable.

Approaches for Ad-Hoc Networking: In their summary "30 years of ad-hoc networking research", Legendre et al. (2011) conclude that "multiple technologies need to be combined to encompass all phases of disaster recovery, while providing differentiated levels of communication services" (p. 7). They elaborated this issue during the initial phase, "delay-tolerant and opportunistic networks have the capacity to provide low-bandwidth data services, while wireless mesh networks have the availability and redundancy to provide limited voice and data services".

Several approaches already deal with information distribution in regions where communication infrastructures fail. Using a *Local Cloud System* (Al-Akkad et al. 2014a), messages can be created and exchanged between mobile devices by creating

Wi-Fi-networks. If a network node provides access to Twitter services, those messages are published and also up-to-date data can be downloaded. The prototype *Twimight* works similarly (Hossmann et al. 2011) and provides a so-called *Disaster Mode,* which gives users the option to publish tweets without requiring Internet connection. Those tweets are uploaded after a successful connection.

The *Help Beacons System* (Al-Akkad et al. 2014b) transfers information directly into the Service Set Identifier (SSID) of a Wi-Fi-network, which then can be found and read by other smartphones. Nishiyama et al. (2014) suggest a concept with permanent Device-to-Device (D2D) communication and present a layer architecture standardizing the use of various devices and radio technologies during communication failures. Furthermore, Reuter (Reuter 2014a) introduces a prototype for displaying crisis-relevant information, which deals with the time period between the cellular network's breakdown and the following overload. Location- and setting-specific information aims to help users getting a better orientation within this timeframe and reducing network load at the same time.

Besides research approaches, applications in the *Google Play Store* support the establishment of ad-hoc networks. Open Garden's *FireChat* enables ad-hoc connections via Bluetooth and Wi-Fi-Direct between various devices, if at least one node has Internet (Shalunov 2013). *Blueeee!* utilizes Bluetooth technology to search for friends online and to chat with them. *The Serval Mesh* is an Android app (Gardner-Stephen 2011) enables ad-hoc connections through Bluetooth and a Wi-Fi ad-hoc mode, which requires rooted devices. Networked people can phone and chat with each other and also up- and download files in a P2P manner.

Research Gap: Our literature study indicates that information from social media is increasingly used for crisis communication (Toriumi et al. 2013). Furthermore, besides the technical possibilities for communicating during power and network breakdowns, direct contact is a highly relevant information source in exceptional situations (Helsloot and Beerens 2009). Kaufhold and Reuter (2016) revealed that location-related data processing with the help of maps can decisively support different actors. A study on Sweden's largest music event outlined the idea for a mobile solution to locate friends without central communication infrastructures (Olofsson et al. 2006).

MANETs can be used in absence of communication infrastructures to establish new and dynamic infrastructures, which offer a certain degree of reliability in the case of multiple connections between different network nodes (Kargl 2003). P2P-systems, which aim to evenly distribute certain tasks in a decentralized manner, can serve as starting point for the design of applications operating in rather inconsistent environments in terms of mobile devices and individuals (Dunkel et al. 2008). Our work aims to combine the areas of (1) ad-hoc networking, (2) provision of social media content, (3) geo-based services respectively map-based representation and (4) direct communication. We are trying to figure out whether and how social media can remain "social" when central communication infrastructures fail. We also tackle technical limitations of existing work by allowing networking and communication over multiple hops (Al-Akkad et al. 2014b; Hossmann et al. 2011) without the need for Internet access (Shalunov 2013) or rooted devices

(Gardner-Stephen 2011). We therefore include further information from our pre-study, serving as a foundation that already addresses certain requirements, to specify an integrated concept.

3 SOMAP 1.0: Social Offline Map Evaluation

SOMAP 1.0, the first version of our mobile Android application, which has already been published (Reuter et al. 2015a), was designed to visualize data from social media on maps. It offers online and offline map functionalities in terms of (a) pro-active loading and storing of potentially needed maps of a respective area as well as (b) the possibility of exchanging information from social media using Bluetooth. Offline maps are downloaded once the application moves to another location, in case an Internet connection is available, and can then be used in areas without network coverage. The application allows adding information from third parties, such as Keyhole Markup Language (KML) layers (Fig. 1). Furthermore, it

Fig. 1 Additional map layer: KML during the European floods

Fig. 2 Integration of social media: information window and cluster marker

provides the option to access Twitter tweets and public Facebook posts via a social media API (Reuter et al. 2015c) and to display them on a map. The user can search for concrete keywords, whereat the search results are filtered by parameters such as location or time (Fig. 2).

The qualitative evaluation of this first version of the application SOMAP aimed to obtain additional information about usability and added value of the application, practical suitability, potentials of enhancement and improvement, and the more specific features of ad-hoc and direct communication. We performed alongside a five-step evaluation guideline to determine the participants' expertise (1) and to introduce a potential scenario of use (2). During the application's general (3) and scenario-specific use test (4), we applied the constructive interaction method (Kahler et al. 2000) and required the users to accomplish both their own and joint tasks; and to maintain a mutual dialogue besides replying to guiding questions.

According to the think-aloud protocol (Nielsen 1993), participants were asked to express frankly their thoughts and their way of proceeding throughout the evaluation. Furthermore, we conducted semi-structured interviews (5) with eight participants of varying age, profession and smartphone experience. The interviews were audio-recorded and transcribed for further analysis according to the categories introduced above.

The evaluation revealed that SOMAP has the potential to support users in emergencies by the combination of an offline map functionality with the opportunity to query and display information from social media. To ensure such support, the application has to be downloaded initially, but it is to be expected that people do not make such provisions before an emergency actually occurs (Lorenz 2010). During the evaluation participants expressed that they would be more likely to download the application if it did not only serve for emergencies, but also had a bearing on everyday life, e.g., for direct communication. Moreover, the handling of the Bluetooth module to exchange social media information was perceived as being complicated.

4 SOMAP 2.0: Mobile Ad-Hoc Networks for Social Media

SOMAP 1.0 facilitates information processing from social media and its geo-based presentation to support during communication infrastructure breakdowns. However, MANET relevant aspects have not been integrated. If users are located in an affected area, new information can only be shared by actors who enter this area from a region with working communication infrastructure. Thus, for direct distribution of information and a more extensive and more user-friendly data exchange, the integrated Bluetooth exchange functionality could be extended by using ad-hoc networking. Direct communication as a further challenge can be derived from certain aspects: On one hand, the subjects asked for the opportunity to create new messages. On the other hand, the literature review shows that direct contact in a region with communication infrastructure breakdowns is a very important source of communication and distribution of information (Helsloot and Beerens 2009). The requirements mentioned above are concretized as derived design challenges, along with further practical requirements, in the following subsections.

R1: **MANET Connection Establishment and Automatized Discovery of Connection Partners**: To enable a direct network, different technologies of radio communication have to be considered: Bluetooth features out with a high degree of platform independence und low energy consumption. It is specified to provide a network range of 50–100 m with up to seven connections to other devices simultaneously. Yet, some devices only provide a range of 1–10 m and the bit rate of 2.1 Mb/s is low compared to Wi-Fi-Direct, which also supports 1:m-connections. It has a high bit rate with 250 Mb/s and a range of 200 m (Wi-Fi Alliance 2014). The connection establishment should be automatized as far as possible and all network participants should get information about entering nodes. During the process of implementation some problems according to Al-Akkad et al. (2014b) occurred when using Wi-Fi-Direct between mobile devices of different manufactures. Thus, merely a Bluetooth service was implemented enabling the operations of accepting, establishing and managing connections, processing data packets and discovering connection partners.

R2: **P2P-Functionality for Social Media Exchange**: Typical P2P-systems as *Gnutella* give notice of themselves with a ping-message to already known network participants when they enter a P2P-network and wait for a pong-message as acknowledgment (Dunkel et al. 2008). This approach is already mapped by entering an ad-hoc network, because the entry is broadcasted to all network participants. The information of social media platforms should directly indicate the clients so that every user has an overview of the existing tweets and posts in an existing ad-hoc network. Additionally, they can be downloaded from the offering clients (Fig. 3).

R3: **Direct Communication and Data Exchange**: The functionality for direct social communication (chat) should incorporate the existing network participants and support interaction besides the exchange of given information. In addition to that, broadcast messages to all clients should be possible. Data exchange formats, respectively a specific protocol, are necessary for the communication over an ad-hoc network. The services communicate with network packets that contain the application data, but do not provide the logical interpretation.

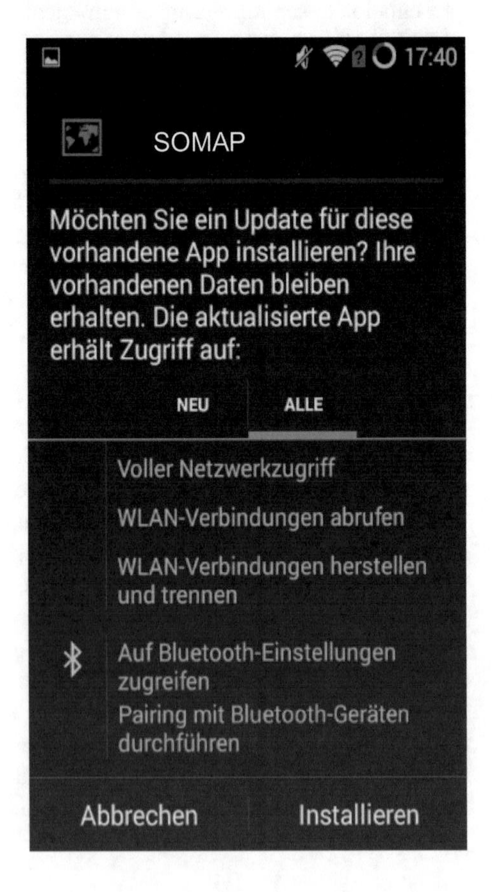

Fig. 3 Download network messages

R4: **Graphical User Interface and Access Control**: Download requests, chat messages and the access control should be illustrated by a graphical user interface (GUI). To split the user interface with regards to content, an Android activity can include fragments that represent different parts of the user interface. Furthermore, break-glass access control was integrated. This kind of access control offers every user access to all information resources. If a resource is requested by a user, it is documented on the providing client revealing who downloaded which information (Stevens and Wulf 2009).

Evaluation Methodology: This iteration (SOMAP 2.0) seeks to evaluate how 'social capabilities', represented with chat and data exchange functions, can be maintained from a rather technical point of view. The philosophy behind the evaluation process was derived from the notion of "situated evaluation" (Twidale et al. 1994) in which qualitative methods are used in order to draw conclusions about real-world use of a technology using domain experts. Furthermore, the evaluation aimed on testing the robustness of the networking functionalities, identifying certain problems and exploring resulting needs for further implementations. The following paragraph describes the setup and conduction of the performance test, concluded with a short interview containing four open-ended questions with eight participants (P1-P8).

The Android application SOMAP 2.0 was deployed on several test devices and was supplied with some social media information. The information (search term *Ukraine*) was gathered within the radius of different and distant towns (*Berlin*, *Hamburg*, *Cologne* and *Munich*), ensuring disjoint datasets on the different devices. After a short introduction of the application (1), the participants performed a connection quality test (2) and were instructed to download the prepared information via P2P (3). Moreover, the participants were asked to send some chat messages over the ad-hoc network (4). Those functional tests were conducted twice, each time with four participants. After this more technical evaluation approach, the evaluation was concluded with four interview questions (5) concerning (a) how participants assess the application's P2P download and chatting functionalities, (b) if they can imagine further applications utilizing ad-hoc networking, (c) their estimation on the limits of the network service and potential enhancements, and (d) whether they want to express other thematic points of interest. The interviews were audio-recorded and transcribed for further analysis in order to enrich the performance test's experiences in a systematic manner. We employed "open" coding (Strauss and Corbin 1998), i.e., gathering data into approximate categories to reflect the issues raised by respondents based on repeated readings of the data and its organization into "similar" statements. These categories reflect the structure of the following results section.

Evaluation Results: The results of the evaluation show that the combination of the four aspects (1) ad-hoc networking, (2) provision of social media information, (3) geo-based services respectively map-based representation and (4) direct communication perform suitable from a technical perspective. The application provides opportunities to download and exchange information from social media via P2P and to discuss certain social media information using the chat component, so that the media is not just exchanged but retains "social" without the need of central communication infrastructure: "I really liked the chat function. Particularly, it is really useful if the Internet breaks down, as you have a connection over such a network, that you can exchange yourself with others over such a chat. The download of messages respectively the tweets, that is certainly nice somehow; the question is what happens, if it is a larger area, if you suddenly find 20000 tweets about one topic" (P2, 30:40).

The participant indicates the requirement to test the application's scalability in greater cases or scenarios. A downside of the current chat implementation is the missing awareness about successful communication: "The problem is if I write a message and wait for a response, I don't know whether the person read the message or has no interest in responding or is involved in a conversation. I cannot check that." (P4, 28:49)

Moreover, participants questioned the suitability of Bluetooth regarding network reach and persistence: "Especially considering emergencies, if we use another technology than Bluetooth that attains a higher network reach, we could deliver the application to a special group of selected users, for instance, emergency services that network among themselves. As an auxiliary network for emergencies services or so" (P5, 27:55).

Therefore, it seems natural to replace Bluetooth with a more modern radio technology, for instance, to offer a reliable auxiliary infrastructure for emergency services in the case of a communication infrastructure breakdown. A higher bandwidth could facilitate the exchange of larger, possibly important data files: "It would be difficult to exchange map data or photos via Bluetooth" (P8, 23:12).

This application aims on providing an auxiliary network infrastructure. Yet, aspects like the P2P-download of social media messages is few mentioned, but the distribution of pictures or additional content is desired. On one hand, this could be the result of the evaluation's focus on the interplay of components; on the other hand, it indicates that further opportunities of improvement have to be considered. To achieve a more concrete assessment of the implemented functionalities and to detect potentials of improvement in different situations of communication infrastructure breakdowns, further appropriate and more comprehensive field research is required.

P8 went about 10 m from the other subjects and left at the same time the space in which the evaluation was conducted. Despite the distance P8 could still successfully download messages. However, the results reveal various connection problems that can be localized in particular on the Samsung Galaxy S devices. When debugging the application, we determined that the application in situations of congestion leads to a CPU overload. Regardless of the crashes during both dates, the download function is largely successfully tested and also most of the chat messages have been sent.

5 Discussion and Conclusion

The aim of our study was to examine whether central characteristics of social media, such as communication and information exchange between different actors, can still be preserved when communication infrastructure fails. Basically we were interested in researching and based on it improving social media resilience during infrastructure breakdowns. These breakdowns might be caused by environmental reasons.

Our literature study has revealed that location-based data and direct communication can play an important role within this context. MANETs are technological concepts to create decentralized and dynamic communication infrastructures. We used it for establishing options of directly downloading the content of Twitter tweets and Facebook posts between individual mobile devices and exchange information from social media.

The evaluation of the extended application SOMAP 2.0 was limited to a small field of participants, but revealed that social media, including the aspects of local reference and direct communication, can remain "social" during communication infrastructure breakdowns, when the above mentioned aspects are considered and state of the art technologies are combined. From a theoretical point of view, the application contributes to resilience by providing an additional layer of communication and collaboration infrastructure for emergent communities during natural or political disasters. Emergent collaboration infrastructures (Reuter 2014b), as a concept to support ad-hoc needs of collaboration, might gain value out of this concept. However, participants questioned the application's performance and its usability in large-scaled scenarios. The evaluation again emphasizes the drawbacks of Bluetooth's low range and low bitrate. Furthermore, because Apple only implements a limited version of the Bluetooth protocol, a native network among Android and iOS devices is not possible. This leads to the requirement of a more suitable network technology like Wi-Fi-Direct to enable a more reliable auxiliary network infrastructure and the exchange of media files. However, technological diversity amongst devices still constitutes a problem here (Al-Akkad et al. 2014b), such as some Wi-Fi-Direct devices only support 1:1-connections.

In comparison to existing approaches, the main advancement of SOMAP 2.0 is the combination of specific aspects, such as the distribution and map-based presentation of geospatial information from social media and the direct communication in combination with MANETs. In particular, compared to Al-Akkad et al. (2014b) or Hossmann et al. (2011), our approach differs in a way that the networking and communication over multiple hops become possible. Our initial approach to consider several radio technologies for networking is in accordance with ideas of Nishiyama et al. (2014). However, the various incompatibilities between individual smartphone manufacturers hampered us implementing Wi-Fi-Direct. In contrast to the application *Open Garden* (Shalunov 2013), within SOMAP 2.0 none of the nodes needs Internet access. Moreover, in contrast to the application *The Serval*

Mesh (Gardner-Stephen 2011), no root access for the use of necessary functions is needed.

Nevertheless, the individual functions should be investigated with regard to usability aspects for improving the links between the map interface and ad-hoc networking interface. Also, the maturity of the approach has to be further examined and improved to provide a robust communication layer during disasters. Thus, a comprehensive functional test is required to gather quantitative and enrich qualitative evaluation results. These are necessary steps to further examine the "social" capabilities of MANET-based applications, like SOMAP 2.0, in large-scale power breakdown scenarios, which—when mature—could (and will) also be integrated in online social networks itself, like Facebook, which are already working in that area (Wiseman et al. 2015).

Acknowledgments The research project 'EmerGent' was funded by a grant of the European Union (FP7 No. 608352). We would like to thank all participants of our empirical study.

References

Al-Akkad, A., Raffelsberger, C., Boden, A., Ramirez, L., Zimmermann, A., & Augustin, S. (2014a). Tweeting "When Online is Off"? opportunistically creating mobile Ad-hoc networks in response to disrupted infrastructure. In S. R. Hiltz, M. S. Pfaff, L. Plotnick, & P. C. Shih (Eds.), *Proceedings of the Information Systems for Crisis Response and Management (ISCRAM)* (pp. 657–666).

Al-Akkad, A., Ramirez, L., Boden, A., Randall, D., Zimmermann, A., & Augustin, S. (2014b). Help beacons: Design and evaluation of an Ad-Hoc lightweight S.O.S. system for smartphones. In *CHI 2014* (pp. 1485–1494).

Brayley, H., Redfern, M. A., & Bo, Z. Q. (2005). The public perception of power blackouts. In *2005 IEEE/PES Transmission and Distribution Conference and Exposition*. Asia Pacific (pp. 1–5).

Dunkel, J., Eberhart, A., Fischer, S., Kleiner, C., & Koschel, A. (2008). *Systemarchitekturen für verteilte Anwendungen*. München: Carl Hanser Verlag.

Gardner-Stephen, P. (2014). *The serval project : Practical wireless Ad-Hoc mobile telecommunications*. http://developer.servalproject.org/files/CWN_Chapter_Serval.pdf.

Helsloot, I., & Beerens, R. (2009). Citizens' response to a large electrical power out-age in the Netherlands in 2007. *Journal of Contingencies and Crisis Management, 17*, 64–68.

Hiete, D. M., Merz, M., & Trinks, C. (2010). Krisenmanagement Stromausfall Kurzfassung—Krisenmanagement bei einer großflächigen Unterbrechung der Stromversorgung am Beispiel Baden-Württemberg., Stuttgart (2010).

Hossmann, T., Legendre, F., Carta, P., Gunningberg, P., & Rohner, C. (2011). Twitter in disaster mode: Opportunistic communication and distribution of sensor data in emergencies. *Proceedings of ExtremeCom* (pp. 1–6). Manaus, Brazil: ACM Press.

Hughes, A. L., Denis, L. A. S., Palen, L., & Anderson, K. M. (2014). Online public communications by police & fire services during the 2012 Hurricane Sandy. *Proceedings of the Conference on Human Factors in Computing Systems (CHI)* (pp. 1505–1514). Toronto, Canada: ACM.

Hunt, A. (2003). Risk and moralization in everyday life. In R. V. Ericson & A. Doyle (Eds.), *Risk and morality* (pp. 165–192). Toronto: University of Toronto Press.

Kahler, H., Kensing, F., & Muller, M. (2000). Methods & tools: Constructive interaction and collab-orative work: Introducing a method for testing collaborative systems. *Interactions, 7*, 27–34.

Kaplan, A. M., & Haenlein, M. (2010). Users of the world, unite! The challenges and opportunities of Social Media. *Business Horizons, 53*, 59–68.

Kargl, F. (2003). Sicherheit in Mobilen Ad-hoc Netzwerken. http://www.tostermann.de/1_public/dissertationMANET.pdf.

Kaufhold, M.-A., & Reuter, C. (2016). The self-organization of digital volunteers across social media: The case of the 2013 European floods in Germany. *J. Homel. Secur. Emerg. Manag., 13*, 137–166.

Legendre, F., Hossmann, T., Sutton, F., & Plattner, B. (2011). 30 Years of Ad Hoc networking research: What about humanitarian and disaster relief solutions? What are we still missing? In *Proceedings of the International Conference on Wireless Technologies for Humanitarian Relief* (pp. 217–217).

Lorenz, D. F. (2010). Kritische Infrastrukturen aus Sicht der Bevölkerung. Forschungsforum Öffentliche Sicherheit der FU Berlin.

Ludwig, T., Reuter, C., Siebigteroth, T., & Pipek, V. (2015). CrowdMonitor: Mobile crowd sensing for assessing physical and digital activities of citizens during emergencies. In *Proceedings of the Conference on Human Factors in Computing Systems (CHI)*. Seoul, Korea: ACM Press.

Nielsen, J. (1993). *Usability engineering*. San Francisco, USA: Morgan Kaufmann.

Nishiyama, H., Ito, M., & Kato, N. (2014). Relay-by-smartphone: Realizing multihop device-to-device communications. *IEEE Communications Magazine, 52*, 56–65.

Olofsson, S., Carlsson, V., & Sjölander, J. (2006). The friend locator: Supporting visitors at large-scale events. *Personal and Ubiquitous Computing, 10*, 84–89.

Reuter, C. (2014a). Communication between power blackout and mobile network overload. *International Journal of Information Systems for Crisis Response and Management (IJISCRAM), 6*, 38–53.

Reuter, C. (2014b). Emergent collaboration infrastructures: Technology design for inter-organizational crisis management (Ph.D. Thesis). Springer Gabler, Siegen, Germany.

Reuter, C., Ludwig, T., Funke, T., & Pipek, V. (2015a). SOMAP: Network independent social-offline-map-mashup. In *Proceedings of the Information Systems for Crisis Response and Management (ISCRAM), Kristiansand, Norway.*

Reuter, C., Ludwig, T., Kaufhold, M.-A., & Pipek, V. (2015b). XHELP: Design of a cross-platform social-media application to support volunteer moderators in disasters. In *Proceedings of the Conference on Human Factors in Computing Systems (CHI)*. Seoul, Korea: ACM Press.

Reuter, C., Ludwig, T., Ritzkatis, M., & Pipek, V. (2015c). Social-QAS: Tailorable quality assessment service for social media content. In *Proceedings of the International Symposium on End-User Development (IS-EUD)*. Lecture Notes in Computer Science.

Reuter, C., Ludwig, T., & Pipek, V. (2016). Kooperative Resilienz—ein soziotechnischer Ansatz durch Kooperationstechnologien im Krisenmanagement. Grup. Interaktion. Organ. Zeitschrift für Angew. Organ.

Shalunov, S. (2013). *Open garden: Multi-hop Wi-Fi offload*. https://opengarden.com/Multi-hop_Wi-Fi_Offload.pdf, (2013).

Stallings, R. A., & Quarantelli, E. L. (1985). Emergent citizen groups and emergency management. *Public Administration Review, 45*, 93–100.

Starbird, K., & Palen, L. (2011). Voluntweeters: Self-organizing by digital volunteers in times of crisis. In *Proceedings of the Conference on Human Factors in Computing Systems (CHI)* (pp. 1071–1080). Vancouver, Canada: ACM-Press.

Stevens, G., & Wulf, V. (2009). Computer-supported access control. *ACM Transactions on Computer-Human Interaction, 16*, 1–26.

Strauss, A. L., & Corbin, J. (1998). *Basics of qualitative research: Techniques and procedures for developing grounded theory*. Sage Publications.

Toriumi, F., Sakaki, T., & Shinoda, K. (2013). Information sharing on Twitter during the 2011 catastrophic earthquake. In *Proceedings of the 22nd International Conference on World Wide Web Companion* (pp. 1025–1028). Rio de Janeiro: ACM.

Twidale, M., Randall, D., & Bentley, R. (1994). *Situated evaluation for cooperative systems Situated evaluation for cooperative systems*. Lancaster, UK: Lancaster University.

Vieweg, S., Hughes, A. L., Starbird, K., & Palen, L. (2010). Microblogging during two natural hazards events: What Twitter may contribute to situational awareness. *Proceedings of the Conference on Human Factors in Computing Systems (CHI)* (pp. 1079–1088). Atlanta, USA: ACM.

Volgger, S., Walch, S., Kumnig, M., & Penz, B. (2006). Kommunikation vor, während und nach der Krise. Amt der Tiroler Landesregierung.

Wi-Fi Alliance: Wi-Fi Direct. http://www.wi-fi.org/discover-wi-fi/wi-fi-direct.

Wilensky, H. (2014). Twitter as a navigator for stranded commuters during the Great East Japan Earthquake. In *Proceedings of the Information Systems for Crisis Response and Management (ISCRAM)* (pp. 695–704).

Wiseman, J., Garcia, D. H., & Toksvig, M. J. M. (2015). Mobile ad hoc networking (US Patent No. US 9,037,653 B2.

Wulf, V., Rohde, M., Pipek, V., & Stevens, G. (2011). Engaging with practices: design case studies as a research framework in CSCW. In *Proceedings of the Conference on Computer Supported Cooperative Work (CSCW)* (pp. 505–512). Hangzhou, China: ACM Press.

Collection and Integration of Multi-spatial and Multi-type Data for Vulnerability Analysis in Emergency Response Plans

Harsha Gwalani, Armin R. Mikler, Suhasini Ramisetty-Mikler and Martin O'Neill

Abstract Public health emergencies, whether natural or manmade, require a timely and well planned response from the local health authorities to mitigate the economic and human loss. Creation of well-defined service areas with Points of Dispensing (POD) facilities for providing medical or other care to the population within these areas is a widely accepted approach. However, not every individual may have equal access to the POD or the resources available at the POD due to various social, behavioral, cultural, and economic or health disparities. Therefore, a realistic and working response plan must provide ways to address these access disparities associated with each POD. The creation of such a response plan is a data intensive problem and requires demographic, spatial, transportation and resource data at different geographic levels. This paper demonstrates the collection of these data from disparate sources and their integration to analyze vulnerabilities in emergency response plans. The variety and volume of the required data and the manipulation process faced challenges of big data.

Keywords Geographic information systems · Response plans · Bio-emergencies · Geospatial data analysis · Vulnerabilities

1 Introduction

Emergency response plans play a crucial part in reducing the threat posed by bio-emergencies like disease outbreaks, natural disasters, terrorist attacks etc. The capacity to reach every person in a community is one of the major goals for emergency

H. Gwalani (✉) · A.R. Mikler · S. Ramisetty-Mikler · M. O'Neill
Center for Computational Epidemiology and Response Analysis (CeCERA),
University Of North Texas, Denton, TX 76203–0250, USA
e-mail: harshagwalani@my.unt.edu

A.R. Mikler
e-mail: mikler@unt.edu

S. Ramisetty-Mikler
e-mail: susie.ramisetty-mikler@unt.edu

M. O'Neill
e-mail: martyo@unt.edu

V. Wohlgemuth et al. (eds.), *Advances and New Trends in Environmental Informatics*, Progress in IS, DOI 10.1007/978-3-319-44711-7_8

preparedness and response (Indrakanti et al. 2016). Dividing the region into well defined service areas with corresponding PODs does not guarantee equal accessibility to the resources for all individuals. Certain vulnerable individuals may be at risk of not receiving critical medical (or other) care because of access disparities. Cutter et al. (2008) define vulnerabilities in the context of response plans as

> the pre-event inherent characteristics or qualities of social systems that create potential barriers increasing the likelihood of inability to receive appropriate communication or required services (e.g., post-exposure prophylaxis) in a given emergency response scenario.

REsponse Plan ANalyzer (RE-PLAN) is a computational framework developed to design response plans and analyze their feasibility in bio-emergencies (O'Neill et al. 2014). The response plan created by RE-PLAN divides a region into service areas and assigns a POD (Point Of Dispensing) location to each service area which acts as the distribution point to dispense medical supplies including prophylactics or vaccines to all the individuals within that service area. The RE-PLAN feasibility analysis includes quantification of transportation, language and age vulnerabilities in each service area. Transportation vulnerabilities arise from inability to travel to a POD to receive resources. The vulnerable individuals are those who lack access to private vehicles or public transportation to reach the assigned POD. Language vulnerabilities arise from the inability to understand the information and follow public health directives. The information must be linguistically accessible to everyone, which is possible only when the planners know about the population subgroups which are not proficient in English (or other official languages). Children and elderly are recognized as the most vulnerable age groups in disaster events. These groups may have special needs that require the assistance of others, therefore age vulnerability needs to be quantified for optimal resource allocation.
Creating service areas in a region is a partitioning problem and we need to identify a smaller geographic area as the base unit for each service area to partition the region. In the UnitedStates, the geographic entities follow a hierarchical structure. Census blocks are at the lowest level, followed by block groups. Next in the hierarchy are census tracts, comprised of block groups. Census tracts align with counties and counties form states (hie 2010). If *census block* is chosen as the base granularity, then each service area is composed of a unique and disjoint set of contiguous *census blocks*. The choice of base granularity is driven primarily by data availability. Section 3.1 discusses the process of selecting a base granularity for RE-PLAN. Collection and integration of population, vulnerability and geo spatial data at the chosen base granularity level is another challenge. Section 3.2 lists the data sources which provide these data at multiple geographic levels. Since the data are accumulated from heterogeneous sources, they may not be consistent. It is possible that the census data list m block groups in a county while the geospatial data source provides data for n block groups for the same county (m and n may not be equal). The inconsistencies occur because of non-synchronous updates in the data sources. Sections 4.1 and 4.2 explore consistent integration and linking of heterogeneous data. Some vulnerabilities cannot be quantified below a certain geographic granularity because the numbers are too small to be considered for analysis at lower levels. Quantification

of these vulnerabilities at each service area is difficult because service areas are constructed at a lower granularity. Section 4.3 discusses the mapping of available data to service areas to overcome this granularity mismatch.

2 Related Work

Aggressive mass treatment is necessary to reduce the number of casualities in a bio-emergency. The local public health authorities are required to prepare response plans to specify how the medications are distributed and dispensed in a timely manner. The Centres for Disease Control and Prevention (CDC) recognize POD based response plans as an effective way to serve populations. RealOpt (2009) provides ways to design customized and efficient POD floor plans and determine optimal resources according to regional needs. However, identification of ad-hoc facilities which can be used as treatment centres during a bio-emergency remains a challenging problem. Data driven computational methods are required for selection of these facilities to ensure optimal and timely resource allocation. The RE-PLAN framework was developed to analyze the feasibility of response plans by selecting locations for these ad-hoc facilities. A set of facilities is considered feasible if its operational efficiency is capable of meeting service requirements (e.g. specific time frames for service completion or proportions of populations to be served) without exceeding available resources (e.g. transportation network capacities or limitations of facility infrastructure) (O'Neill et al. 2014).
The tragedy of hurricane Katrina in USA showed the repercussions of not including the communities needs in preparedness plans. Areas most damaged by Katrina were largely populated by low-income African Americans, many living in substandard housing and lacking access to personal transportation for evacuation. Latinos and Asian Americans faced similar barriers during Katrina, compounded by issues of language, culture, and their status as undocumented or uninsured residents (vul 2006). Government reports (kat 2006; Senate 2006) also emphasize on the importance of community preparedness and access disparities to minimize disastrous effects of such events. Flanagan et al. (2011) compute a social vulnerability index value for states at the census tract level ranking social vulnerabilities in terms of socioeconomic status, household composition, minority status or language and housing or transportation to identify the most socially vulnerable locations in a region. Indrakanti et al. (2016) explore data driven methods to quantify language and transportation vulnerabilities in the context of response plans to establish a balance between the demand and availability of resources in a region. The novel contribution of this paper is that it explains the data collection and integration processes needed for quantifying vulnerabilities. These processes can be also used to integrate geospatial and demographic data for applications other than response planning such as Medical GIS, crime mapping etc.

3 Data Collection

A response plan for a region includes locations of POD facilities and the respective non overlapping and non empty sub-regions served by the PODs referred to as service areas (O'Neill et al. 2014). The location and design of POD facilities are established with respect to the population composition of the sub-regions served by the PODs. A response plan r, for a region R can be represented as a one to one mapping $r =< P, S >$, where P is the set of PODs and S is the set of service areas contained in the region R.

3.1 *Base Geographical Granularity*

The set of service areas S, is created by combining a lower level geographic region contained in R. The lowest level geographic granularity is the census blocks and creating service area by combining census blocks will result in high resolution service areas. The decennial census conducted by the United States Census Bureau (USCB) tabulates the population at the census block level every 10 years and the geospatial data are also available at this level. But census block is not an appropriate choice as the base granularity because:-

1. The decennial census data are the least current demographic data as they are updated only every 10 years. The geographic features of a region are subject to change and the associated population counts for that region may vary because of changes in boundaries or changes in regional population over time. Therefore, the decennial data are not the best representation of the real time scenario.
2. All the data (for example, disability status and language proficiency) are not available at the census block level due to time constraints.

RE-PLAN, currently uses block groups as the base granularity as most of the data are available at this level and they are more frequently updated than the decennial census data. The region R, is defined as a set of contiguous non overlapping block groups b such that $\cup\, b \in B = R$. Each block group, $b \in B$ is represented by its centroid *centroid*(b) and has a population, *pop*(b), such that,

$$\bigcup b \in B, \sum pop(b) = pop(R).$$

A response plan is created by computing a partitioning S of R, such that, $|S| = k$ and each $s_i \in S$ represents a service area with,

$$s_i = B_i \text{ where,}$$
$$B_i \subseteq B \text{ and}$$
$$B_i \bigcap B_j = \phi \forall B_i, B_j \subseteq B \text{ and}$$
$$s_i \bigcap s_j = \phi \forall s_i, s_j \in S \text{ for } i, j = 1, 2, 3...k$$

3.2 *Data Sources and Data Format(s)*

The sources of the data used in RE-PLAN, along with the lowest granularity levels at which the data are extracted are listed in Table 1. The American Community Survey (ACS) conducted by the United States Census Bureau provides 1-year, 3-year, and 5-year estimates for several demographic variables. The ACS, 5-year estimates were chosen because they cover almost all the variables required for the analysis and are more accurate than the ACS 1-year or ACS 3-year estimates. The ACS, 1-year or 3-year estimates though, are updated more frequently than the 5-year estimates but the sample size used to estimate these is narrow therefore the standard error is high. Also the 1-year and 3-year estimates are available only for regions with larger populations (acs 2010). The increased precision of the 5-year estimates has greater utility when examining smaller populations and geographies (Ramisetty-Mikler et al. 2015).

Population data

The ACS 5 year estimates table *B00003*: "Total Population" tabulates the population count at the block group level. This data are used for equal population partitioning and POD performance analysis in RE-PLAN (Jimenez et al. 2012; O'Neill et al. 2014).

Geo-spatial data

Topologically Integrated Geographic Encoding and Referencing (TIGER) provides geospatial data in the shapefile format which is the Environmental Systems Research Institute (ESRI) regulated standard format for representing non topological data. The shape file describes shapes (points, lines or polygons) using a list of vertices (tig 1995).

Table 1 Data sources for data used in RE-PLAN

Requirement	Data source	Owner	Granularity
Population data	ACS, 5 year estimates	US Census Bureau	Block group
Geospatial data	Topologically Integrated Geographic Encoding and Referencing (TIGER)	US Census Bureau	Block group
Public transport data	General Transit Feed Specification (GTFS)	Local transit authorities	County
Personal transport data	American Community Survey, 5 year estimates.	US Census Bureau	Census tract
Language vulnerability data	ACS, 5 year estimates.	US Census Bureau	Census tract
Age vulnerability	ACS, 5 year estimates.	US Census Bureau	Block group

Public transport data

The General Transit Feed Specification (GTFS) defines a common format for public transportation schedules and associated geographic information (Exchange 2016). RE-PLAN uses the transit stops and transit routes data to calculate the number of individuals who may have access to the POD via public transit and the shape data to draw the routes on the maps.

Personal transport data

The ACS table *B08201* provides the data for "Household Size by Vehicles Available" at the census tract level. It lists the number of 1-person, 2-person, 3-person and 4 or more-person households with respect to the number of vehicles (0, 1, 2, 3 and 4 or more), the households own. The number of people with no vehicle access can be estimated by using this data.

Language data

RE-PLAN uses 2 ACS tables to quantify language vulnerability as the number of individuals with limited ability to speak English. Table *B16002* titled "Household Language By Household Limited English Speaking Status" provides information about the number of households in which English is not spoken very well at the block group level. RE-PLAN uses this table to estimate the number of individuals with limited English speaking abilities for each block group. But these data are not enough to allocate translators as the specific language spoken by individuals is still not known. Table *B16004* titled "Language Spoken At Home By Ability To Speak English For The Population 5 Years and Over" is used to quantify specific language vulnerabilities. It is available at the census tract level. It lists the number of people above 5 years in age who do not speak English very well by the language they speak.

Age data

The ACS 5 year estimates, maintain a table *B01001*: "Sex by Age" at the block group level. The vulnerable age groups are quantified by adding the male and female data for the specific age range from this table.

The transportation, language and age vulnerability data are collected from the sources discussed above and stored in a *postgres* database along with the geospatial data.

4 Data Integration

The vulnerability data are collected from heterogeneous sources and they can vary in terms of three factors: time of collection, type of data and the geographic level at which they were collected. It is important that all the data are integrated without any redundancy or inconsistency. The following subsections discuss consistent integration with respect to time of release, type of data and geographic granularity.

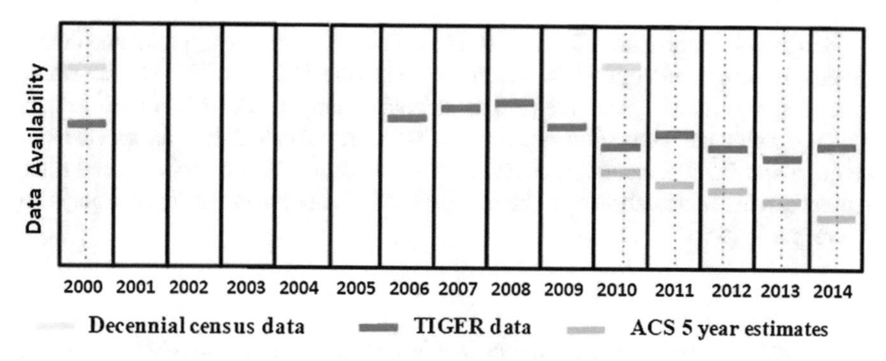

Fig. 1 Decennial census data, ACS 5 year estimates and TIGER data availability from 2000 to 2014. The *red dotted line* indicates data consistency with respect to geographic boundaries

4.1 Consistency with Respect to Time

Ensuring consistency with respect to spatial granularity of data and time of release across data sources is important. While mapping statistical data to the TIGER data, it is important to make sure that, the geographic boundaries, that were in effect at the time the statistical data were tabulated, are used. The census bureau releases updated versions of the TIGER shape files annually. Tabulating the statistical data is a more complex process for the census bureau than listing geographic boundaries, therefore, the most recent geographic boundary files available are about a year ahead of the most recent statistical data files available. Figure 1 shows the data releases which are consistent with one another. The ACS data launched in 2010 are the 5 year estimates for 2006 to 2010, similarly the decennial census data released in 2010 tabulates the population data for 2001–2010. RE-PLAN, currently uses the TIGER13 data integrated with the ACS 2009–2013 estimates.

4.2 Consistency with Respect to Type of Data

Since the geospatial and demographic data come from different sources, linking them is not a trivial task. The geospatial and the statistical data have a geographic region identifier (geoid) to map the statistics for a region with its geographic features, *but this id may not be consistent or unique* (tig. 1995). Also, geoid is a multiple character string value created by concatenating other attributes like the Federal Information Processing Standard (FIPS) code (5 digits), state id, county id etc., which makes it difficult to use as a primary key in the database. RE-PLAN creates its own integral unique identifier for each geographic region which is used to map the statistical data with the geographic area.

The *postgres* rank() function is used to create a unique key that maps all the records in the geospatial tables (TIGER) to all the records in demographic tables (ACS).

Geographic Information Systems (GIS) are needed to manage spatial and demographic information required for RE-PLAN (McLafferty 2003). The data are stored in a *postgres* database with the *postgis* extension support. Postgis is used to handle geographic and geometric calculations like centroid calculation for service area creation and POD placements, checking if a POD lies within a service area or not, closest distance calculation etc. The *GeoTools* API is used to create layered maps for the response plans.

4.3 Consistency with Respect to Geographical Granularity

ALGORITHM 1: Mapping service areas at the block group level to service areas at the census tract level

Input → $S = \{s_1, s_2, s_3, ...s_k\}$,
/* S is the initial set of all service areas */
$s_i = B_i$
Output → $S' = \{s'_1, s'_2, s'_3, ...s'_n\}$ where $n \leq k$
/* S' is the final set of service areas */
$s'_j = C_j$
where, $B_i \subseteq B$ and $C_j \subseteq C$
$\forall\, i = \{1, 2, 3, ...k\}$ and $\forall\, j = \{1, 2, 3, ...n\}$
/* B is the set of all block groups and C is the set of all census tracts in R*/
$L = C$
/* L is a list of census tracts initialized to C */
$l = 1$
foreach $s_i \in S$
  while $L.has(c_j)$
    if s_i contains $centroid(c_j)$
      Add c_j to s'_l
      $L = L - c_j$
    End if
  End while
  $l = l + 1$
End For
Return S'

RE-PLAN creates the service area set, S by combining block groups in B, so the total number of vulnerable individuals $V(i, j)$, at a service area s_i, with vulnerability v_j is given by :

$$V(i,j) = \sum_{k=1}^{p} (d(k,j)) \tag{1}$$

where service area $s_i = B_i$, $|B_i| = p$ and $d(k,j)$ is the number of vulnerable individuals with vulnerability v_j at block group b_k for $b_k \in B_i$ and $B_i \subseteq B$. B is the set of all block groups in R and C is the set of all census tracts in R.

However, for some vulnerabilities the data are not available at the block group level. Therefore, these vulnerabilities cannot be quantified at the block group level. RE-PLAN uses the next higher geographic level, census tracts (Fig. 1) to quantify these vulnerabilities. ACS provides the data at the census tract level. The service areas need to be recreated at this level for accurate analysis. RE-PLAN uses Algorithm 1 to create service areas (S') at the census tract level by using the initial set of service areas (S) (block group level).

5 Case Study

Transportation vulnerable individuals

First, the at-risk individuals were identified as the number of people who do not have access to private vehicles without taking the public transit network into account. Figure 4a demonstrates the at-risk households at the census tract level. There are around 114,117 people in Dallas County who do not have access to a private vehicle and may not be able to make it to the assigned POD. If, it is assumed that the people can walk at most a distance of 2 km to the POD, the at-risk population reduces to 55,736. Figure 4b shows the coverage when a walkable distance (dw) of 2 km is taken into account. If public transit feed is taken into account and individuals are allowed to walk at most 2 km either to the POD directly or to the nearest public transit stop to reach the POD, the number of at-risk individuals reduces to 42,488 (Figs. 2 and 3). Figure 4c demonstrates the coverage in Dallas County when public transit is also considered.

Language vulnerable individuals

Vulnerable individuals in the Dallas County were identified as the number of people who have no or limited proficiency in English but are proficient in another language. For example, there are around 372,266 Spanish speaking individuals above the age of 5 years in Dallas county who have limited proficiency in English. Figure 4 shows the distribution of these individuals in the 70 service areas.

Age vulnerable individuals

The individuals below the age of 5 years and the individuals above the age of 65 years were identified as the two vulnerable age groups. There are 193,525 under-age individuals and 217,693 elderlies in the county. Figure 5 shows the distribution of individuals under 5 years in the service areas in Dallas County (Fig. 3).

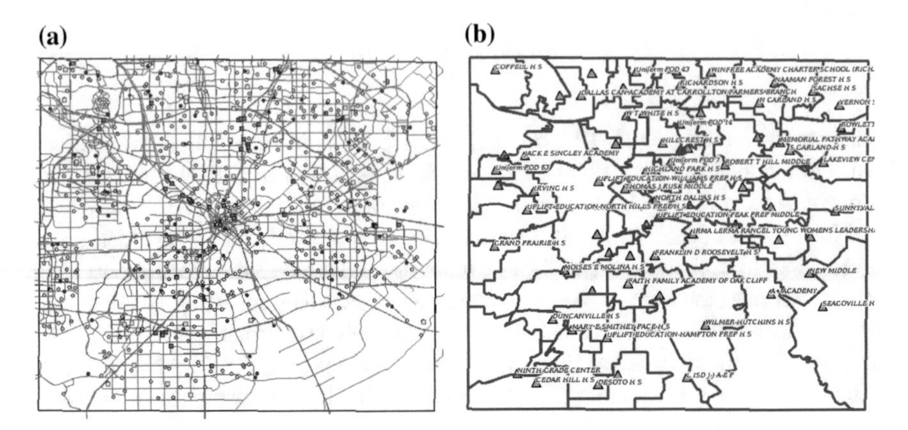

Fig. 2 Dallas County. **a** Dallas County map, the *red line* represents the road network and the *dots* represent the candidate POD locations. **b** Dallas county partitioned into 70 service areas using an equal population based partitioning. The *green triangles* depict the assigned PODs

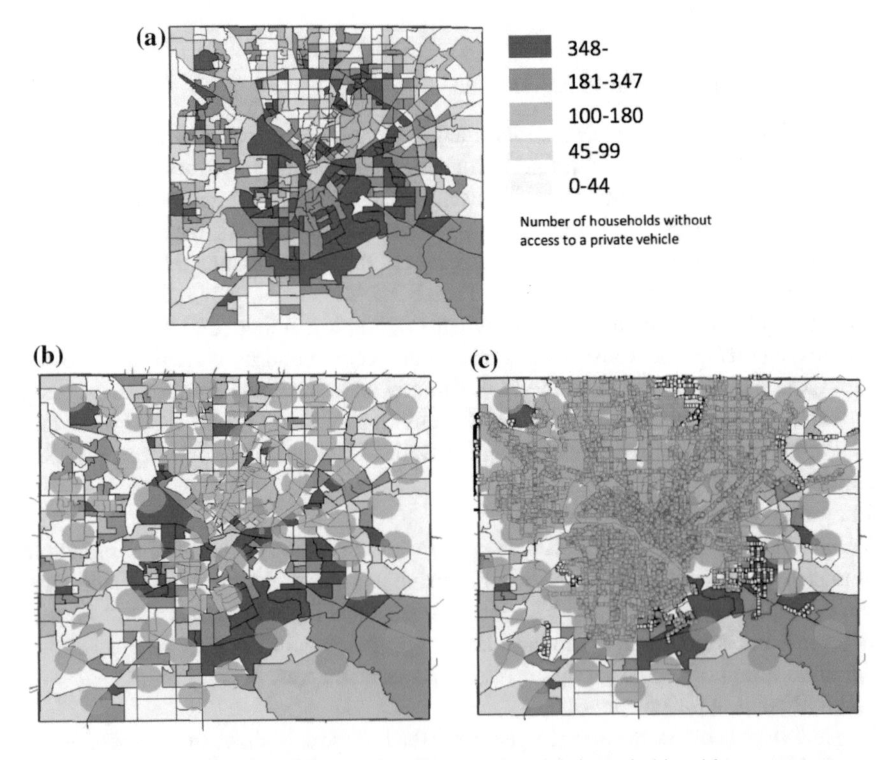

Fig. 3 Transportation vulnerability. **a** Distribution of at-risk households with respect to transportation vulnerability at the census tract level in Dallas County. **b** Coverage of at-risk individuals depicted by *green circles* when a walkable distance of 2 km is considered. **c** Coverage of at-risk individuals when a walkable distance of 2 km and public transit are taken into account

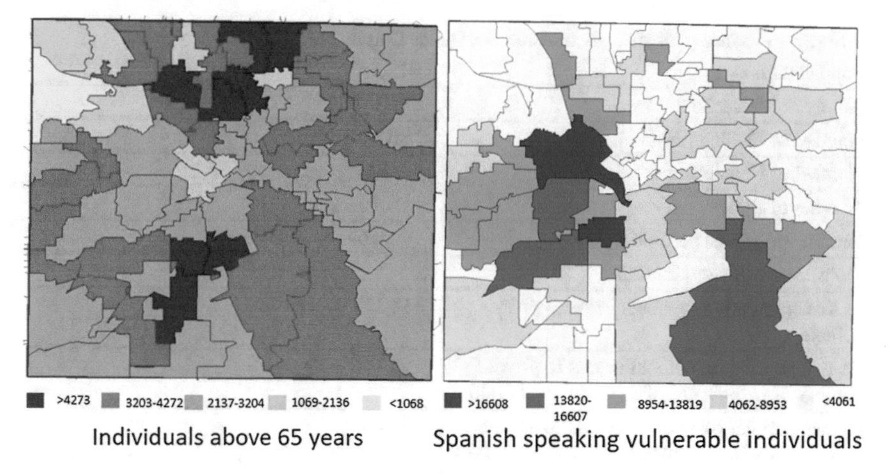

Fig. 4 Age and language vulnerability distribution in 70 service areas

6 Conclusion

In conclusion, this paper demonstrates the collection of demographic and geospatial data from disparate sources and their consistent integration to analyze vulnerabilities in emergency response plans. The variety and volume of the data required make the acquisition and manipulation process a big data problem. The methods described above have been integrated into the RE-PLAN Framework which has been deployed for practical use at state and county offices (O'Neill et al. 2014). The quantification of language, transportation and age vulnerability highlights the need to devise specific methodologies to address the access disparities in response plans. Transportation vulnerability can be addressed by maximizing the reach of transportation vulnerable populations to the PODs. The reach of transportation vulnerable populations can be maximized by selecting POD locations which are located in densely populated areas and/or are easily reachable via public transit. The effects of language vulnerability can be reduced by targeted allocation of resources to the PODs in each service area. Quantification of age vulnerability can be used to determine the dosage requirements at each POD for children and special infrastructure requirements for the elderly population. Resources may have to be shared between PODs due to constraints with respect to temporal availability or scarcity, which necessitates scheduling of resources. RE-PLAN uses resource allocation resources to allow temporal and spatial sharing of resources to minimize vulnerable populations whose needs are not addressed.

Table 2 Response plan analysis statistics for Dallas County

Population count	2,412,481
Number of households	858,332
Number of block groups	1669
Number of Census tracts	529
Number of PODs	70
Number of booths per POD	10
Transportation vulnerabilities	
At-risk population without access to private vehicle	114,117
At-risk population without access to private vehicle allowing a walkable distance of 2000m	55,736
At-risk population without access to private vehicle allowing a walkable distance of 2000 m and taking into account public transit	42,488
Language vulnerabilities	
Spanish	372,266
Vietnamese	15,504
Chinese	6,046
Other Asian languages	5,625
African languages	4,580
Age vulnerabilities	
Individuals below 5 years	193,525
Individuals above 65 years	217,693

The future goals for this research include broadening the set of indicators of vulnerabilities (include disability, homelessness, refugee status etc.) for response planning and developing methodologies to minimize the number of vulnerable individuals by co-locating the resources and at risk individuals (Table 2).

References

acs., (2010). *A Compass for Understanding and Using American Community Survey Data*. Department of Commerce Economics and Statistics Administration: U.S.

Cutter, S. L., Barnes, L., Berry, M., Burton, C., Evans, E., Tate, E., et al. (2008). A place-based model for understanding community resilience to natural disasters. *Global Environmental Change*, *18*(4), 598–606.

Exchange, G. D. (2016). General transit feed specification reference.

Flanagan, B. E., Gregory, E. W., Hallisey, E. J., Heitgerd, J. L., & Lewis, B. (2011). A social vulnerability index for disaster management. *Journal of Homeland Security and Emergency Management*, *8*(3).

hierarchy. (2010). Standard Hierarchy of Census Geographic Entities. US Census Bureau. http://www2.census.gov/.
Indrakanti, S., Mikler, A. R., Neill, M. O., & Tiwary, C. (2016). Quantifying access disparities in response plans. *Plos One*. journal.pone.0146350.
Jimenez, T., Mikler, A. R., & Tiwari, C. (2012). A novel space partitioning algorithm to improve current practices in facility placement. *IEEE Transactions on Systems, Man and Cybernetics-Part A: Systems and Humans*, *42*(5), 1194–1205.
kat. (2006). *A Failure of Initiative, Final Report of the Select Bipartisan Committee to Investigate the Preparation for and Reponse to Hurricane Katrina*. US House of Representatives.
Lee, E. K., Chen, C. H., Pietz, F., & Benecke, B. (2009). Modelling and optimizing the public health infrastructure for emergency response. *Security and Emergency Management*, *8*(3).
McLafferty, S. L. (2003). Gis and health care. *Annual Review of Public Health*, *24*, 25–42.
O'Neill, M., Mikler, A. R., Indrakanti, S., Tiwari, C., & Jimenez, T. (2014). Re-plan: An extensible software architecture to facilitate disaster response planning. *IEEE Transactions on Systems, Man, and Cybernetics: Systems*, *44*(12), 1569–1583.
Ramisetty-Mikler, S., Mikler, A. R., O'Neill, M., & Komatz, J. (2015). Conceptual framework and quantification of population vulnerability for effective emergency response planning. *ncbi PubMed journal*, (13(3)).
Senate, U. (2006). A nation still unprepared. Special report of the Committee on Homeland Security and Governmental Affairs.
tig. (1995). ESRI Shapefile Technical Description. Environmental Systems Research Institute Inc. https://www.esri.com/library/
vul. (2006). The federal response to hurricane katrina, lessons learned.

EPISECC Common Information Space: Defining Data Ownership in Disaster Management

Gerhard Zuba, Lina Jasmontaite, Uberto Delprato, Georg Neubauer and Alexander Preinerstorfer

> *Disaster-affected people need information as much as water, food, medicine or shelter: accurate, timely information can save lives. The right information helps aid organizations to understand better the needs of affected communities and ways to meet those needs. Today's information technology presents new possibilities, but has not been fully exploited by humanitarian organizations. Lack of information can make people victims of disaster.*
>
> World Disaster Report (2013) International Federation of Red Cross.

Abstract This paper provides a summary of the EPISECC consortium decisions taken when developing a common information space with respect to data sharing in the field of public protection and disaster relief. After explaining why defining data flows and design process in the EPISECC is important, the paper introduces a high-level overview of the EPISECC Common Information Space (CIS). Additionally, the paper explores the CIS user requirements and software architecture. To attain the main objective of this paper and to provide an operational overview of the

G. Zuba
Frequentis AG, Vienna, Austria
e-mail: gerhard.zuba@frequentis.com

L. Jasmontaite (✉)
KU Leuven – Centre for IT and IP Law – iMinds, Leuven, Belgium
e-mail: lina.jasmontaite@kuleuven.be

U. Delprato
IES Solutions, Rome, Italy
e-mail: u.delprato@iessolutions.eu

G. Neubauer · A. Preinerstorfer
AIT, Vienna, Austria
e-mail: Georg.Neubauer@ait.ac.at

A. Preinerstorfer
e-mail: Alexander.Preinerstorfer@ait.ac.at

V. Wohlgemuth et al. (eds.), *Advances and New Trends in Environmental Informatics*, Progress in IS, DOI 10.1007/978-3-319-44711-7_9

EPISECC CIS, the paper defines a methodology to map data flows within the CIS and examines the basic functionality of the proposed system.

Keywords Common Information Space • Disaster relief • Information sharing • Public protection

1 Introduction

Advancements in the Information Communication Technology (ICT) have led to the growth of data that can facilitate the decision-making process in various sectors, including the public protection and disaster relief (PPDR). For the PPDR organisations, ICT tools can provide access to necessary information in a particular situation (e.g. a plan of the suburb in a case of fire). Timely availability of data can not only enhance situational awareness of first responders but also lead to better decisions when responding to crises. Yet many ICT tools used for Incident Command Systems for the PPDR primarily focus on needs of organisations on the operational level. This often results in a lack of interoperability of different information systems used by different PPDR organisations. This is of a great concern in transboundary emergency events, including natural or man-made disasters and environmental crises, where PPDR organisations need to cooperate and exchange information.

The number of information sharing platforms on the EU level has been growing. Usually these platforms serve only one purpose and focus on one particular issue. For example, the Critical Infrastructure Warning Information Network (CIWIN) allows to exchange critical infrastructure protection related information and the Common Emergency Communication and Information System (CECIS) contains info of Member States resources that could be used to respond to emergencies.[1] In addition to a limited purpose, these platforms are often based on voluntary participation and thus consequently have limited impact on enhancing coordination, cooperation and recognition of technical standards or specifications among organisations working in a specific field, such as PPDR.

To improve the current situation and address the needs of Public Protection and Disaster Relief organisations, the "Establish Pan-European information space to Enhance seCurity of Citizens" (EPISECC) project has been developing a Common Information Space (CIS).[2] This paper will provide an overview of various decisions taken when developing a common information space with respect to data sharing. The paper will aim at introducing the EPISECC Common Information Space (CIS). Consequently, the paper will explore the CIS user requirements and software

[1]Critical Infrastructure Warning Information Network (CIWIN), Weblink: http://ec.europa.eu/dgs/home-affairs/what-we-do/networks/critical_infrastructure_warning_information_network/index_en.htm, accessed 03.06.2016.

[2]EPISECC is a Collaborative Project which will Establish a Pan-European Information Space to Enhance seCurity of Citizens, funded by the EU, grant agreement no. 607078.

architecture. The paper will also define a methodology to map data flows within the CIS and examine the basic functionality of the system. The underlying objective of this paper is to provide an operational overview of the EPISECC CIS.

2 Defining Data Flows and Design Process in the EPISECC: Why Is It Important?

In response to the changing nature of emergency management, the EPISECC consortium aims at developing an architecture which would facilitate information exchange between and among different types of first responders and PPDR stakeholders ranging from public authorities to private entities and NGOs contributing to relief actions in emergency events. Before the architecture has been defined, it has been important to engage in a discussion with a wider community about the EPISECC approach and decisions made throughout the system design phase. Explaining the step by step approach is valuable as it can help to improve acceptance of the CIS by the end users and allows identifying weak points of the proposed framework. In other words, sharing the EPISECC experience to a larger public, in particular to the end-user community, can enhance ethical acceptability, social value and scientific validity of the research and innovation efforts of the consortium members. Even more so, by provoking discussions it may allow to attain an acceptable ratio of potential benefits to risks or harm that may occur from the CIS to vulnerable populations.

At the same time, defining data flows within the CIS is important for the legal (sometimes referred to as regulatory) compliance purposes. Legal compliance is a quality aspect of ICT tools for PPDR organisations. It is reasonable to expect that only tools that were designed to conform to applicable regulatory measures, such as standards and laws, can be adopted by PPDR organisations.

Finally, data exchange and processing in PPDR include various types of data. Some of the data, such as personal data or data subjected to the intellectual property rights, is subject to a particular set legal requirements. Provided the highly fragmented nature of technology regulatory, ensuring regulatory compliance of ICT tools for PPDR prove to be very difficult. Numerous technical, operational and cultural standards exist in the field of the PPDR.[3] Therefore, defining and visualising data flows is valuable as it can help to identify applicable regulatory frameworks. Within the EPISECC project, legal compliance was focusing on access to information held by the public sector, intellectual property rights and protection of

[3]Examples of technical standards include: EMTEL TS 102 181 (Requirements for communication between authorities/organizations during emergencies), TS 102 182 (Requirements for communications from authorities/organizations to individuals, groups or the general public during emergencies) and TS 102 410 (Basis of requirements for communications between individuals and between individuals and authorities whilst emergencies are in progress); CEN-CENELEC. CEN/TC391 (Societal and Citizen Security).

personal data. For the latter domain, defining data flows helps not only to carry out legal compliance assessments and define applicable data protection requirements, but also to conduct Data Protection Impact Assessment that evaluates the project against the needs and expectations, concerns of the relevant stakeholders.

To ensure the quality of information it is recommended to clarify the responsibilities of the engaged actors/organizations. The practice shows that better information quality is ensured if actors own the data they generate and create (e.g. the automotive industry). The attribution of data ownership also may help to answer questions about the control of the information flow, the cost of information, and the value of information.

Finally, defining data flows is of relevance to software developers as it can assist when developing the information governance strategy for a particular system as well and identify security flaws which can result from information exchange as well as anticipate functionality and security drawbacks. For example, it can ensure more precise understanding of data and complex system and define the chain of responsibility.

3 Information Inventory and a Data Mapping Exercise

To ensure compliance with the relevant regulatory frameworks and the expectations of organisations assumed to share their information, developers of CIS need to understand and determine what types of data are being processed and what types of data are stored in the system. For these purposes, the developers are suggested to pin down all information that could potentially be made available on the EPISECC CIS. Within the ESPISECC project this was done in Deliverable 6.1 "Proof of concept design", where a selected list of communications tools was analysed in greater detail. The overall objective of the information inventory is to determine:

- types of data producers (organizations that create, compile, aggregate, package, and provide information to be inserted into an information sharing system);
- what types of data are processed (e.g., weather forecast, personal information, structured, semi-structured, and unstructured information);
- purposes of information consumption (e.g., information about situational awareness, decision-making process);
- flows of information; and
- status or sensitivity of information (confidential, sensitive personal data).

While the data mapping exercise may allow to understand better what kind of tools can be connected to the CIS, it should be noted that it hardly ever leads to a complete and final document. Data mapping exercise continues through the lifecycle of the CIS and needs to be updated on a regular basis as different tools could be connected.

Certainly creating such an inventory contributes to the quality of the IT system. Yet "producing good software designs requires a considerable amount of practice, with a great deal of feedback on the quality of the designs with the resulting software products influencing the design of the next software system."[4] In other words, it requires human effort and is costly. Within the scope of the EPISECC project, the advisory board is regularly consulted about the CIS architecture.

4 Common Information Space: EPISECC Approach

Common Information Spaces have been deployed for complex governance in both the public and private sectors with complex governance structures to improve efficiency of activities carried out by an entity. Some attempts on the national level have been made with respect to developing information systems for PPDR, yet such attempts have not received much attention on the international level. To aid the current situation and enhance interoperability of different tools used for the PPDR, the EPISECC CIS will be based on the distributed processing implementation by peer to peer architecture. This means that there will be no dedicated servers and clients, instead all processing responsibilities will be allocated among all the machines that that are considered to be peers. To become a peer an entity needs to register.

This architecture ensures that there is no centralised data storage or information gateway which could become a single point of failure or a target of cyber-attacks, leaking sensitive or protected data. In fact, the information to be shared stays in the domain of the data owner and the implemented authorisation and data protection concept guarantees that every piece of information is only accessible by authorised participants. Apart from security related aspects, an additional advantage of the architecture is to ensure that messages are sent to the right address in shorter time. The possibility of selection of wrong email addresses and loss of time due to late decisions is considerably reduced.

5 Common Information Space and Its Objectives

The overarching aim of the EC funded project, titled "Establish Pan-European information space to Enhance seCurity of Citizens" (EPISECC), is to enhance security of citizens by improving data management practices and information sharing capabilities between the various parties involved in responding a disaster situation. The main objective of the CIS that will be developed within the EPISECC project is to serve as a tool enabling interaction and optimising coordination of civil protection assistance in disasters of various scale, including disasters that prompt activation of the EU Civil Protection Mechanism.

[4]Leach, R. J. *Introduction to software engineering* (2nd ed.). Chapman and Hall/CRC, 2016. Vital Source Bookshelf Online.

5.1 Defining User Requirements for the CIS

The first step in developing the architecture for the EPISECC CIS entailed a detailed analysis of the management practices of past disasters. For this purpose, a questionnaire was developed that served as an interface for the interviewees that represent the majority of the EU countries. In this context, more than 40 interviews with European crisis managers were performed. The interviews covered multiple types of disasters such as flooding (hydrological disasters), earthquakes (geological disasters) or complex events such as the management of the currently ongoing refugee flow. Taking all types of events into account, the predominant request was related to interoperability (34 % of all requests). Out of these requirements, 30 % were related to improved information exchange on tactical and or operational level, 26 % of the requests were dealing with information exchange on political/strategic level and finally 26 % were related to technical interoperability requests.[5] The majority of examined disasters were predominantly events managed on a national level, the request on improved interoperability was therefore mainly related to the national level. In detail, those statements include predominantly requests on an improved operational picture, lack of shared information as well as requests on tools allowing improved information exchange including adequate interfaces for the crisis management tools of stakeholders. Other important requirements focus on the need of a common taxonomy as well as the problem of language barriers in case of border crossing crisis managements. Another set of requirement is focusing on request on technical solutions such as a general lack of adequate tools for communication and information exchange or missing resilience of communication tools. It was also pointed out, that main problems arise due to communication problems or political challenges. In addition, data protection requirements have to be integrated in any future solution.

Taken together, the requirements from stakeholders clearly demonstrate the need of an information space that allows interoperability of the very heterogeneous landscape of crisis management tools used by European stakeholders, including proprietary solutions. Following on the end-users' recommendations to integrate existing standards and taxonomies in the development of new platforms and to develop decentralized platforms instead of centralized approaches in order to avoid hosting and funding problems for the system to be installed in case of a disaster, the EPISECC CIS has been designed.

A CIS concept has multiple advantages compared to the current information exchange approaches that are based on direct connection of all systems of the stakeholders (see Fig. 1). Automatic sharing of information ensures use of correct addresses and saves time, assuring that measures can be taken faster in time critical situations. Moreover, the requested technical efforts are reduced in a long term, because it is only necessary to design one adaptor for each system connected to the

[5]EPISECC, Deliverable 3.4 "Pan-European inventory of disasters and business models for emergency management services", available at: https://www.episecc.eu.

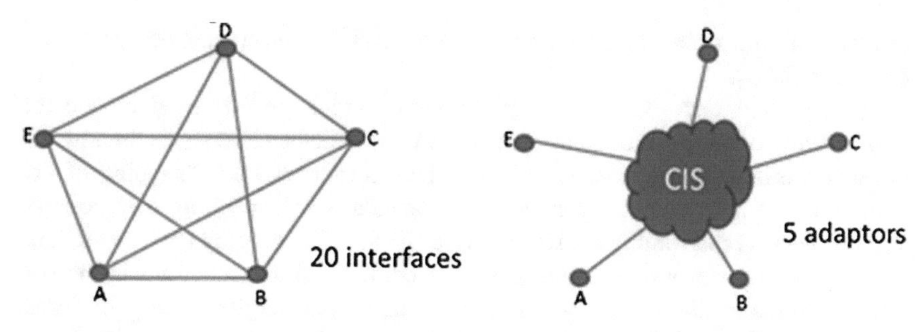

Fig. 1 Connecting 5 communication tools without/with CIS

CIS only once. There is not need to develop new interfaces each time a new system is included in a communication process.[6]

5.2 *Information Sharing*

For efficient response to a disaster, access to necessary information, seamless communications between rescuers and stakeholders as well as the coordinated availability of resources are key factors. Additional challenges particularly in cross border events include language barriers, organizational and technical barriers hindering communication and information exchange. To improve the collaboration in these situations, new concepts like a common information space are necessary.

The Common Information Space represents an architecture for sharing information in an easy way between the IT systems of organisations engaged in disaster relief. Beyond pure data exchange, it also includes concepts how to bridge the different terminologies making the available data understandable, and dynamic authorisation for safe data access dependent on needs in a given situation. The basic idea is to share information automatically between tools of different organisations that do not have dedicated interfaces for daily cooperation. Instead of developing specific interfaces for every (potential) pair of partners, CIS provides standardised interfaces that need to be implemented only once per tool (CIS adaptor).

The shared information is transported within the CIS as standard messages, using both standard formats (e.g. CAP, EMSI), and uniform EPISECC taxonomy for key terms which are not predefined by the used standard definition and namespace. While creating/interpreting the standard messages sent/received via CIS, the CIS adaptor has to transform proprietary interface formats of the connected tool to the CIS standard syntax and semantic. That means the adaptor has firstly to transform the parameters of the tool used by the sending organisation to the

[6]Could interoperability have prevented the metro bombing in Brussels?, Weblink: http://www.iessolutions.eu/en/interoperability-metro-bombing-brussels/, accessed 03.06.2016.

elements of the used standard and backwards at the receiver side (*syntactical interoperability*).

As a second step, it has to map the terms and keywords used by the tool (organisation owning a tool connected to the CIS) with standard terms used in CIS, in other words *semantic interoperability*. This comprises both, mapping of proprietary terms to key values defined by the standard, and mapping of proprietary terms to values representing the standardised EPISECC taxonomy (free text will not be translated by semantic mapping). To overcome possible limitations in the coverage of terms in the taxonomy, the project has introduced the concept of "broad match", according to which, each term will be available on the receiving end either as an "exact match" or flagged as having the closest meaning (taxonomy). The more the taxonomy will be used and enriched, the more accurate matches will be offered to the users. The transmitted information will always contain both, the originally sent message providing accuracy and the semantic annotations providing an easier understanding for the receiver.

All business logic and the ownership of information stay with the tools participating in information sharing. The common information space itself does not create, own or process the data. It is not a central data repository, but just an information broker that distributes information that is released for sharing with defined partners (Fig. 2).

The standard-based adapter concept ensures information exchange without the need of knowing the interfaces and terminologies of other partners. That allows ad hoc integration of new tools or services without modifying existing adaptors.

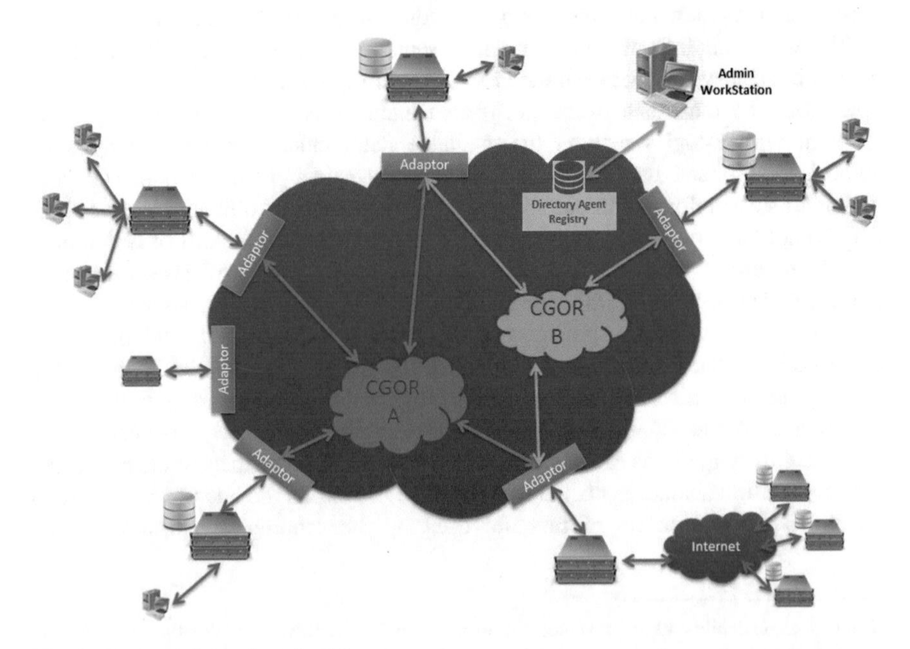

Fig. 2 Sharing information via CIS

Beyond the technical capability for automated data sharing, trust between the partners and confidence in data security and integrity is key for the acceptance and willingness of organisations to share their information. A context aware and configurable security concept is integrated in the CIS architecture. The trust policy requires the registration and certification of the CIS participants. Communication Groups and Online Rooms (CGOR) can be established dynamically according to current communication needs, and only the trusted and accepted participants are able to read or publish the information circulated in a CGOR. Data wrapping and encryption related to the CGOR prohibit unauthorised use of the shared information and minimises security risks. While the CIS concept provides appropriate tools for configuring safe communication groups, the responsibility for the classification of information and the lawful handling of sensitive date remains with the participating organisations and their tools.

6 Conclusion

It is important to note that the EPISECC project does not intend to replace existing and successful operational procedures of first responders. It rather aims at creating an information platform that would assist PPDR organisations with additional information and with the capability for cooperation. Simple but effective collaboration procedures shall be added in order to achieve acceptance within the stakeholder community. The operational capacity of CIS will be first tested in the proof of concept of the EPISECC scenario. This scenario will demonstrate the extent to which the CIS architecture is of relevance, actuality and completeness of shared data, when it comes to information sharing between different types of first responders acting jointly in transboundary emergency events and crises.

A prototype implemented for the proof of concept exercise will comprise the CIS distribution mechanism including the data protection components and adapters for tools that the project partners bring in, as well as samples of taxonomy from the participating end-user organisations.

Acknowledgments This paper was made possible thanks to the funding from the European Union's Seventh Framework Programme for research, technological development and demonstration under the EPISECC project (Establish Pan-European information space to Enhance seCurity of Citizens), grant no. 607078.

References

Could interoperability have prevented the metro bombing in Brussels? Weblink: Retrieved July 03, 2016, from http://www.iessolutions.eu/en/interoperability-metro-bombing-brussels/.

Critical Infrastructure Warning Information Network (CIWIN), Weblink: Retrieved July 03, 2016, from http://ec.europa.eu/dgs/home-affairs/what-we-do/networks/critical_infrastructure_warning_information_network/index_en.html.

EPISECC, Deliverable 3.4. *Pan-European inventory of disasters and business models for emergency management services*. https://www.episecc.eu.

Leach, R. J. (2016). *Introduction to software engineering* (2nd ed.). Chapman and Hall/CRC, 2016. VitalSource Bookshelf Online.

World Disaster Report 2013, International Federation of Red Cross, Weblink: Retrieved July 03, 2016, from http://worlddisastersreport.org/en/chapter-3/index.html.

Part III
Energy Systems

Integrating Social Acceptance of Electricity Grid Expansion into Energy System Modeling: A Methodological Approach for Germany

Karoline A. Mester, Marion Christ, Melanie Degel and Wolf-Dieter Bunke

Abstract Present energy system models are mainly based on techno-economical input parameters whereas social or political factors are neglected. This paper presents an approach to include social acceptance in energy system modeling, focusing on electricity transmission grid expansion projects in Germany. Qualitative as well as quantitative research techniques were applied: An analysis to quantify social acceptance was developed and implemented for 19 German districts (*Landkreise*) and acceptance-based delay assumptions for all German districts were derived. The dimension of social acceptance was integrated through years of delay. On the basis of assumed delays, different electricity grid expansion scenarios could be created. The results show that low delays can only be expected in regard to a few projects in Schleswig-Holstein. Moreover, the findings emphasize the importance of a commitment to grid expansion projects on the part of regional governments.

Keywords Electricity transmission grid expansion · Energy system modeling · Scenarios · Social acceptance · Socioeconomic factors

K.A. Mester (✉) · M. Degel
IZT, Schopenhauerstr. 26, 14129 Berlin, Germany
e-mail: k.mester@izt.de

M. Degel
e-mail: m.degel@izt.de

M. Christ · W.-D. Bunke
Europa-Universität ZNES Flensburg, Munketoft 3b, 24937 Flensburg, Germany
e-mail: marion.christ@uni-flensburg.de

W.-D. Bunke
e-mail: wolf-dieter.bunke@uni-flensburg.de

V. Wohlgemuth et al. (eds.), *Advances and New Trends in Environmental Informatics*, Progress in IS, DOI 10.1007/978-3-319-44711-7_10

1 Introduction

The transformation of the energy system towards a low carbon future with secure energy supply around the globe represents one of the major challenges of the 21st century. In order to illustrate pathways of integrating larger shares of renewable energies, various energy system models have been developed in recent years (Connolly et al. 2010). Although the energy transition is regarded as a "social contract for sustainability" (Wissenschaftlicher Beirat der Bundesregierung Globale Umweltveränderungen 2011), present energy models used to generate scenarios of preferable futures focus on technical and economic input parameters. They rarely consider social or political aspects (Mai et al. 2013; Fortes et al. 2015). The need for "integrative" (Grunwald and Schippl 2013) energy system modeling can be explained in the context of the German *Energiewende*: Despite a general acceptance of the promotion of renewable energies by the German population (Agentur für Erneuerbare Energien e.V. 2016), infrastructure measures affecting the landscape are facing increasing public oppositions at local levels (Wüstenhagen et al. 2007; Grunwald and Schippl 2013). This is evident especially in the case of the electricity transmission grid expansion.

In energy system modeling, the electricity grid development is considered by specifying the particular year when planned transmission capacities will be installed (also known as commissioning year). In Germany the four transmission system operators (TSOs) identify the required grid expansion projects and its associated transmission capacities on a yearly basis. Since 2012, they have been in charge of designing the Network Development Plan (*Netzentwicklungsplan—NEP*). Every *NEP* is based on a scenario framework representing presumptions of energy generation and consumption for the next ten (and 20) years. In line with the *NEP* 2015, preferential requirement is accorded to 22 projects based on the Power Grid Expansion Act 2009 (*EnLAG*) and to 43 projects corresponding to the Federal Requirement Plan Act (*BBPlG* in combination with *NABEG* 2013, *EnWG* 2011) (50 HzTransmission GmbH et al. 2016). Predominantly, these projects were designed for transmitting the electricity generated by wind turbines (and conventional power plants) in the north of Germany to major power consumption areas in the south (Gailing et al. 2013). In contrast to this political target the implementation is proceeding slowly: Only about one third of the total length of planned *EnLAG* projects has been completed until today (Bundesnetzagentur 2016). Besides other obstacles like the cooperation of federal authorities or the modification of legal rules, problems of social acceptance are increasingly causing delays (Deutsche Energie-Agentur GmbH 2012). That is why in the last few years numerous studies had been published, referring e.g. to characteristics of protests of particular projects or recommendations to increase social acceptance of grid expansion projects in Germany (Deutsche Umwelthilfe e.V. 2010; Rau and Zoellner 2010; Kurth 2011; Schnelle and Voigt 2012; Zimmer et al. 2012; Deutsche Umwelthilfe e.V. 2013; Menges and Beyer 2014; Lienert et al. 2015). More than 26,000 comments from the public on the *NEP* 2014 underline the actual relevance of this issue in Germany

(50 Hz Transmission GmbH et al. 2014). Thus, it is argued that commissioning years of projects can be delayed due to social acceptance issues. As a consequence, social acceptance may become a constraining factor for the success of the German energy transition.

The objective of this study is to develop an approach to include social acceptance in energy system modeling. Thereby the understanding of social acceptance comprises not only the occurrence of resistance against but also the commitment to grid expansion projects (Rau et al. 2009). The study focuses on planned *EnLAG* and *BBPlG* projects whereas the determination of future or alternative power lines was excluded. The approach presented in this paper is part of the *VerNetzen* research project that aims for the technical, economical, social and ecological modeling of future pathways towards a 100 % renewable electricity supply in Germany until 2050 (Bunke et al. 2015; Degel and Christ 2015; Christ et al. 2016). For the subsequent simulation the open source model *renpass* (renewable energy pathways simulation system), developed at the University (*Europa-Universität*) of Flensburg (Wiese 2015), is used.

The next section presents the developed methodological framework, followed by results and the discussion explained in Sects. 3 and 4, completed by concluding remarks.

2 Methodology

With regard to the original commissioning year of grid expansion projects, the authors decided to indicate the dimension of social acceptance through years of delay. In the course of reverting to the original commissioning year and consulting the assumed delays of each *EnLAG* or *BBPlG* project, new commissioning years of planned transmission capacities could be calculated.

The methodology consists of two main sections: (i) the development and implementation of an approach to quantify social acceptance in selected German districts (*Landkreise*); (ii) the integration of the results into energy system modeling by deriving acceptance-based delay assumptions for Germany and creating electricity grid expansion scenarios. The research design of this study followed the objective to integrate qualitative data into a quantitative model. Consequently, it was composed of a mix of qualitative and quantitative techniques, referring to the approach of mixed methods (Fakis et al. 2014; Kuckartz 2014).

2.1 Social Acceptance in Selected Districts

First attempts to systematically evaluate the social acceptance of grid expansion projects were made by Neukirch (2014). His approach served as a basis for this study. However, it did not comprise the examination of commitments to grid expansion projects, and the evaluation of protests remained on a qualitative, narrative level. In order to develop a measurement for the resistance against and the commitment to electricity grid expansion projects in this approach, eleven sub-projects of *EnLAG 1-6* and two sub-projects of *BBPlG 8* had been analyzed, representing overall 19 German districts. The selection process was mostly determined by activities of the environmental and consumer protection non-governmental organization (NGO), called *Deutsche Umwelthilfe e.V.* (*DUH*).[1] The *DUH* gained considerable experience in regard to social acceptance of grid expansion projects since it organized stakeholder information and dialogue events, workshops and discussions on-site of the above-named projects (Becker and Grünert 2015).

The findings derived from these proceedings referred to investigated projects: reasons, regional diffusion and actors of protests or applied participatory methods to involve the public. This qualitative data was converted into spreadsheets, containing information about the investigated project, the district where a categorized concern had been expressed by which stakeholder (e.g. citizens' initiative) and the applied participatory process. Overall 356 entries could be collected. It has to be pointed out that these findings were based on subjective expert experience of the *DUH*. Furthermore, the number of entries differed among districts. As a second step, the project-specific data was structured by defining (sub-) categories for the subjects 'contents and actors of protest' and 'degree of public participation' so that protests and participation processes could be analyzed and characterized for each examined district. For example, in terms of 'contents of protest' the following categories (representing at the same time reasons for resistance) were created: 'technical planning', 'location/development of the region', 'distributive and procedural justice' and 'grid expansion requirement'.

Moreover, the *DUH* determined success factors for realizing grid expansion projects: consistent energy transition policy, commitment to grid expansion by federal state and district governments or extensive and early public participation (e.g. affording citizens to comment on projects before the official planning approval proceedings) (Becker and Grünert 2015). In summary, the input of the *DUH* was used to develop two systems of indicators, each of them consisting of seven indicators (Table 1):

- The rate of resistance against a project defines the extent of activities that could lead to a delay of the initially intended commissioning year. The indicators relate to: contents (width, type, status of regional diffusion), actors (width, type, operating level) and effectively implemented actions (e.g. legal complaints) of protest. Six out of seven indicators relied on project-specified data of the *DUH*.
- The rate of commitment to a project describes the extent of activities that could lead to meet the initially intended commissioning year. The indicators are related to activities by: federal states (promotion of renewable energies and commitment to grid expansion), districts (promotion of renewable energies and

[1]The *DUH* is considered as *VerNetzen* research project partner.

Table 1 Indicators and weighing of the rate of resistance and the rate of commitment

Field	Group	Indicator	Weighing
Resistance	Contents of protest	Width[a]	0.125
		Type[a]	0.125
		Status of regional diffusion[a]	0.125
	Actors of protest	Width[a]	0.125
		Type[a]	0.125
		Operating level[a]	0.125
	Actions of protest	Effectively implemented actions	0.250
Commitment	Governmental activities of federal state	Promotion of renewable energies	0.125
		Commitment to grid expansion	0.125
	Governmental activities of district	Promotion of renewable energies	0.125
		Commitment to grid expansion	0.125
	Activities of TSO	Degree and time of public participation[a]	0.300
		Preparation of information	0.100
		Conversational efforts	0.100

[a]Based on project-specific data of the DUH

commitment to grid expansion) and TSOs (degree and time of public participation, preparation of information and conversational efforts). One of these indicators was based on the project-specific data of the *DUH*.

In general, the development of the named indicators followed this process: Based on project-related data of the *DUH*, literature reviews or online research, the first step was to create qualitative variables (representing nominal characteristic values) for every single indicator. Secondly, these variables were rated in accordance with the degree of resistance and commitment (the more points, the stronger the degree). Hence, the minimum and maximum achievable number of points were established for every indicator. After all, indicators had been standardized on a scale from zero to one, weighted (Table 1) and subsumed to one rate. Since a result of these rates could not be regarded as an absolute value, the authors utilized the approach of typification (Kelle and Kluge 2010) in order to indicate little, medium and strong resistance and commitment. Four types had been generated: best, mid-best, mid-worst, worst. In this context every variable was attached to one of these types. Also, the types were adjusted to the meaning of application: In the case of resistance, 'best' implied a low level of resistance. In the case of commitment, 'best' represented a strong commitment. The evaluation was conducted on the level of German districts. For detailed information about the developed approach to quantify social acceptance, see Mester (2016).

2.2 Integration into Energy System Modeling

The results of the first part were used to derive acceptance-based delay assumptions by opposing the evaluated degree of social acceptance to delays of regarded projects in a crosstabulation: To begin with, every investigated district was assigned to its identified combination of the type of resistance and commitment. Secondly, the districts among the registered combinations were compared with occurred actual delays (in years) of the sub-projects. It should be mentioned that at the time of this survey one of the investigated projects had been completed. The delays within districts that were associated with a specific combination of the type of resistance and commitment were calculated as statistical averages: by adding the number of years of actual delay of projects within districts that were assigned to a certain combination divided by the number of assigned districts. The examination of statistical spreads of delays supported the analysis. On the basis of occurred resistance, commitment and delays in regarded districts the authors derived certain levels of delay for different degrees of resistance and commitment.

Thereafter, the evaluation of resistance and commitment had to be proceeded for every German district (in total 395, out of 402 districts seven were excluded). This step proved to be challenging, as—in contrast to the rate of commitment—the approach of the rate of resistance could not be used to define the type of resistance in more than the already investigated districts. Because the additional data was not only difficult to access, but the recording of e.g. contents and actors of protest in every single German district would have also implied great research efforts, this analysis had been regarded as not feasible within the limits of the study.

In order to define regions where high delays of grid expansion projects could be expected, a quantitative data analysis was applied. This procedure was based on the assumption that districts with high delays could at the same time represent regions with strong resistance. Consequently a multiple regression analysis was implemented. The objective was to transfer delays of the overall 66 *EnLAG* sub-projects to future projects, by using statistical regional information. The values of utilized delays (in years) were based on the *EnLAG* monitoring report in 2014 (second quarter) (Bunke 2015). At that time, less than 25 % of all sub-projects were completed.

For the multiple regression analysis, Germany was divided into a georeferenced 400 × 400 m raster grid. The information about the delay of transmission grid projects was transferred to each raster field. The results of the regression model could be displayed either on the level of the raster fields, or aggregated for each German district. For the regression, a georeferenced database was used, containing information about population densities, renewable energy installations, existing railways and power lines, tourism, land use and settlements, economic factors and nature reserves. Based on the information in the database, 29 variables regarding their retarding effect on the examined *EnLAG* projects were used for the regression. The result of the optimal multiple regression model (out of 117 models) with almost 30 % of empirical variance ($R^2 = 0.29$) used ten variables to describe the retarding effect of the examined projects. These are the following (listed hierarchically according to their influence on the model): GDP and income per capita, existence of electricity grids, share of biosphere reserves, installed wind capacity per municipality, amount of touristic beds per municipality, share of forest area, share of nature parks, motorways and share of nature reserves. In this manner, a delay in years was generated for each raster field (for detailed information about the proceeding of the regression analysis, see Bunke 2015).

The results from the regression model were combined and adjusted with the findings of the qualitative analysis. The combination process is explained in Sect. 3 as it depended on further results. In order to transform the delay assumptions per district to the particular projects, an algorithm was developed to find the shortest connection with a minimum delay between the starting and ending point of the planned grid expansion projects. The final delay assumptions (combination of quantitative and qualitative analysis) per district were converted to the 400 × 400 m raster grid for the path-finding calculation. Additional to the starting and ending point, the planned commissioning year and a corridor for each project, defining the range where the future power line might be, was given as input for the path finding algorithm. With the algorithm a new commissioning year was defined for every planned grid expansion project.

The delay assumptions per district were not specified (by an exact number of years) but provided as ranges, representing low, medium and high delays (e.g. low: 0–2 a). In this context, three electricity grid expansion scenarios were created: By considering only the lowest value of the appropriate range which had been assigned to each district, the scenario 'low' was generated. Medium values were associated with scenario 'mid' and the upper boundary of the range was used as scenario 'high'. Depending on the chosen scenario, the resulting commissioning years of transmission capacities could be used as assumptions for energy system models.

3 Results

This section illustrates the results of the developed approach. It introduces not only the degree of social acceptance and occurred delays in selected districts, but also delay assumptions for Germany and the subsequent grid expansion scenarios.

3.1 Acceptance and Delay in Selected Districts

The systematic qualitative evaluation of acceptance according to the rate of resistance and commitment revealed that most of the regarded districts could be related to medium types of resistance (n = 16) and commitment (n = 17) (Table 2). Districts with medium resistance types had average delays of three to four years, varying from zero to eight years. Three districts were associated with resistance

Table 2 Occurred actual delays, resistances and commitments in 19 selected German districts

District	Project no. (EnLAG/BBPlG)	Actual delay of projects[b]	Rate of resistance	Type of resistance	Rate of commitment	Type of commitment
Dithmarschen	8_3[a]	0a	0.79	Mid-worst	0.93	Best
Nordfriesland	8_3/8_4[a]	0a	0.83	Worst	0.93	Best
Borken	5_2	1a	0.59	Mid-worst	0.57	Mid-best
Peine	6_1	3a	0.74	Mid-worst	0.53	Mid-worst
Hildesheim	6_1/6_2	3–5a	0.74	Mid-worst	0.57	Mid-best
Emsland	5_7/5_8	4a	0.57	Mid-worst	0.63	Mid-best
Steinfurt	5_7	4a	0.48	Mid-best	0.57	Mid-best
Hersfeld-Rotenburg	6_4	4a	0.67	Mid-worst	0.64	Mid-best
Kassel	6_4	4a	0.67	Mid-worst	0.64	Mid-best
Pinneberg	1_2	5a	0.56	Mid-worst	0.70	Mid-best
Segeberg	1_2	5a	0.56	Mid-worst	0.66	Mid-best
Göttingen	6_3	5a	0.67	Mid-worst	0.57	Mid-best
Northeim	6_2/6_3	5a	0.74	Mid-worst	0.57	Mid-best
Coburg	4_5	6a	0.70	Mid-worst	0.46	Mid-worst
Lichtenfels	4_5	6a	0.70	Mid-worst	0.46	Mid-worst
Märkisch-Oderland	3_1	7a	0.87	Worst	0.51	Mid-worst
Uckermark	3_1	7a	0.87	Worst	0.59	Mid-best
Diepholz	2_3	8a	0.65	Mid-worst	0.45	Mid-worst
Oldenburg	2_3	8a	0.65	Mid-worst	0.45	Mid-worst

[a]Investigated BBPlG projects
[b]Calculation based on Bundesnetzagentur 2011 and Bundes-netzagentur 2015, a = delay in years

type 'worst', referring to delays between zero and seven years and an average delay of five years. No district could be assigned to little protests ('best'). The opposite was valid for detected commitments: No district was related to the type 'worst'. Two districts were defined as type 'best', each of them with no actual delay. The delays in districts with a medium type of commitment ranged from one to eight years. Hence, medium commitment efforts could also imply high delays. In this context the results suggested that both, resistance and commitment seemed to appear more in medium and occasionally in strong degrees. The resulting types of resistance can be explained (amongst others) by the fact that in almost all of the districts, actors of protest submitted objections (n = 12) or even legal complaints (n = 4) against a project. In terms of commitments, it was evident that TSOs made considerable efforts to prepare information and to communicate with the public. Nevertheless, prevailing public participations hardly started before the official planning approval proceedings. Moreover, governmental activities to commit to grid expansion projects were neglected in several federal states and districts.

The next step was to derive delay assumptions from occurred delays and examined levels of acceptance (represented by the particular combinations of medium and strong resistance and commitment). In order to simplify the process, three categories of delay were established, representing little (0–2 a), medium (2–5 a) and high delays (5–9 a). On the basis of 19 investigated districts the following could be assumed: The combination of strong resistance and medium commitment was attached to high delays. Medium resistance and medium commitment were associated with medium delay. The central assumption was that with a medium or strong degree of resistance, only strong commitment could lead to little delays.

3.2 Delay Assumptions for Germany and Scenario Development

As stated in Sect. 2, the qualitative approach to evaluate resistance could not be applied to all German districts. However, the delay assumptions based on 19 selected districts should be transferred. In order to do so, the authors had to differentiate between regions with medium or strong commitment respectively medium or strong resistance. So far, the evaluation of commitment could be made for Germany: Only Dithmarschen and Nordfriesland, belonging to the federal state of Schleswig-Holstein, disposed of a strong commitment to electricity grid expansion. All other German districts (n = 393) referred to medium types of commitment. Equal to the already investigated 19 districts, a lack of commitment was noticed above all on the part of several federal states and districts.

In terms of estimating the degree of resistance, the question of where to expect medium or strong resistance could only be answered by reverting to statistical data. Considering the result that in districts with strong resistance on average five years of delay had been observed, the multiple regression analysis had been used to distinguish between districts with expected delays of less than five years and districts equal to five years of delay or above. The application of the regression model revealed that delays equal to five years or above could be expected in 49 German districts. Most of them were situated in the eastern part of Germany. All other districts (n = 346) were associated with delays of less than five years. In addition, districts with less than five years of delay were regarded as regions where medium resistances could occur. In the same way the authors assumed that districts with a prospective delay equal to five years or more reflect regions with the potential for strong resistances.

By combining the findings of the quantitative regression model (representing the results for resistance) with the actual qualitative assessment of commitment for all German districts and by referring to the delay assumptions derived from 19 investigated districts (Sect. 3.1), it was evident that the two districts Dithmarschen and Nordfriesland were assigned to low delays. This was caused by the assumption that with a medium or strong degree of resistance only strong commitments could result in low delays. For the majority of districts (n = 344) medium delays were assumed, based on the derived delay assumptions and the result that most German districts have a combination of medium resistance and medium commitment. For the 49 districts, where strong resistance and medium commitment was identified, high delays were assumed. In order to demonstrate the scope of the developed approach it was assumed that future public participation will be realized according to the type 'best' of the corresponding indicator of the rate of commitment (equal to an early public participation with citizens in an active role) in all German districts. This modification resulted in three additional districts with low delays (Ostholstein, Pinneberg and Steinburg, also located in Schleswig-Holstein). Presuming that future participation will be undertaken to that extent, the modified delay assumptions were adopted for the process of creating electricity grid expansion scenarios. Figure 1 shows the results of scenario 'high'. In each of the defined scenarios ('low', 'mid', 'high'), the only power lines with delays ranging from zero (scenario 'low') to two years (scenario 'high') were the projects *BBPlG 8* and *33* (Fig. 1), located in Dithmarschen and Nordfriesland. Maximum delays were defined for projects in the eastern part of Germany (e.g. *BBPlG 11*). Average delays of all planned *EnLAG* and *BBPlG* power lines varied among scenarios: from two (scenario 'low'), four (scenario 'mid') to five years (scenario 'high').

The calculated commissioning years were integrated into the simulation model *renpass* by defining the available transmission capacities between each German dispatch region (21 German dispatch regions and exchange to neighboring countries) for every year. With regard to the simulation of the German electricity system in 2050, the developed grid scenarios can be used to analyze possible consequences of social acceptance issues for the German energy transition.

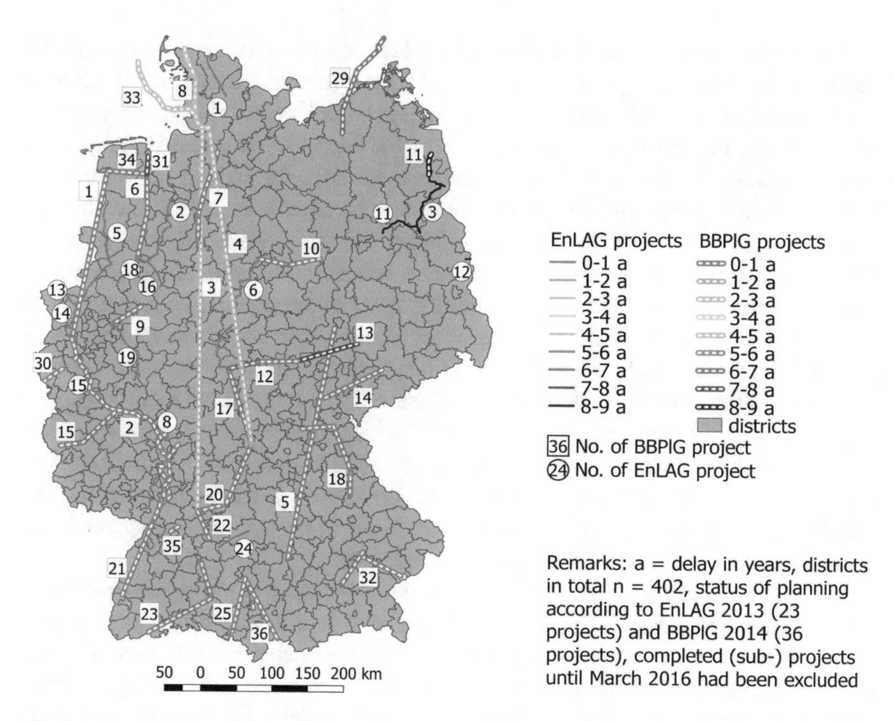

Fig. 1 Expected delays of planned *EnLAG* and *BBPlG* projects according to the electricity grid expansion scenario 'high'

4 Discussion and Conclusions

In this section the developed methodological approach and its implications for the energy system modeling will be critically reviewed.

In accordance with the derived delay assumptions and the result that most of the German districts were classified as regions where medium resistances can occur, it is argued that it takes at least medium commitments to avoid high delays. Only the districts Dithmarschen and Nordfriesland could be related to low delays. These two districts are part of the grid development initiative (*Netzentwicklungs-initiative*), together with the federal state of Schleswig-Holstein, other affected districts, TSOs and additional stakeholders (Hirschfeld and Heidrich 2013). The participation in this initiative demonstrates the outstanding position of the mentioned districts, and it suggests that such an initiative could help to prevent delays. Moreover, it was evident that for most of the German districts, the modification of the assumed degree of public participation did not result in the type of commitment 'best'. This implies that a more extensive and early participation process alone is not sufficient to lower implementation delays. As the analysis shows that on the part of the TSOs, considerable efforts have already been made, it is governments of federal states and districts that need to increase their activities.

It has to be pointed out that during the described analysis most of the investigated projects had not yet been completed. The difference between the planned commissioning year and the actual or anticipated completion of the power lines was set according to the information available at the time of this research. The actual delays of these projects, which were used as a basis for the scenario development, can therefore increase further. In the same way it should be mentioned that the variables used to describe the retarding effect of the examined *EnLAG* projects rely on the location of these projects. Since only a few examined *EnLAG* projects were situated in the south of Germany, the results of the regression model indicated low delays, e.g. in Bavaria. Comparing the delay assumptions with present acceptance problems in these regions, it can be concluded that even higher delays are likely to appear in reality than in the developed scenarios. The results of the methodological approach strongly depend on available data and the developed valuations systems (rate of resistance and commitment, regression model). Therefore, a high level of transparency is necessary for the implementation as well as for the interpretation of the described approach. The delays in years have to be regarded as approximate scenario values, and not used as precise projections.

All in all, this paper demonstrates how social acceptance could be integrated into energy system modeling. The developed approach is a first step towards an integrative research concerning the transformation of the energy system, including not only techno-economical but also social or political factors. In this context, it can be applied to expand the results of present energy models. Besides, the approach allows for an analysis of social acceptance at local levels so that it can help to detect and recommend (political) needs for action: for example in critical regions where planned projects that have to be completed at the initially intended year (with regard to security of electricity supply) are confronted with strong resistances and (only) medium commitments. Again, it is underlined that the approach is used to analyze possible implications of social acceptance and delays for the energy system and should not be regarded as a planning instrument for the grid development because its results might not be socially optimal: Routing grids by avoiding districts with strong resistance and high delays would lead to an exclusion of these regions on the expense of 'poorer' regions (where resistance is less likely) that would have to accept higher shares of grid expansion projects.

Finally, the approach depends on further developments: It requires an extended (with special regard to the data base of 19 districts for the survey of resistance and derived delay assumptions) and continuing input of data in order to improve the quality of assumptions. In particular, the investigation of commitments should be pursued, e.g. to determine impacts of enlarged participation processes. Further research is also attached to the potential to advance the predominant use of open data and open source software: By providing not only the calculated commissioning years but also the assessed (qualitative) data in terms of resistance and commitment per district with an open source database, (regional) governments could directly benefit from specific examinations of particular regions without depending on general recommendations.

Acknowledgments The developed approach is a result of the research activities of the Institute for Futures Studies and Technology Assessment (IZT), the *Europa-Universität* Flensburg and the *DUH* within the *VerNetzen* project. The authors acknowledge Judith Grünert and Liv Anne Becker (*DUH*) for their contributions to the project. The presented results are based on quantitative analyses by Wolf-Dieter Bunke (2015) and the combination with qualitative investigations by Karoline Mester, see Mester (2016).

References

50 Hertz Transmission GmbH, Amprion GmbH, TenneT TSO GmbH, TransnetBW GmbH (2014) Netzentwicklungsplan Strom 2014: Zweiter Entwurf der Übertragungsnetzbetreiber. Retrieved November 9, 2015, from http://www.netzentwicklungsplan.de/netzentwicklungsplaene/2014.

50 Hertz Transmission GmbH, Amprion GmbH, TenneT TSO GmbH, TransnetBW GmbH (2016) Netzentwicklungsplan Strom 2025, Version 2015: Zweiter Entwurf der Übertragungsnetzbetreiber. Retrieved May 20, 2016, from http://www.netzentwicklungsplan.de/netzentwicklungsplan-2025-version-2015-zweiter-entwurf.

Agentur für Erneuerbare Energien e.V. (2016). Akzeptanzumfrage 2014: 92 Prozent der Deutschen unterstützen den Ausbau Erneuerbarer Energien. Retrieved April 13, 2016, from https://www.unendlich-viel-energie.de/themen/akzeptanz2/akzeptanz-umfrage/akzeptanzumfrage-2014.

Becker, L., & Grünert, J. (2015). Basispapier: Die politisch-rechtlichen Rahmenbedingungen des Stromnetzausbaus und ihre Auswirkungen auf die Akzeptanz: Entwurf. Forschungsarbeiten der DUH im Rahmen des Projekts VerNetzen: Sozial-ökologische, technische und ökonomische Modellierung von Entwicklungspfaden der Energiewende (Arbeitspaket 2). Deutsche Umwelthilfe e.V. (DUH), Berlin.

Bundesnetzagentur. (2011). Bericht gemäß § 63 Abs. 4a ENWG zur Auswertung der Netzzustands- und Netzausbauberichte der deutschen Elektrizitätsübertragungsnetzbetreiber. Retrieved January 25, 2016, from http://www.bundesnetzagentur.de/SharedDocs/Downloads/DE/Sachgebiete/Energie/Unternehmen_Institutionen/Versorgungssicherheit/Berichte_Fallanalysen/Bericht_6.pdf?__blob=publicationFile&v=2.

Bundesnetzagentur. (2015). EnLAG-Monitoring: Stand des Ausbaus nach dem Energieleitungsausbaugesetz (EnLAG) zum zweiten Quartal 2015. Retrieved January 18, 2016, from http://www.netzausbau.de/leitungsvorhaben/de.html.

Bundesnetzagentur. (2016). Leitungsvorhaben. Retrieved April 13, 2016, from http://www.netzausbau.de/leitungsvorhaben/de.html.

Bunke, W. (2015). Entwicklung einer Methodik zur Erstellung von Netzausbauszenarien in Abhängigkeit von Akzeptanzannahmen. Abschlussarbeit im Rahmen des Projekts VerNetzen: Sozial-ökologische, technische und ökonomische Modellierung von Entwicklungspfaden der Energiewende; Masterthesis im Masterstudiengang Energy and Environmental Management, Europa-Universität Flensburg.

Bunke, W., Christ, M., & Degel, M. (2015). VerNetzen: Sozial-ökologische und technisch-ökonomische Modellierung von Entwicklungspfaden der Energiewende. In: Bundesnetzagentur (Ed.), *Wissenschaftsdialog 2014: Technologie, Landschaft und Kommunikation, Wirtschaft* (pp. 107–128).

Christ, M., Soethe, M., & Degel, M. (2016). Windausbauszenarien unter Berücksichtigung gesellschaftlicher und ökologischer Rahmenbedingungen für die Simulation des deutschen Energiesystems bis 2050 Tagungsband: EnInnov2016. 14. Symposium Energieinnovation, pp. 110–111.

Connolly, D., Lund, H., Mathiesen, B. V., & Leahy, M. (2010). A review of computer tools for analysing the integration of renewable energy into various energy systems. *Applied Energy, 87* (4), 1059–1082. doi:10.1016/j.apenergy.2009.09.026.

Degel, M., & Christ, M. (2015). Sozial-ökologische und technisch-ökonomische Modellierung von Entwicklungspfaden der Energiewende Tagungsband: FONA Statuskonferenz des BMBF zur Fördermaßnahme Umwelt- und gesellschaftsverträgliche Transformation des Energiesystems, pp. 197–202.

Deutsche Energie-Agentur GmbH. (2012). Stand der Genehmigungsverfahren der Netzausbauprojekte nach EnLAG. Retrieved April 13, 2016, from http://www.dena.de/fileadmin/user_upload/Projekte/Energiedienstleistungen/Dokumente/Stand_EnLAG-Trassen.pdf.

Deutsche Umwelthilfe e.V. (Ed). (2010). Plan N: Handlungsempfehlungen an die Politik. Forum Netzintegration Erneuerbare Energien.

Deutsche Umwelthilfe e.V. (Ed). (2013). Plan N 2.0: Politikempfehlungen zum Um- und Ausbau der Stromnetze. Forum Netzintegration Erneuerbare Energien.

Fakis, A., Hilliam, R., Stoneley, H., & Townend, M. (2014). Quantitative analysis of qualitative information from interviews: A systematic literature review. *Journal of Mixed Methods Research, 8*(2), 139–161.

Fortes, P., Alvarenga, A., Seixas, J., & Rodrigues, S. (2015). Long-term energy scenarios: Bridging the gap between socio-economic storylines and energy modeling. *Technological Forecasting and Social Change, 91*, 161–178. doi:10.1016/j.techfore.2014.02.006.

Gailing, L., Hüesker, F., Kern, K., & Röhring, A. (2013). Die räumliche Gestaltung der Energiewende zwischen Zentralität und Dezentralität: Explorative Anwendung einer Forschungsheuristik. Retrieved April 29, 2015, from http://www.irs-net.de/download/wp_energiewende_raum_zentral_dezentral.pdf.

Grunwald, A., & Schippl, J. (2013). Die Transformation des Energiesystems als gesellschaftliche und technische Herausforderung: Zur Notwendigkeit integrativer Energieforschung. In J. Radtke, & B. Hennig (Eds.), *Die deutsche „Energiewende" nach Fukushima: Der wissenschaftliche Diskurs zwischen Atomausstieg und Wachstumsdebatte* (1st ed. Beiträge zur sozialwissenschaftlichen Nachhaltigkeitsforschung, Vol. 8, pp. 21–35). Metropolis, Weimar (Lahn).

Hirschfeld, M., & Heidrich, B. (2013). Die Bedeutung regionaler Governance-Prozesse für den Ausbau des Höchstspannungsnetzes. In B. Klagge & C. Arbach (Eds.), *Governance-Prozesse für erneuerbare Energien. Arbeitsberichte der ARL* (Vol. 5, pp. 94–113). Hannover: Akademie für Raumforschung und Landesplanung.

Kelle, U., & Kluge, S. (2010). *Vom Einzelfall zum Typus: Fallvergleich und Fallkontrastierung in der qualitativen Sozialforschung* (2nd ed.). Qualitative Sozialforschung. Wiesbaden: VS Verlag für Sozialwissenschaften.

Kuckartz, U. (2014). *Mixed Methods: Methodologie, Forschungsdesigns und Analyseverfahren* (1st ed.). Lehrbuch: VS Verlag für Sozialwissenschaften, Wiesbaden.

Kurth, M. (2011). Was verzögert den Bau von Stromleitungen? Thesenpapier des Präsidenten der Bundesnetzagentur. Retrieved April 13, 2016, from https://www.wirtschaftsrat.de/wirtschaftsrat.nsf/id/4B8B8604F67E7894C1257841005090E1/$file/kurth_thesenpapier_klausurtagung_energie_umwelt_25_02_11.pdf.

Lienert, P., Suetterlin, B., & Siegrist, M. (2015). Public acceptance of the expansion and modification of high-voltage power lines in the context of the energy transition. *Energy Policy, 87*, 573–583. doi:10.1016/j.enpol.2015.09.023.

Mai, T., Logan, J., Blair, N., Sullivan, P., & Bazilian, M. (2013). *RE-ASSUME—A decision maker's guide to evaluating energy scenarios, modeling, and assumptions*. IEA—Renewable Energy Technology Deployment.

Menges, R., & Beyer, G. (2014). Underground cables versus overhead lines: Do cables increase social acceptance of grid development? Results of a contingent valuation survey in Germany. *International Journal of Sustainable Energy Planning and Management*(03), 33–48.

Mester, K. A. (2016). Entwicklung einer Methodik zur Integration gesellschaftlicher Akzeptanz in die Strommarkt-Modellierung am Beispiel ausgewählter Netzausbauvorhaben in Deutschland. Abschlussarbeit im Rahmen des Projekts VerNetzen: Sozial-ökologische, technische und ökonomische Modellierung von Entwicklungspfaden der Energiewende; Masterarbeit im

Studiengang Sustainability Economics and Management, Carl von Ossietzky Universität Oldenburg.

Neukirch, M. (2014). Konflikte um den Ausbau der Stromnetze: Status und Entwicklung heterogener Protestkonstellationen. Retrieved April 29, 2015, from http://www.uni-stuttgart.de/soz/oi/publikationen/soi_2014_1_Neukirch_Konflikte_um_den_Ausbau_der_Stromnetze.pdf.

Rau, I., Schweizer-Ries, P., & Zoellner, J. (2009). Akzeptanz Erneuerbarer Energien und sozialwissenschaftliche Fragen. Projektabschlussbericht. Retrieved April 28, 2015, from http://edok01.tib.uni-hannover.de/edoks/e01fb09/612638286.pdf.

Rau, I., & Zoellner, J. (2010). Abschlussbericht: „Umweltpsychologische Untersuchung der Akzeptanz von Maßnahmen zur Netzintegration Erneuerbarer Energien in der Region Wahle - Mecklar (Niedersachsen und Hessen)". Retrieved April 13, 2016, from http://www.fg-umwelt.de/assets/files/Akzeptanz%20Netzausbau/Abschlussbericht_Akzeptanz_Netzausbau_Juni2010.pdf.

Schnelle, K., & Voigt, M. (2012). Energiewende und Bürgerbeteiligung: Öffentliche Akzeptanz von Infrastrukturprojekten am Beispiel der „Thüringer Strombrücke". Retrieved November 9, 2015, from http://germanwatch.org/fr/download/4135.pdf.

Wiese, F. (2015). renpass—Renewable Energy Pathways Simulation System—Open Source as an approach to meet challenges in energy modeling. Doctoral thesis, Europa-Universität Flensburg.

Wissenschaftlicher Beirat der Bundesregierung Globale Umweltveränderungen. (2011). Welt im Wandel: Gesellschaftsvertrag für eine Große Transformation. Hauptgutachten. Retrieved April 29, 2015, from http://www.wbgu.de/fileadmin/templates/dateien/veroeffentlichungen/hauptgutachten/jg2011/wbgu_jg2011.pdf.

Wüstenhagen, R., Wolsink, M., & Bürer, M. J. (2007). Social acceptance of renewable energy innovation: An introduction to the concept. *Energy Policy, 35*(5), 2683–2691. doi:10.1016/j.enpol.2006.12.001.

Zimmer, R., Kloke, S., & Gaedtke, M. (2012). Der Streit um die Uckermarkleitung—Eine Diskursanalyse: Studie im Rahmen des UfU-Schwerpunktes „Erneuerbare Energien im Konflikt". Retrieved November 9, 2015, from http://www.ufu.de/media/content/files/Fachgebiete/Ressourcenschutz/Publikationen/Streit%20um%20die%20Uckermarkleitung.pdf.

Dynamic Portfolio Optimization for Distributed Energy Resources in Virtual Power Plants

Stephan Balduin, Dierk Brauer, Lars Elend, Stefanie Holly, Jan Korte, Carsten Krüger, Almuth Meier, Frauke Oest, Immo Sanders-Sjuts, Torben Sauer, Marco Schnieders, Robert Zilke, Christian Hinrichs and Michael Sonnenschein

Abstract The aggregation of distributed energy resources in virtual power plants (VPPs) is a feasible approach to overcome entry barriers for energy markets like, e.g., the European Power Exchange SE (EPEX SPOT SE). An increasing number of energy supply companies offer the integration of decentralized units in VPPs aiming to achieve the maximum profit by trading the power produced by the VPP at energy markets. However, the coordination of the generation units' operational modes (operation schedule) within a VPP as well as the selection of offered market products (product portfolio) are optimization problems that are mutually dependent. In this contribution a method is proposed that allows automating both the optimized composition of the product portfolio and the determination of the matching operation schedule for the VPP, in terms of profit maximization. Application example of the method is the EPEX SPOT SE day-ahead market. The concept of the approach can be roughly described as follows: First of all, machine learning techniques are used to predict the market prices for the trading day. Then, the market forecast is used in combination with feasible schedule samples from the generation units as input for the optimization process. During the optimization a hybrid approach comprising heuristic algorithms, such as simulated annealing or tabu search, and linear optimization is applied. The approach is evaluated using historical market data and intricate simulation models of generation units.

Keywords Virtual power plant · Energy market · Product portfolio optimization · Operation schedule optimization

S. Balduin (✉) · D. Brauer · L. Elend · S. Holly · J. Korte · C. Krüger · A. Meier · F. Oest · I. Sanders-Sjuts · T. Sauer · M. Schnieders · R. Zilke · C. Hinrichs · M. Sonnenschein
Carl von Ossietzky University of Oldenburg, 26111 Oldenburg, Germany

V. Wohlgemuth et al. (eds.), *Advances and New Trends in Environmental Informatics*, Progress in IS, DOI 10.1007/978-3-319-44711-7_11

1 Introduction

For some years, an increasing number of energy supply companies has employed VPPs in order to enable the politically intended market integration of renewable energy resources. Especially in Germany political decision-makers desire this integration since the previous direct feed-in of renewable energy resources into the power grid is expensive due to subsidies and affects the grid stability because of the limited predictability of, e.g., the produced wind power. A VPP is a logical aggregation of multiple distributed energy resources, such as photo-voltaic facilities, cogeneration units or wind power plants (Awerbuch and Preston 1997). VPPs integrate both company-owned and privately owned units. By regarding these aggregated resources as a single virtual entity, fluctuating generation can be balanced effectively. On the one hand, this compensates for uncertainties in power generation from individual energy units. On the other hand, the aggregation allows to overcome market entry barriers, so that trading the generated power in energy markets such

e-mail: Stephan.Balduin@uni-oldenburg.de; pg-vk@informatik.uni-oldenburg.de

D. Brauer
e-mail: Dierk.Brauer@uni-oldenburg.de; pg-vk@informatik.uni-oldenburg.de

L. Elend
e-mail: Lars.Elend@uni-oldenburg.de; pg-vk@informatik.uni-oldenburg.de

S. Holly
e-mail: Stefanie.Holly@uni-oldenburg.de; pg-vk@informatik.uni-oldenburg.de

J. Korte
e-mail: Jan.Korte@uni-oldenburg.de; pg-vk@informatik.uni-oldenburg.de

C. Krüger
e-mail: Carsten.Krueger@uni-oldenburg.de; pg-vk@informatik.uni-oldenburg.de

A. Meier
e-mail: Almuth.Meier@uni-oldenburg.de; pg-vk@informatik.uni-oldenburg.de

F. Oest
e-mail: Frauke.Oest@uni-oldenburg.de; pg-vk@informatik.uni-oldenburg.de

I. Sanders-Sjuts
e-mail: Immo.Sanders-Sjuts@uni-oldenburg.de; pg-vk@informatik.uni-oldenburg.de

T. Sauer
e-mail: Torben.Sauer@uni-oldenburg.de; pg-vk@informatik.uni-oldenburg.de

M. Schnieders
e-mail: Marco.Schnieders@uni-oldenburg.de; pg-vk@informatik.uni-oldenburg.de

R. Zilke
e-mail: Robert.Zilke@uni-oldenburg.de; pg-vk@informatik.uni-oldenburg.de

C. Hinrichs
e-mail: Christian.Hinrichs@uni-oldenburg.de; pg-vk@informatik.uni-oldenburg.de

M. Sonnenschein
e-mail: Michael.Sonnenschein@uni-oldenburg.de; pg-vk@informatik.uni-oldenburg.de

as the European Power Exchange SE (EPEX SPOT SE) is enabled. In this market, orders for hourly quantities of power (i.e., both supply and demand) are matched day ahead in an auction process.

A selection of the market products (product portfolio) that are to be offered in the market has to be made by the VPP operator. An optimal product portfolio enables the VPP operator and thus also the plant owners to achieve the maximum profit. The optimal solution of this product portfolio optimization problem depends on the operational modes (schedules) of the energy units in the VPP. A schedule specifies for each quarter-hour interval of the day which energy amount a unit has to produce. Vice versa, as many energy units are flexible in their operation to some degree, the optimal selection of a schedule for each unit depends on the target product portfolio. Nowadays, the selection of a product portfolio is usually still solved manually by the VPP operator. With an increasing number of units within the VPP, it becomes too complex to manually find a solution considering the flexibilities of each plant. In this paper a hybrid approach is presented that allows automatically solving these mutually dependent optimization problems in combination. Previous approaches take only one of these optimization problems into account (von Oehsen 2012; Sonnenschein et al. 2015) or are not applicable directly to a VPP (Gröwe-Kuska et al. 2003; Bagemihl 2003).

In the following, Sect. 2 discusses the two optimization problems and the proposed hybrid solver in detail. Afterwards the approach is evaluated in a simulation study in Sect. 3. Finally, Sect. 4 concludes the paper.

2 Optimization

As outlined in the introduction, the dynamic product portfolio optimization in VPPs comprises two mutually dependent optimization problems. A solution algorithm has to coordinate the schedules of the energy resources in order to obtain a feasible operation schedule for the VPP. In addition it has to select the market products which the VPP operator should offer at the energy market. In the optimization process, the amount of produced power and the amount of sold power have to be synchronized: If too much power is sold, the demand cannot be accommodated by the generation of the VPP and contractual penalties become due. If too much power is produced, the excess power cannot be sold on the market. Hence no profit is gained, while production costs still occur.

Both optimization problems rely on a defined power profile (i.e., the intended power generation for the course of one day). On the one hand, to determine which market products can be chosen requires information about the available power levels for each time period of the considered day. On the other hand, as many generation units are flexible to some degree, their schedules must be optimized towards realizing a given target power profile (that in turn depends on the chosen product portfolio). Thus the product portfolio optimization depends on fixed schedules for all plants, whereas the schedule optimization depends on a given target product portfolio. To resolve this dilemma, a bootstrap method is applied that allows solving one of the problems independently in order to establish feasible starting conditions. The result of this optimization can then be taken as input for the other optimization process.

This can be achieved by using a heuristic algorithm that generates random solutions in a structured way and evaluates their quality afterwards. For the computation of the solution quality the overall optimization result, including the product portfolio and the operation schedule, is needed. Accordingly, the optimization problems are split into an outer and an inner optimizer. The outer optimization uses a heuristic and generates a selection of random solutions. For each of these solutions, an inner optimization is performed to obtain the respective counterpart solution from the dependent optimization problem. Afterwards, the overall profit for each combined optimization result can be calculated. These are then again used by the outer heuristic to generate further optimized solutions to be fed to the inner optimizer. This iterative procedure repeats until a predefined termination criterion (e.g., a certain amount of elapsed time) is met.

Heuristics can be adapted to solve arbitrary optimization problems, but they cannot guarantee to find the optimal solution. Therefore it is preferable to use a non-heuristic algorithm for the inner optimization in order to improve the global result. To decide, which optimization is suitable as outer and which as inner optimization, both problems have to be regarded in depth and individually. This is done in the following subsections.

2.1 Product Portfolio Optimization (PPO)

Following the motivating use case of trading power in the EPEX SPOT SE day-ahead market, we focus on product types that are available in this market. The market rules permit selling individual hours or block products. A block comprises a sequence of hour products and is offered on an all-or-none basis. This means that all hour products within a block can only be sold collectively or not at all. There is a variety of predefined blocks such as "Sun-Peak" or "Rush Hour" (European Power Exchange 2016).

In order to be able to determine promising market products for the VPP, a precise prediction of the market prices of the following day is required. Furthermore, to trade a specific product, the amount of power and the offer price have to be chosen for this product. In the presented approach, this selection process (i.e., choosing products, their volumes and their offer prices) relies on a market forecast, which is performed before the product portfolio optimization. Historical market data, which is available freely from EPEX SPOT SE, and a meteorological weather forecast for the predicted day form the basis of this computation. For the calculation of the forecast, the popular Weka framework of the University of Waikato was employed, implementing various machine learning algorithms (Hall et al. 2009). Weka learns from the market and weather data of the last 110 days and predicts the price of each EPEX SPOT SE market product.

Table 1 Identifiers for the formulation of the PPO

i	Identifies a product within a portfolio, with $i \in \{0, 1, \ldots, n\}$, n = portfolio size
v_i	Hourly volume of product i in MW
$price_i$	Offer price of product i in $\frac{€}{\text{MWh}}$ (selected during the market forecast)
d_i	Duration of product i in hours (predefined from the EPEX SPOT SE)
h	Hour of the day, with $h \in \{0, 1, \ldots, 23\}$
$p(h)$	Amount of available power during hour h
$V(h)$	Set of products which are currently applicable in hour h

During the product portfolio optimization, only the amount of power for each product is variable. It is limited by the generated power of the VPP. For reasons of simplification it is assumed that the volume of blocks has to stay at a constant level for their whole duration, in order to satisfy acceptance conditions of the EPEX SPOT SE, which are contained in (European Power Exchange 2016). Using the identifiers given in Table 1, the objective of the PPO, to find the best distribution of the produced power on the market products, can then be formulated as follows:

$$\max\left(\sum_{i=0}^{n} price_i \cdot d_i \cdot v_i\right) \tag{1}$$

The product portfolio with the highest profit shall be chosen. $price_i \cdot d_i \cdot v_i$ represents the sales for product i if the volume v_i is chosen. In addition at no point of time more power must be sold, than power is produced:

$$\text{s.t.} \sum_{v_i \in V(h)} v_i \leq p(h) \tag{2}$$

Furthermore, by restriction of the EPEX SPOT SE, the volume of a product v_i has to be a multiple of 0.1 MW and within the permitted range:

$$0\text{ MW} \leq v_i \leq 600\text{ MW} \tag{3}$$

We consider two different methods for solving this optimization problem. As the problem is formulated in a set of (in)equalities and only discrete values can be chosen, it is an integer linear optimization problem (Dantzig and Thapa 1997). Therefore linear programming solvers can be applied. As an alternative, we propose RIPPO (Recursive-Interval-Product-Portfolio-Optimizer), which is a tailor-made solver for the above optimization problem.

In RIPPO, a given profile of available power is first split into rectangular regions of maximal width and height from bottom to top, as illustrated in Fig. 1 for three layers. For each region, the optimal combination of products is then calculated as follows. Let $R = (u, v, p)$ be the currently considered rectangular region with height

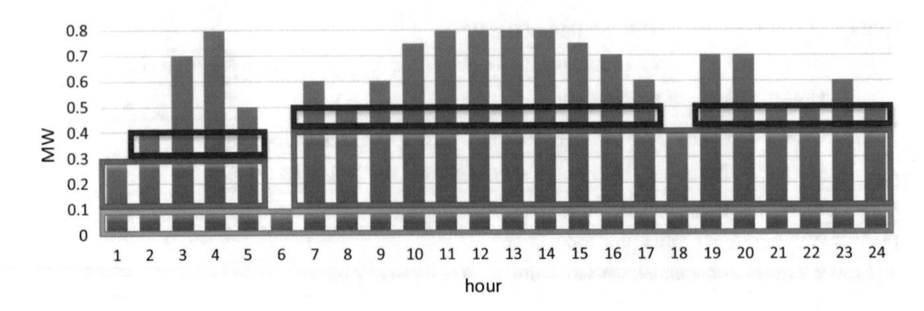

Fig. 1 Recursive interval partitioning

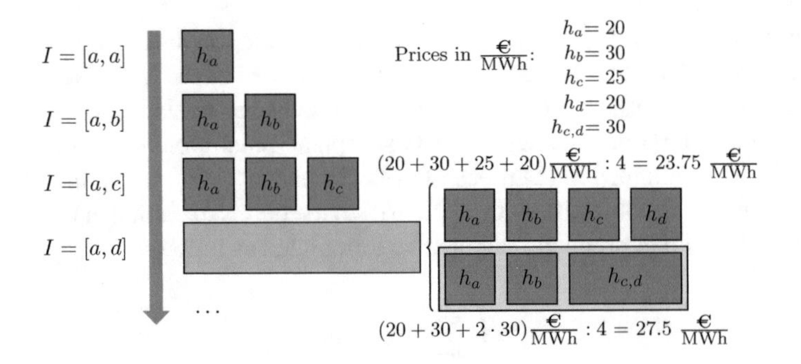

Fig. 2 Compositions of products in increasing intervals

p (given in MW), spanning hours u to v of the day. Similar to a dynamic programming approach (Bradley et al. 1977), optimal product combinations are now determined for increasing sections of R subsequently. For each subinterval $I = [u, i]$ with $u \leq i \leq v$ and increasing values of i, all possible market products of volume p are determined which start somewhere in I but end exactly at hour i. From these products, the optimal one with respect to previous (smaller) subintervals is selected. This way, the optimal combination of market products for R is constructed step by step. Figure 2 illustrates this process with an example of an arbitrary region $R = (a, d, p)$ with $d = a + 3$. In the first step, the subinterval $I = [a, a]$ is considered. As it spans just one hour, the only possible market product is the one-hour product h_a, which is set as initial solution for the subsequent steps. Following, subinterval $I = [a, b]$ is considered. Here, the one-hour product h_b is the only possible product that ends at hour b, so the result from the previous subinterval, h_a, is merged with the current product, h_b, to form the optimal product combination (h_a, h_b) so far. The same situation arises in $I = [a, c]$, resulting in (h_a, h_b, h_c). In $I = [a, d]$, however, we find two possible market products that end at hour d: the one-hour product h_d and the block-product $b_{c,d}$. This results in two solution candidates at this point in time: (h_a, h_b, h_c, h_d) and $(h_a, h_b, b_{c,d})$. Comparing these, the latter yields a higher profit and thus forms the optimal combination of market products in this example. After processing all regions this way, the resulting list of optimal market products can finally be truncated by merging similar products spanning the exact same hours of the day. For instance, having products h'_a with volume p' and h''_a with volume p'' in the result set, we combine these to form the product h_a with volume $p = p' + p''$.

In summary, both a linear solver and RIPPO yield optimal solutions for the PPO, but RIPPO has a considerably lower runtime as will be shown in the evaluation study (see Sect. 3.1). As both require accurate information about available power per time interval, the PPO qualifies as inner optimization only and needs the operation schedule as input from the outer optimizer.

2.2 Operation Schedule Optimization (OSO)

For each generation unit within a VPP, a schedule has to be selected in accordance with the target product portfolio on the one hand, and the inherent restrictions of the individual energy unit on the other hand. The set of selected schedules for all VPP members is called operation schedule (of the VPP). The involved restrictions differ among unit types and even for similar units due to varying operational modes. For demonstration, we consider photo-voltaic facilities (PV plants) and cogeneration units in the following. The flexibility of PV plants is severely limited. Their performance relies on external circumstances such as solar radiation and other factors that cannot be influenced. While the power generation can be confined in principle, this doesn't provide any benefits in our motivating use case. Instead, this could be a measure to maintain grid stability, which is not of concern in this paper. Hence the planning of the overall output of the VPP involves the estimation of the power generation of the PV plants, based on the weather forecast. These predictions are taken as assigned schedules for the PV plants. Following, solving the OSO focuses on selecting schedules for the cogeneration units. For each cogeneration unit, a feasible schedule has to be chosen. Within the scope of their device-specific constraints, each cogeneration unit can be operated in many different ways. In order to reduce the problem size only a fixed number of feasible schedules for each unit is considered.

In most cases where this type of optimization is performed, the pursued energy amount for each hour of the day is given. In this the case it is possible to solve the problem with a linear equation system. The formalization is presented below, using the identifiers in Table 2.

The objective is to minimize the costs, deducting the returns of the chosen schedules. This can also be formulated as a maximization problem:

$$max\left(\sum_{a=0}^{n}\sum_{j=0}^{m_a}\left(z(f_{a_j}) - c(f_{a_j})\right)\cdot x_{a_j}\right) \tag{4}$$

where $z(f_{a_j}) - c(f_{a_j})$ is the total sales of the schedule f_{a_j}. $z(f_{a_j})$ and $c(f_{a_j})$ are fixed values whereas the x_{a_j} (i.e., the selection of schedules) has to be optimized. Additionally the following constraint must be considered which guarantees that at least as much power is produced as is required by the lower threshold $t(h)$:

Table 2 Identifiers for the formulation of the OSO

a	Identifies a single generation unit within the VPP, with $a \in \{0, 1, \ldots, n\}$ and n = number of all generation units of the VPP
j	Single schedule of a generation unit, with $j \in \{0, 1, \ldots, m_a\}$ and m_a = number of schedules of unit a
f_{a_j}	Schedule j of generation unit a
x_{a_j}	Boolean indicator that is 1 if schedule f_{a_j} is selected and 0 otherwise
$c(f_{a_j})$	Costs for schedule f_{a_j} in $\frac{€}{\text{MWh}}$
$z(f_{a_j})$	Returns for schedule f_{a_j} due to subsidies (based on German law) in $\frac{€}{\text{MWh}}$
$l_{h_k}(f_{a_j})$	Power of schedule f_{a_j} in hour h and quarter-hour interval k in MW
$t(h)$	Lower threshold that has to be generated during hour h in MW

$$s.t. \quad t(h) \leq \sum_{a=0}^{n} \sum_{j=0}^{m_a} l_{h_k}(f_{a_j}) \cdot x_{a_j}, \quad h \in \{0, \ldots, 23\}, \quad k \in \{0, \ldots, 3\} \tag{5}$$

The threshold has a constant value for the whole hour. The schedules have a granularity of quarter-hour intervals. On this account the sum of power for all selected schedules has to exceed the threshold of the hour for each quarter-hour interval. Another constraint specifies that exactly one schedule has to be selected per unit:

$$\sum_{j=0}^{m_a} x_{a_j} = 1 \quad \forall a \in \{0, 1, \ldots, n\} \tag{6}$$

In contrast to the PPO, however, the OSO problem is much more complex with respect to the number of units and possible schedules for each unit, so that optimal solution methods are not feasible. Therefore the OSO is solved with a heuristic as the outer optimization process. There is a variety of heuristic algorithms that can be applied to the problem. We focus on simulated annealing and tabu search in the following.

Simulated annealing is a specific form of evolutionary algorithms, with only one individual in both, the parent and the child generation. Additionally, a cooling factor determines the probability that an inferior child generation will nevertheless be chosen as a new parent (Kirkpatrick et al. 1983). This cooling factor decreases over time. The purpose is to perform a conversion from exploration to exploitation.

In the OSO an individual represents an operation schedule, comprising one schedule per unit. To calculate the fitness of the individuals, the matching product portfolio is required. Therefore this counterpart has to be determined for the initial operation schedule as well as for all following generations. This is done by using integer linear optimization or the RIPPO. The fitness is the achieved profit of the combined operation schedule and product portfolio. The mutation operation within the process of creating offspring generations alters schedules of the operation schedule: Depending on the mutation rate, a proportion of the power levels in the quarter-hour intervals of each schedule is replaced by random values. To ensure the feasibility of the schedules a decoder is applied. The decoder takes the mutated schedule as input and returns a feasible schedule, that is as similar as possible to the original, cf. (Bremer and Sonnenschein 2013). Through the use of the decoder, the algorithm becomes independent of knowledge about unit models, their characteristics and operational state.

Tabu search relies on the definition of a neighborhood relation between solution candidates. Its specific feature is the use of a tabu list to prevent getting stuck in local optima (Gendreau 2003). Let the current solution x be a candidate operation schedule. For this operation schedule a number of neighboring solutions is generated by mutating one schedule at a time. This means that if the VPP contains n units, n neighboring solutions are created. Again the use of a decoder is required to assure the feasibility of the schedules. For each of the neighboring solutions the matching product portfolio and the estimated profit is calculated to evaluate the fitness, similar to the simulated annealing approach. The best of the neighboring solutions x' is then compared to the best global solution. If x' yields a better fitness and is not contained in the tabu list, it replaces the current solution ($x \leftarrow x'$) and is put on the tabu list. If x' already is in the tabu list, this comparison is repeated for the second best neighboring solution, and so on. If the termination criterion is not fulfilled, a new set of neighboring solutions for x is created afterwards, and the process starts from the beginning.

Both algorithms have the anytime characteristic and therefore provide feasible solutions at every iteration. Their termination criterion depends on elapsed time.

3 Evaluation

For both described optimization problems two different solution approaches, respectively, were presented in the previous section. Against this background, the evaluation aims to determine which combination of optimizers achieves the best quality in terms of fitness and runtime. In the evaluation process, first the inner optimizers (i.e., the PPO solvers) were examined. Afterwards the best inner optimizer was used for the evaluation of the outer optimizers (i.e., the OSO solvers). In the following sections the results for the evaluation of the inner and outer optimizers are presented.

3.1 Evaluation of the Inner Optimizers

Ten experiments were assembled to evaluate the proposed PPO solvers, each of them with another overall power of the VPP. For each experiment an individual arbitrary operation schedule was generated as input for the PPO solvers. The operation schedule was based on real schedules of simulated unit models. Since a new operation

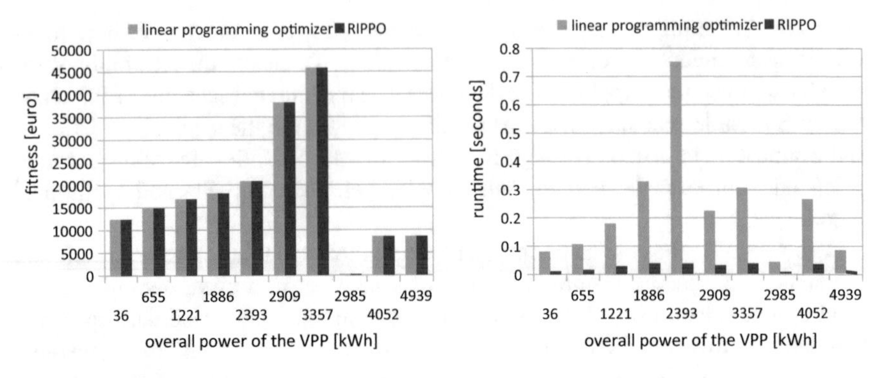

Fig. 3 Fitness and runtime of the inner optimizers

schedule was created for each experiment, the relation between overall power and fitness or runtime was not examined. Only the comparison of the optimizers results is important for the evaluation.

For solving the integer linear problem formulation as stated in Sect. 2.1, the commercial mathematical programming solver gurobi has been used. Further information about gurobi can be found on the company site (Gurobi Home Page 2016). The comparison between RIPPO and gurobi shows that both optimizers always have the same fitness, whereas the runtime of the linear programming optimizer exceeds the runtime of RIPPO by a multiple, see Fig. 3. The runtime difference is lower than one second, but in the context of the overall optimization this is a considerable difference as the outer optimizer calls the inner optimizer thousands of times, depending on the available runtime. For this reason, RIPPO was employed for the subsequent evaluation of the outer optimizers.

3.2 *Evaluation of the Outer Optimizers*

For the design of the OSO evaluation experiments the Latin Hypercube Sampling strategy was applied. A subset from all possible combinations of values (levels) of the system parameters (factors) is chosen randomly, but with a reasonably good cover of the data space (Mckay et al. 2000). Therefore the ranges of the factors are divided into the same specified number of intervals. Then from each interval a value for one evaluation run (experiment) is chosen. The other experiments are constructed in the same way with the remaining intervals.

The ranges for the factor levels of the assembled design can be found in Table 3. The design was repeated four times, each time with another amount of available runtime as termination criterion for the optimizers: 2, 15, 35 and 300 min. These scenarios (see Fig. 4) show that the simulated annealing optimizer always, with one exception, has a better fitness than the tabu search optimizer. The tabu search optimizer reaches only 94–97 % of the fitness of the simulated annealing optimizer. Even

Table 3 Latin hypercube design for evaluating the OSO solvers

Number of plants	$\in \{1, \dots, 5000\}$
Overall power of the VPP [in kWh]	$\in \{1, \dots, 5000\}$
Ratio of power from cogeneration plants to PV power	$\in \{0.0, \dots, 1.0\}$
Number of schedules per cogeneration plant	$\in \{1, \dots, 300\}$

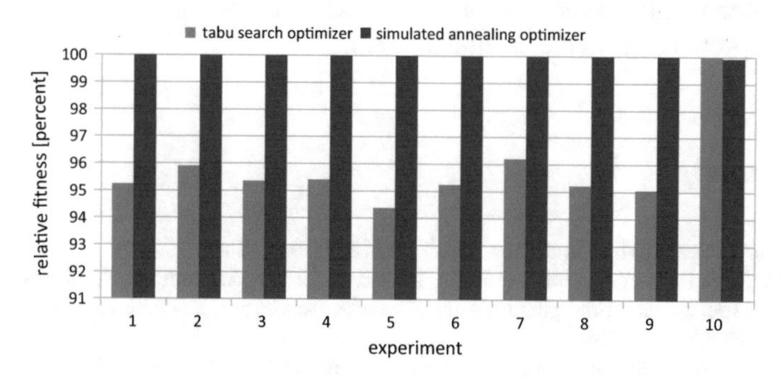

Fig. 4 Relative fitness with 2 min runtime

a longer runtime does not lead to significant changes. An increase of runtime from 2 to 300 minutes results in an improvement of less than 1 % point.

Additionally, in order to scrutinize the correctness of the heuristic approaches in solving the OSO, a combined integer linear optimizer was implemented, which solves the OSO and the PPO in a single linear (in)equation system by combining the problem formulations from Sects. 2.1 and 2.2. While this optimizer always yields the overall optimum for the combined optimization problem, it requires extensive runtime. It does not qualify for production use and is employed for validation only. Thus a limited set of experiments comprising few plants and schedules was executed. The results show acceptable small fitness differences of up to 10 %.

Summing up all evaluation results, the simulated annealing optimizer with RIPPO is the best optimizer combination, due to the better fitness of the simulated annealing optimizer and the lower runtime of RIPPO. Since the available runtime of the optimizers is a user-defined tuning parameter, they are suitable for calculating product portfolios and operation schedules for the day-ahead market.

4 Conclusion

The subject of this paper was to find an automated approach for the composition of product portfolios and their matching operation schedules within VPPs for trading at the EPEX SPOT SE day-ahead market. For each of the subproblems two possible solution methods were suggested, respectively, and an overall solution concept was presented. The evaluation shows that the proposed approach leads to sufficient results in terms of runtime and solution quality even for large VPPs comprising thousands of energy units. Furthermore the results comply with restrictions of real world generation units and market regulations. Thus the automated composition of the product portfolio for trading at the EPEX SPOT SE day-ahead market and the matching operation schedule for the VPP has been successfully implemented. Therefore, the presented approach strengthens the market integration of renewable energy resources and allows the VPP operator and plant owners to operate the VPP or the generation plants, respectively, with the maximum profit.

References

Gurobi Home Page (2016). http://www.gurobi.com. Accessed 12 April 2016

Awerbuch, S., & Preston, A. (Eds.) (1997). The Virtual Utility: Accounting, Technology & Competitive Aspects of the Emerging Industry, *Topics in Regulatory Economics and Policy*, vol. 26. Kluwer Academic Publishers

Bagemihl, J. (2003). Optimierung eines Portfolios mit hydro-thermischem Kraftwerkspark im börslichen Strom-und Gasterminmarkt. Ph.D. thesis, Universität Stuttgart. http://elib.uni-stuttgart.de/handle/11682/1614. Accessed 11 April 2016

Bradley, S., Hax, A., & Magnanti, T. (1977). Applied Mathematical Programming. Addison-Wesley. http://web.mit.edu/15.053/www/. Accessed 4 April 2016

Bremer, J., & Sonnenschein, M. (2013). Constraint-handling for Optimization with Support Vector Surrogate Models—A Novel Decoder Approach. In J. Filipe & A. L. N. Fred (Eds.), *ICAART 2013—Proceedings of the 5th International Conference on Agents and Artificial Intelligence* (Vol. 2, pp. 91–100). Barcelona, Spain: SciTePress.

Dantzig, G. B., & Thapa, M. N. (1997). *Linear Programming 1: Introduction*. Springer.

European Power Exchange (EPEX SPOT SE) (2016). EPEX SPOT Operational Rules. http://www.epexspot.com/en/extras/download-center. Accessed 17 March 2016

Gendreau, M. (2003). An introduction to tabu search. In F. Glover & G. A. Kochenberger (Eds.), *Handbook of Metaheuristics* (pp. 37–54). Kluwer Academic Publishers.

Gröwe-Kuska, N., Liebscher, A., Lucht, M., Römisch, W., Spangardt, G., & Wegner, I. (2003). *Mittelfristige risikoorientierte Optimierung von Strombeschaffungs-Portfolios kleinerer Markteilnehmer*. Humboldt-Universität zu: Berlin. http://www.mathematik.hu-berlin.de/~romisch/papers/drewag7new2.pdf. Accessed 11 April 2016

Hall, M., Frank, E., Holmes, G., Pfahringer, B., Reutemann, P., & Witten, I.H. (2009). The WEKA Data Mining Software: An Update. *SIGKDD Explorations*, *11*(1), 10–18. www.cms.waikato.ac.nz/~ml/publications/2009/weka_update.pdf. Accessed 11 April 2016

Kirkpatrick, S., Gelatt, C. D., & Vecchi, M. P. (1983). Optimization by simulated annealing. *Science*, *220*(4598), 671–680. doi:10.1126/science.220.4598.671.

Mckay, M. D., Beckman, R. J., & Conover, W. J. (2000). A Comparison of Three Methods for Selecting Values of Input Variables in the Analysis of Output from a Computer Code. *Technometrics*, *42*(1), 55–61. http://www.jstor.org/stable/1271432. Accessed 11 April 2016

von Oehsen, A. (2012). Entwicklung und Anwendung einer Kraftwerks- und Speichereinsatzoptimierung für die Untersuchung von Energieversorgungsszenarien mit hohem Anteil erneuerbarer Energien in Deutschland. Ph.D. thesis, Universität Kassel). http://nbn-resolving.de/urn:nbn:de:hebis:34-2013050742706. Accessed 11 April 2016

Sonnenschein, M., Appelrath, H. J., Canders, W. R., Henke, M., Uslar, M., Beer, S., Bremer, J., Lünsdorf, O., Nieße, A., Psola, J. H., & Rosinger, C. (2015). Decentralized Provision of Active Power. In: L. Hofmann & M. Sonnenschein (Eds.) *Smart Nord—Final Report*. Hanover

Distributed Power Management of Renewable Energy Resources for Grid Stabilization

Bengt Lüers, Marita Blank and Sebastian Lehnhoff

Abstract To increase the share of *distributed energy resources* (DER), they need to provide grid supporting ancillary services. In order to fulfill large energy products, *virtual power plants* (VPP) aggregate many small DER. In this paper we present an agent-based approach to the optimization problem of scheduling the DER of a VPP. We compare our approach which uses a evolutionary algorithm to one that uses a mathematical solver. When comparing the two approaches we found that our approach approximates the optimal solution well. The central benefit of our approach is that it scales better wrt. the VPP size. The increased scalability opens the ancillary services market to VPPs that aggregate more and smaller DERs.

Keywords Distributed power generation · Genetic algorithms · Multi-agent systems · Self-adaption · Scheduling · Unit commitment

1 Introduction

In recent years, the share of renewable energy has increased rapidly. Until 2020, the European Union's *Renewables Directive* (2009/28/EC) requires 20 % of the energy consumed in the European Union to be renewable. Beyond that, renewable power stations will need to fully replace conventional power stations.

Current power products are designed for large-scale conventional power plants. To replace conventional power stations, renewable power stations need to provide the same power supply tasks. This includes the provision of *ancillary services* (AS). AS have grid-supporting functions and are necessary for a stable power supply. They

B. Lüers · M. Blank (✉)
OFFIS - Institute for Information Technology, Oldenburg, Germany
e-mail: marita.blank@offis.de

B. Lüers
e-mail: bengt.luers@offis.de

S. Lehnhoff
Department of Information Technology, University of Oldenburg, Oldenburg, Germany
e-mail: sebastian.lehnhoff@offis.de

V. Wohlgemuth et al. (eds.), *Advances and New Trends in Environmental Informatics*, Progress in IS, DOI 10.1007/978-3-319-44711-7_12

have to be provided with high reliability. However, renewable power stations are often small-scale, *distributed energy resources* (DER) and far less predictable than usual conventional power plants. Therefore, DER cannot provide AS products on their own. To overcome market barriers, *virtual power plants* (VPPs) have been used Bitsch et al. (2002). These execute a coordination algorithm to aggregate power from multiple DER. The coordinating algorithm of VPPs can be central or distributed. A central algorithm can in theory exploit global information for greater social welfare. However, there are practical reasons, why a distributed scheduler is preferable. Among the most important is that distributed scheduling algorithms ensure privacy for the unit owner's confidential information Setiawan (2007). A programming paradigm commonly used for distributed control algorithms are *multi-agent systems* (MAS). Here, each agent focuses on its own interests and can negotiate with other agents. As shown in Hernandez et al. (2013), VPPs can be modelled as MAS. In that case, VPPs are also called coalitions.

In Lehnhoff et al. (2013), a central approach has been suggested to support the AS of frequency containment control by providing *primary control reserves* (PCR) by VPPs of DER. To be competitive, DER need to be scheduled as economic as possible. With rising numbers of DER, the scalability of the scheduling method becomes an important factor. A viable compromise might be to trade some economy for scalability. In this paper we quantified, how economic and how scalable a evolutionary algorithm can schedule DER for providing PCR, thus investigating how this approach can support the integration of DER into the power system. To that end, we developed a *self-adaptive evolutionary strategy* (SAES), specific for scheduling DER represented by a MAS and compared it to an analytical approach to the same optimization problem.

This paper is organized as follows: Sect. 2 describes requirements and common concepts of the optimization. In Sect. 3 our proposed approach—a *self-adaptive evolutionary strategy* (SAES)—to the problem of scheduling the units of a VPP is introduced. Section 4 analyses and compares experimental results. Section 5 ends the paper with conclusions and outlook.

2 Concept and Optimization Problem

The focus on this paper lies on the direct comparison of the central and distributed approach for providing PCR by a VPP or coalition of DERs. In the following, DERs are also called units. The requirements for providing PCR and the common concepts of the central and distributed solution are given as well as the underlying optimization problem. Details on the central concept can be found in Lehnhoff et al. (2013).

PCR are used for stabilizing the frequency in the power system and corresponding products are tendered at dedicated markets. A coalition committed to PCR must provide the stipulated amount of power throughout a specified product horizon. The power is partially provided by the units in the VPP and they must activate their share automatically in case of frequency deviations. This leaves no time to execute the control algorithm of a coalition. Furthermore, coalitions must offer PCR with a certain reliability. In Lehnhoff et al. (2013) reliability has been defined as the probability with which the reserves can be provided for the offered time. This estimation is done given short-term forecasts as renewable power stations such as wind turbines and photovoltaics are supply-dependent. Furthermore, they fluctuate in their possible power output with the weather conditions. In a first approximation, the reliability of a coalition as the product of its unit reliabilities. Note that with a growing coalition lifespan the power prediction error grows.

The core idea is to schedule units within the coalition such that the stipulated reserves with the required reliability throughout the whole product horizon are guaranteed. To this end, subsets of the coalition are formed for shorter time intervals within the product horizon such that within this *lifespan* requirements for power reserves and reliability are fulfilled. This means, that only the units in the subset must activate reserves during the subset's lifespan where the sum of individual reserves amount to the offered product. To contribute to a coalition, units must quantify their ability to provide a given power for a given time based on forecasts. The scheduling for the next subset must have terminated before the end of the current subset's lifespan. The advantages of this approach are that reserves can be automatically activated within the active subset and current short-term forecasts can be included for planning that have smaller prediction errors leading to a higher reliability. Furthermore, prime costs are taken into account. Cost of the VPP are considered to be the product of each unit's prime cost and its provided reserves. To minimize costs, VPPs must schedule the most cost-efficient units to contribute during a lifespan. The most cost-efficient DER in terms of primary costs, such as wind and photovoltaics, are usually the least reliable. The less reliable the DERs of a coalition are, the shorter its lifespan is. Thus, longevity and cost efficiency of a coalition are conflicting objectives.

The model presented in Lehnhoff et al. (2013) can be formulated as

$$cost_c(T_C) = t_c \cdot \Sigma_{i=1}^{n} e_{cont,a_i}(T_C) \cdot cost_{a_i} \tag{1}$$

In this model each unit agent a_i has a fix prime cost $cost_{a_i}$. An agents prime cost multiplied by the power it contributes e_{cont,a_i} in the coalition interval T_C the length of which is the coalition's lifespan t_C. gives the cost of each agent in a single time step. The costs of all agents yield the coalition cost.

A straightforward approach to optimizing this mathematical model uses the solver IBM ILOG CPLEX.[1] This approach calculates pareto-optimal solutions in two steps. Both are under the constraint that their solution must at least match the reliability and power given in the product. In a first step, the coalition lifespan is maximized. In a second step, the coalition cost is minimized. The second step is under the additional constraint that the solution must match the maximum lifespan, as determined in the first step Lehnhoff et al. (2013).

[1] http://www.ibm.com/software/commerce/optimization/cplex-optimizer/.

Whereas central scheduling using a solver guarantees a pareto-optimal solution upon termination, it has the drawback of not scaling well with the number of units and time steps. Here scalability concerns primarily, how long it takes for every additional unit or time step to find a solution. As a corollary, scalability also concerns, for how many units and time steps a valid solution can be found at all. As we will discuss in detail later, the scalability of this approach can be considered insufficient for scheduling VPPs consisting of many renewable power stations for many time steps.

3 Distributed Scheduling—the SAES Approach

We employed a multi-agent system where each unit is represented by an agent. We chose to use JADE as a multi-agent platform and agent framework.[2] On top of JADE's agent model, we implemented initiator and responder behaviors. Initiator and responder are the two roles of the Contract Net protocol, which defines interaction between producer and consumer of a good or service FIPA (2002). To allow repeated negotiations, we extended our agent implementation with iteration of the contracting process, as defined in FIPA (2002). In each iteration, an agent evaluates a call for proposal for the reliability with which it can provide the requested power for the requested lifespan. Once it gets proposals from all responders, the initiator agent calculates the coalition reliability as the product of the proposal reliabilities. In effect, an iteration of the Contract Net protocol is a generation of the evolutionary strategy.

Evolutionary algorithms are meta-heuristics that have proven to be robust methods of approximating global optima (Kramer 2009, p. 13). We employed a *self-adaptive* $(\mu/\rho + \lambda)$ *evolutionary strategy* (SAES), as given in (Kramer 2009, p. 16). With them, a population of individuals, which are created by crossover, get modified by mutation and selected for survival and reproduction based upon their fitness. In our approach, the genes of an individual represent the contributions of units in a coalition. In turn, a population is a list of possible coalitions. Our evolutionary strategy considers the coalition reliability as the individual fitness. Being self-adaptive means here that individuals carry a standard deviation σ for each of their gene elements, which gets influenced and propagated along with them. Being a $(\mu/\rho + \lambda)$ evolution strategy means here that ρ parent individuals get selected from the union of an offspring population of λ individuals and (+) a current population of μ individuals. The ρ parent individuals get overcrossed to produce λ offspring individuals, which get added to the population of which μ individuals survive to the next generation. In our experience from testing the evolutionary algorithm while developing it, the best parameters for the evolutionary strategy are a population size μ of 12, a number of parent individuals ρ of 3 and an offspring population size λ of 9.

Note that the evolution strategy is not distributed but runs on the initiator. However, the evaluation of the partial fitness functions runs on the responders. Cost and

[2]http://jade.tilab.com/.

reliability calculation are privacy-sensitive to a unit operator and computationally expensive. They are combined to the coalition's fitness on the initiator through multiplication, which consumes only negligible time. This approach has properties of interest for the application to real units. On the one hand, the error models and predictions do not leave the unit's agent, thereby a unit operator does not need to expose its confidential information to the aggregator or other unit operators. On the other hand, the coalition reliability also never gets known to any of the responders, which means that an aggregator, which operates a VPP, would also be able to keep its confidential information.

Our evolutionary algorithm follows the methods given in the literature, e.g. (Kramer 2009, Fig. 2.1). It initializes its population with λ individuals which each contain the power required by the product divided evenly among the contributions. It iterates until either all three optimization objectives are satisfied or time runs out. In each iteration, it selects, overcrosses, repairs and evaluates individuals. The selection of ρ individuals for reproduction is done in a fitness-proportionate way. To improve performance, the evolutionary reproduction process gets repeated until it yields a promising child candidate. The overcrossing of the selected parent individuals uses an intermediate overcrosser on the gene elements as well as the lifespan. The mutation of the candidate generated by overcrossing is done in a self-adaptive way, which firstly mutates the standard deviations using log-normal mutation and then mutates the genes using Gaussian mutation (Kramer 2010, Eqs. 13 and 14). The repairing makes the mutated candidate's powers sum up to the total power requested by the product. Our repairer corrects deviations by distributing them equally among the powers. The evaluation of the child candidate's reliability is executed on the agent system. For each offspring that it gets added to the population, the worst individual gets removed.

Our fitness function ranks the population's individuals by how well they meet the multiple objectives and derives a fitness value. This is similar to pareto-ranking approaches given in the literature, e.g. Konak et al. (2006). To that end, we implemented a comparator that decides, which of two given individuals is better.[3] It defines a total order on the individuals and can thus be used for sorting. When the goal reliability, lifespan or cost is reached, further improvements in their dimensions is considered bad. That is because optimizing in one dimension hinders optimization in the other directions. Since there are always three dimensions to be optimized, over-optimization in one dimension is harmful for both other dimensions. The fitness function f of an individual uses the sorted population and is defined as $f(x) = N - i(x)/N$, where N the number of individuals in the population and $i(x)$ the rank of the individual x in the population. Therefore, the lower the rank of an individual is the higher its fitness is, limited by $1/N$ (for $i(x) = N - 1$ and 1 (for $i = 0$).

[3]https://docs.oracle.com/javase/8/docs/api/java/lang/Comparable.html.

4 Evaluation

We exploratively tested the CPLEX optimizer on problem instances (domains), to determine which could be solved within a given time. The CPLEX optimizer could not find solutions for domains with more than 7 time steps or 100 units in under 15 min. Within these bounds we created a problem instance for every combination of unit count and time step count. Our domain generator initializes the units with predictions of available power, error models about these predictions and costs of produced power.

4.1 Conduction

To evaluate how well our SAES performs, we compared it with the preexisting CPLEX approach. All tests were run on a desktop computer with an AMD Phenom 2 X6 1090 T processor. When running exploratory tests with the domains, we found that beyond the limits shown in Table 1a, CPLEX consumed more than the maximum amount of time of 15 min.

Firstly, we ran the CPLEX optimizer on all applicable domains. The CPLEX optimizer stops its execution when it finds a pareto-optimal solution wrt. coalition cost and lifespan. The execution time and the coalition lifespan of the CPLEX optimizer could therefore serve as a target lifespan and allowed time for the SAES optimizer. This left only the coalition's cost as a dependent variable, which allows us to compare the two optimizers directly. Since the platform takes some time to set up and to spawn agents, the SAES optimizer was not capable of solving any domain in under 3 s. We excluded domains from being run, where the CPLEX optimizer had found a solution faster than possible for the SAES. Table 1b lists these lower limits of the SAES's applicability for comparison with CPLEX. There, a 0 denotes that the SAES was quick enough to generate solutions for unit counts of 1 or more. Secondly, we ran the SAES optimizer for each of its applicable domains under the same conditions as CPLEX. Finally, we ran the SAES optimizer on all domains, to explore its robustness against domains of varying difficulty. For these runs, we set a time constraint of 60 s, because that seemed a viable compromise between execution time per domain and total execution time.

Table 1 Unit limits for both optimizers

(a) Upper unit limits of CPLEX (better: bigger)

Steps	1	2	3	4	5	6	7	8	9
Units	100	81	60	50	43	37	31	28	26

(b) Lower unit limits for SAES (better: smaller)

Steps	1	2	3	4	5	6	7	8	9
Units	36	21	17	11	10	9	6	0	0

4.2 Results

This section presents the results of comparing the CPLEX to the SAES approach.

Cost results: Figure 1 compares the costs of the solutions generated by CPLEX and the SAES optimizer using boxplots. The upper boxplot visualizes the solution distribution wrt. cost of the CPLEX optimizer on all applicable domains. The middle boxplot visualizes the run of the SAES optimizer under the same conditions as the CPLEX optimizer. The SAES optimizer was run on the same domains as CPLEX, because we had lifespan and runtime limits for those, only. The lower boxplot visualizes the SAES running on all domains under a time constraint of 60 s. When running under the same conditions as CPLEX, the SAES optimizer generates solutions with less spread wrt. cost than the CPLEX optimizer. When running on all domains for 60 s, the SAES optimizer generates solutions with more spread wrt. cost than when run under the same conditions as CPLEX. On average CPLEX generates solutions that are 30 % more cost-efficient than those of the SAES. This is to be expected, since CPLEX finds a pareto-optimal solutions wrt. cost and the SAES only approximates such solutions. While higher cost is the main weakness of our SAES in comparison to the CPLEX-based approach, the results with regard to cost are still comparable and within a reasonable margin for an application-relevant meta-heuristic approach. Additionally, the SAES solutions show less spread wrt. cost, which means that they are more predictable than the CPLEX solutions.

Reliability results: Figure 2 compares the reliabilities of the solutions generated by CPLEX and the SAES optimizer using boxplots. Again, the upper boxplot represents

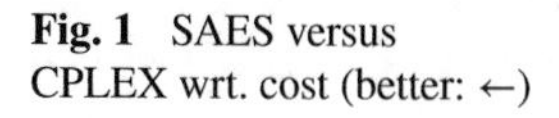

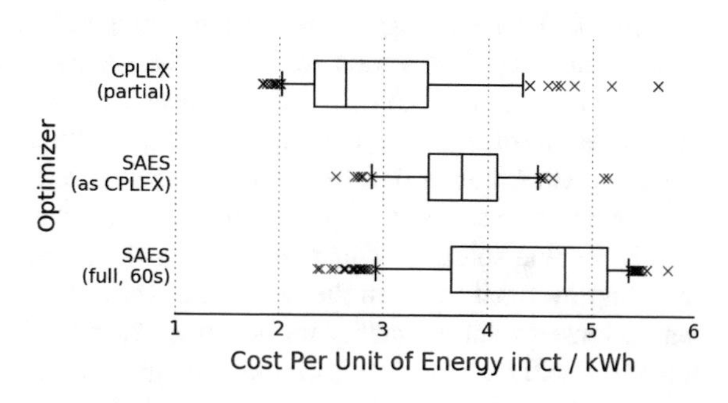

Fig. 1 SAES versus CPLEX wrt. cost (better: ←)

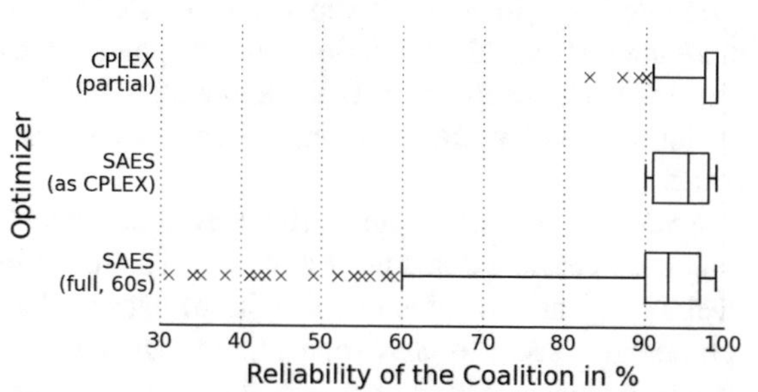

Fig. 2 SAES versus CPLEX wrt. Reliability (better: →)

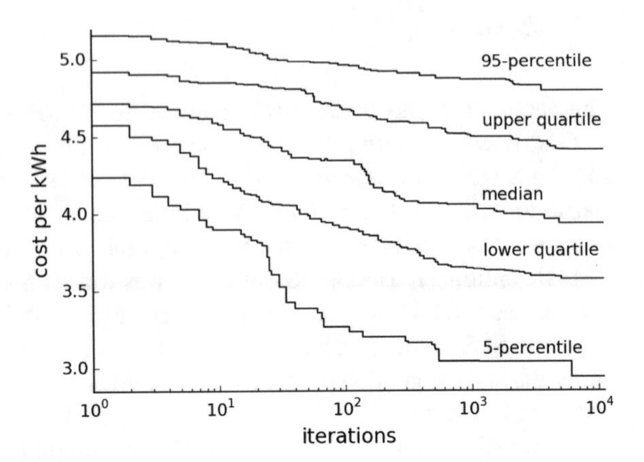

Fig. 3 Cost of SAES over iterations (better: ↓)

CPLEX run on its own terms, the middle boxplot represents the SAES run as CPLEX did and the lower boxplot represents the SAES run on all domains for 60 s. When running under the same conditions as CPLEX, the SAES optimizer solutions show more spread wrt. reliability than those of the CPLEX optimizer. When running on all domains for 60 s, the SAES optimizer generates solutions with the same amount of spread wrt. reliability as when run under the same conditions as CPLEX. The box plot whiskers seem to indicate a bigger spread, but they mark the inter quantile range of 90 % of the values ($QR_{.05}$). The SAES finds reliable solutions to all domains that it and CPLEX can solve at all. However, the SAES solutions show more spread than those of CPLEX, which underlines that they only approximate optimal solutions.

Runtime behavior: Figure 3 shows the aggregated runtime behavior of the SAES when running applicable domains under the same conditions as CPLEX. The five step plots depict the 95th percentile, upper quartile, median, lower quartile and 5th percentile of the costs per kWh on each iteration. Note that the x-axis is logarithmic, which means that later improvements take much more iterations than earlier improvements. Some runs spend the first few iterations on finding a first sufficiently reliable solution. This explains why the upper quartile and median exhibit little improvement between the second and third iteration. The SAES enables trading decreased solution quality for decreased runtime. The more units there are, the less likely will the mutation favor the best unit. So domains with smaller unit counts see more improvement. This means that CPLEX finds solutions for domains with more units so quickly that it does not compensate the difficulty of additional search space dimensions for the SAES. On the other hand, CPLEX runs that much longer on domains with more steps that it overcompensates the increased difficulty for the SAES.

Scalability results: Figure 4 shows solution reliabilities of the SAES when running each possible domain for a minute. The reliability axis is inverted to increase visibility of the low reliability area in the back of the graph. As a general tendency, the domains seem to get exponentially harder with increasing unit and time step counts, as one would expect from the curse of dimensionality Sniedovich (2010).

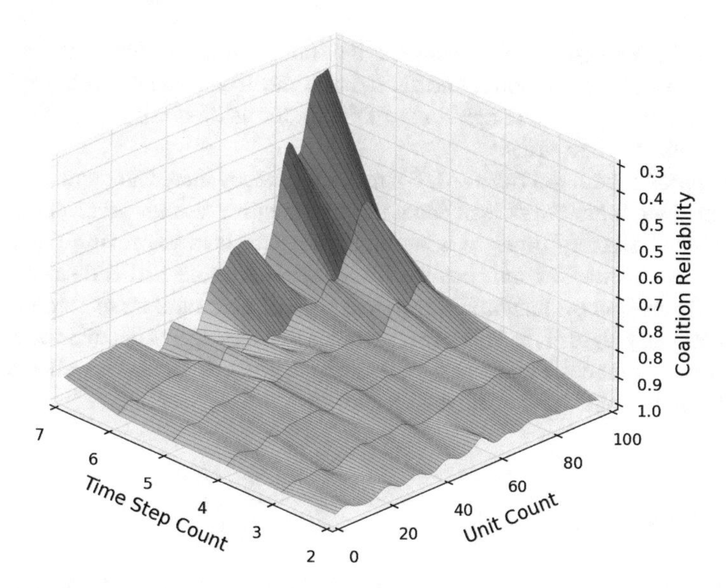

Fig. 4 Reliability of SAES over 1 min runs (better: ↓)

Table 2 Unit limits of SAES in 1 min (better: bigger)

Time step count	2	3	4	5	6	7
Maximum unit count with a reliable solution	100	100	100	87	87	80
Minimum unit count with an unreliable solution	–	–	–	82	63	43

Table 2 lists the found reliability limits of the SAES for runs at all domains for 60 s. These outline the practical limits of using the SAES to get sufficiently reliable solutions. Compared to the limits we found for the CPLEX-based approach as shown in Table 1a, our SAES-based approach can find solutions to more domains. This is despite the fact that our approach only ran for 1 min on each domain, while CPLEX was allowed up to 15 min of run time. This goes to show that our SAES is more scalable than CPLEX for the time step count and also for the unit count.

5 Conclusion and Future Work

This paper has presented an agent-based meta-heuristic approach that implements the SAES—an evolutionary strategy—for VPP unit scheduling. It is capable of providing PCR for frequency containment control. We compared it to an analytical implementation using synthetic data and found clear experimental results. Our system performs comparably well to the analytical one. The results are not optimal, i.e. they are more expensive, as it is a meta-heuristic approach but they approximate to a reasonable margin. Our system is a more universal tool as it was able to find reliable solutions to more problem instances. The main novelty of our system is that it scales well with high unit counts. Another benefit of our approach is that it allows to trade increased cost of the solution for decreased runtime. Consequently, the usage of our approach can support the integration of DER in terms of substituting conventional power plants and their tasks.

Our approach suffered from JADE's message latency increasing with every message sent. Our future work will include implementing a high-performance MAS which lessens start up times to a negligible amount thus improving the scalability of our SAES for low unit counts. In our approach, the SAES distributes only the fitness function evaluation. Future systems will distribute more components of the evolutionary algorithm. This paper assumes that unit agents act upon an accurate prediction error model. Future work will integrate grid constraints as well as a notion of trust that accounts for deviating behavior.

References

Bitsch, R., Feldmann, W., & Aumayr, G. (2002). Virtuelle Kraftwerke—Einbindung dezentraler Energieerzeugungsanlagen, etz-Jg (vol. 123, pp. 2–9).

FIPA, (2002). *FIPA Contract Net Interaction Protocol Specification*. Retrieved November 23, 2014, from http://www.fipa.org/specs/fipa00029/.

FIPA, (2002). *FIPA Iterated Contract Net Interaction Protocol Specification*. Retrieved November 24, 2014, from http://www.fipa.org/specs/fipa00030/.

Hernandez, L., et al. (2013). A multi-agent system architecture for smart grid management and forecasting of energy demand in virtual power plants. *IEEE Communications Magazine*, *51*(1), 106–113.

Konak, A., Coit, D. W., & Smith, A. E. (2006). Multi-objective optimization using genetic algorithms: A tutorial. *Reliability Engineering & System Safety*, *91*(9), 992–1007.

Kramer, O. (2010). Evolutionary self-adaptation: A survey of operators and strategy parameters. *Evolutionary Intelligence*, *3*(2), 51–65.

Kramer, O. (2009). *Computational intelligence*. Springer-Verlag.

Lehnhoff, S., Klingenberg, T., Blank, M., Calabria, M., & Schumacher, W. (2013). Distributed coalitions for reliable and stable provision of frequency response reserve. *IEEE*, 11–18.

Setiawan, E. A. (2007). *Concept and controllability of virtual power plant*. Kassel University Press GmbH.

Sniedovich, M. (2010). *Dynamic programming: foundations and principles*. CRC Press.

Proposing an Hourly Dynamic Wind Signal as an Environmental Incentive for Demand Response

Anders Nilsson and Nils Brandt

Abstract Demand Response (DR) is expected to play a crucial role in balancing supply and demand in future smart grids with increased proportion of electricity from renewable sources. However, previous studies on price-based DR programs have shown that there is a substantial need to strengthen the incentive models in order to achieve sufficient end-user response. In addition, recent studies are starting to explore alternative incentives based on environmental performance as a support to dynamic pricing tariffs. In this paper, we investigate in the potential of using a dynamic wind signal, reflecting the hourly variations in wind power generation, as an environmental incentive for load shift in DR programs. A wind signal is constructed based on Swedish electricity generation data for 2014, and intraday and seasonally patterns of wind power generation are analyzed with respect to hourly electricity spot prices. The results show that a wind signal is supportive to the economic incentive of a dynamic price signal to stimulate intraday load shift by end-use customers; shifting electricity consumption from hours of high price and low wind power generation to hours of low price and high wind power generation, leading to both consumer cost-savings and reduced climate impact in the long term.

Keywords Wind power · Electricity spot price · Demand response · Smart grids · Renewable energy

A. Nilsson (✉) · N. Brandt
Department of Sustainable Development, Environmental Science and Engineering, School of Architecture and the Built Environment, Royal Institute of Technology (KTH), 100-44 Stockholm, Sweden
e-mail: anders.nilsson@abe.kth.se

N. Brandt
e-mail: nils.brandt@abe.kth.se

V. Wohlgemuth et al. (eds.), *Advances and New Trends in Environmental Informatics*, Progress in IS, DOI 10.1007/978-3-319-44711-7_13

1 Introduction

1.1 Background

Increasing energy demand, the threats posed by climate change, and ambitious global emission reduction targets have resulted in a need for radical changes in the conventional approach of producing and consuming electricity (Ipakchi and Albuyeh 2009; IEA 2014). Development of smart grids, symbolizing the transformation from single-directional and non-intelligent electricity grids into interconnected, automatic-intelligent, and flexible electricity network systems, is expected to play a game changing role for future sustainable energy systems, and opens up for a broad spectra of new technologies and strategies to increase the efficiency and reliability of the electricity grid (Amin and Wollenberg 2005; Farhangi 2010; Rahimi and Ipakchi 2010). Although the arguments for smart grid programs often are considered to be of technical and economic character, including market and infrastructure improvements allowing increased reliability, optimization, and security of supply, smart grids also offers potential benefits of environmental character (Amin and Wollenberg 2005; Farhangi 2010). The environmental advantages of smart grids are usually discussed in terms of reduced electricity demand leading to a reduction in carbon emissions, but improved system flexibility will also allow for an increased integration of electricity from intermittent renewable sources (Gottwalt et al. 2011; Jacobson and Delucchi 2011). However, due to the unpredictable nature of intermittent renewable energy, without the ability to control or require generation at a certain time, increased penetration of renewables requires increased demand flexibility; active and reliable demand side participation to maintain the balance between supply and demand in the power system (Ilic 2007; Xie et al. 2009).

1.2 Demand Side Management and Demand Response

Demand Side Management (DSM), defined as the planning, implementation, and monitoring of those utility activities that designed to influence customers use of electricity in ways that will produce desired changes in the utility's load shape (Gellings 1985), and Demand Response (DR), a key DSM tool, are expected to play an important role in balancing a future grid with increased non-controllability (Albadi and El-Saadany 2008; Strbac 2008; Palensky and Dietrich 2011). DR refers to all intentional adjustments in electricity consumption pattern by end-use customers that are intended to modify the timing, level of instantaneous demand, or total electricity consumption (IEA 2003; Albadi and El-Saadany 2008) Although there are several definitions of DR, the U.S. Department of Energy defines it as:

"Changes in electric usage by end-customers from their normal consumption pattern in response to changes in the price of electricity over time, or to incentive payments designed to induce lower electricity use at times of high wholesale market prices or when system reliability is jeopardized" (U.S. Department of Energy 2006).

DR programs may be deployed with a wide range of different features, however, there are two main categories; (1) incentive-based, and (2) price-based programs (Tan and Kirschen 2007; U.S. Department of Energy 2006; Siano 2014). The former category are designed by system operators and include Direct Load Control, Interruptible service, Demand Bidding, Emergency DR, and Capacity Market, and the customers receive payments as a bill credit or discount rate for their participation. Depending on the type, incentive-based DR programs may also include economic penalties to customers that do not respond in a sufficient way at end-use critical times (Albadi and El-Saadany 2008; Siano 2014). The latter category are often fully voluntary and encourages to a change in electricity consumption behavior by end-use customers in response to dynamic price tariffs; either by a general decrease in overall electricity consumption, or by shifting load from peak demand hours to off-peak hours, which may be considered as a direct supportive action to maintain the balance between supply and demand in the grid (Albadi and El-Saadany 2008; Conejo et al. 2010; Xu et al. 2011). As price-based DR programs, opposed to incentive-based programs, are free from binding agreements between system operators and end-use customers, the success of these types of programs are heavily dependent on the willingness of adoption and motivation of high activeness among customers. Consequently, the positive-incentive value for end-use customers to change electricity consumption behavior must exceed the sacrifices in comfort and daily routines that such a change might entail.

1.3 Previous Research

Previous studies on deployed price-based DR programs indicate that there is a significant challenge to develop strong incentive models that ensure a sufficient response from end-use customers (Darby and McKenna 2012; Torriti et al. 2010). Several studies have evaluated the residential response to dynamic price tariffs and the results vary significantly from modest to substantial, where the potential has proven to be closely related to the scale of the economic incentive, the availability of automation technologies, and study design. An early review by Darby (2006), compiling results from real-time feedback studies in Europe and U.S., suggests an average saving in demand of 5–15 %. Similar levels were suggested by a state-wide pilot experiment in California (Faruqui and George 2005; Faruqui et al. 2009) using time-of-use tariffs and critical peak pricing, with a residential response of 5 % and 13 % respectively. Two following reviews, including the fourteen (Farugui and Sergici 2009) and fifteen (Farugui and Sergici 2008, 2009, 2012) most recent pricing experiments in U.S. at the time, concluded that time-of-use tariffs lead to reduced peak demand of 5 % and 3–6 % respectively, and critical peak pricing tariffs to reductions of 15–20 % and 13–20 % respectively. An extensive review of research on experiences with DR programs, covering most of the EU-15 countries, Slovenia, Australia, U.S., and Canada (Chardon et al. 2008), suggests that the average peak demand reduction varies from 20–50 %, where the latter represents studies involving enabling, or "smart", technologies and thus including automated reductions in peak demand. The Nordic countries have similar deregulated market designs, which means that distribution system operators and retailers are legally separate entities and thus, customers are charged separately for electricity distribution and consumption. A large-scale Norwegian study (Grande and Sæle 2005), applying real-time rates in retail, a time-of-use distribution tariff, and a combination of the two, suggests peak load reduction varying from 0.5–0.18 kWh/h, at a price difference of 0.15 €/kWh. Another Norwegian study (Stokke et al. 2010), combining a variable energy rate (NOK/kWh) and a variable demand charge (NOK/kW/year), showed on an average peak load reduction of 0.37 kWh/h, corresponding to 5 %.

1.4 Environmental Incentives

Although the focus of previous studies has been on economic incentives, dynamic pricing tariffs, almost exclusively, recent Swedish studies have started to investigate in the possibility, and the potential, to introduce alternative incentives based on environmental performance to strengthen the overall incentive model in price-based DR programs (Stoll et al. 2013; Nilsson et al. 2015). A main reason to the increasing interest of environmental incentives is due to the need to compensating for the declining strength in the economic incentive, which is a result of decreasing levels and volatility in electricity spot market prices the recent years (Kirschen and Strbac 2004; Darby and McKenna 2012). When the intraday dynamics in the price levels decrease, consequently, so do the economic benefit for end-use customers to shift load from peak hours to off-peak hours. Another factor of importance is the increasing environmental awareness and interests among Swedish electricity consumer, for instance illustrated by the increased demand of green electricity contracts, ensuring fossil-free electricity supply (SEA 2013), opening up for novel approaches targeting customers that are willing to participate in DR programs of environmental reasons. Taking such motivational factors into account, it is reasonable to argue that a combination of environmental and economic incentives for electricity consumption behavioral change may form a stronger total positive-incentive value compared to each aspect individually, as it would appeal to a wider range on electricity consumers with varying preferences on economic and environmental issues.

1.5 Aim and Objectives

Given this background, with (1) smart grids as flexible systems that supports increased integration of electricity generated by intermittent renewable sources but whose environmental benefits often are neglected in the literature, (2) the emerging need to boost the incentive models for demand side participation in DR programs, and (3) the increased environmental awareness among electricity end-use customers, there is a significant need to investigate in the possibility to link these challenges to each other; analyzing the potential of using information on variations in intermittent renewable electricity generation as an environmental incentive for demand response. The rationale of such an approach is to provide a motive for end-use customers to shift electricity consumption from hours of low intermittent renewable generation, to hours of higher intermittent renewable generation, in order to offset alternative, more carbon-intensive, generation.

In this paper we investigate in the potential of using a wind signal, reflecting the hourly variations in domestic wind power generation, as an environmental incentive for load shift in residential price-based DR programs. The approach is based on the hypothesis that such a signal, provided as a supportive incentive to a price-signal, would strengthen the overall incentive model for end-use customers to adjust their daily electricity consumption patterns, shifting load from hours of high price and low wind power generation to hours of low price and high wind power generation. In order to examine the relation between the economic and environmental incentive, an hourly dynamic wind signal is constructed based on Swedish electricity generation for 2014, and seasonally and intraday patterns of the signal are analyzed with respect to hourly electricity spot prices of the Swedish electricity market.

The study contributes to increased knowledge on how real-time electricity generation data can be utilized to develop novel incentive models for demand response strategies, addressing the significant need to highlight the environmental benefits of smart grid systems and boost demand side participation. Further, the study examines the relation between intermittent renewable power generation and electricity market prices, providing valuable insights to the planning, design, and implementation of alternative incentive models for future DR programs.

2 Methods

2.1 Construction of an Hourly Dynamic Wind Signal

The rationale of an environmental incentive in a residential DR context is to change electricity consumption behavior for environmental benefits. In this study, the environmental incentive is represented by a hourly dynamic signal showing the variations in wind power generation, which means that end-use customers are assumed to respond to the signal by shifting electricity consumption from hours of low wind power generation (wind off-peak hours), to hours of high wind power generation (wind peak hours).

The reasoning behind choosing wind power as the only renewable intermittent power source to base the analysis on is due to the conditions of the Swedish electricity system, where solar power accounts for less than 0.01 % on yearly basis, considered as negligible. Although wind power represented 8 % only of the total generation during 2014 (Svenska Kraftnät (SvK) 2015), the level of variations in wind power generation is considered as sufficient to base the study on. Wind power generation capacity in Sweden also increases continuously (The Swedish Energy Agency (SEA) 2015), strengthening the future relevance of such an approach further. In additional favor, hourly prognosis data of wind power generation for the Swedish electricity market for the next 24 h are accessible since 2014 (Nordpool Spot 2015), allowing a construction of an hourly dynamic signal reflecting the day-ahead variations in wind power generation. This might be considered as a requirement for implementation in order to allow end-use customers to plan and adjust their electricity consumption according to the signal.

Two approaches were considered for the construction of an hourly dynamic wind signal; (1) reflecting the quote of wind power generation in relation to the total power generation in Sweden, and (2) reflecting the quote of wind power generation in relation to the total installed wind power capacity in Sweden. Given that the total generation is a function of total demand, the risk with the former approach is that the value of the signal would decrease with increasing demand, regardless of the level of wind power generation, which makes the latter approach more suitable for the purpose of the analysis.

Hence, the wind signal, expressed as the quote of generated wind power of total installed wind power capacity in Sweden, are calculated for all hours of 2014 as

$$Wind\,signal\,[\%] = \frac{Wind\,power\,generation\,[MW]}{Total\,installed\,capacity\,of\,wind\,power\,generation\,[MW]} \quad (1)$$

Data of hourly wind power generation in Sweden for 2014 was obtained from the Swedish electricity grid authority (Svenska Kraftnät (SvK) 2015), and the total installed capacity of wind power generation in Sweden at the end of 2014 was 5 097 MW, distributed on a total of 2 961 wind power plants (The Swedish Energy Agency (SEA) 2015).

2.2 *Correlation of Wind Power Generation and Electricity Spot Price*

Using Eq. (1), wind signal values are calculated for all hours of 2014. These values, and the hourly electricity spot price obtained from (Nord Pool 2015), are then used to study the correlation between the wind signal and the spot price. As the relation is assumed to be non-parametric, the analysis is based on Spearman Rank Correlation, where the sign of the Spearman Rank Correlation coefficient, ρ, indicates the direction of association between the wind signal values and spot price; an absolute positive correlation corresponds to $\rho = 1$, and an absolute negative correlation corresponds to $\rho = -1$. Choosing the level of significance to $p = 0.05$, and $n > 30$, the critical t-value for rejection of the null hypothesis is $t = 1.96$. This means that the null hypothesis of no correlation can be rejected for all observed periods with an absolute t-value of 1.96 or higher.

3 Results

The calculated wind signal values for all hours during 2014, with monthly mean values plotted, are illustrated in Fig. 1. As shown, wind power generation varies significantly, from a maximum hourly value of 91 % of the total installed capacity in December, to a minimum hourly value of 1 % in July. On a total annual basis, the hourly mean wind signal value is 26 %, indicating that the total wind power generation capacity in Sweden during 2014 was utilized to approximately one fourth on an average basis. In addition, some seasonally variations in wind power generation can be noticed. A slightly increase is shown during the winter season, with a maximum monthly mean value in December of 43 %, followed by October and January of 35 % and 32 %, respectively. Contrary, the wind power generation decreases during the summer months, with July as the month with the lowest mean value of 14 %, followed by June and May of 17 % and 18 %, respectively.

Figure 2 illustrate the hourly mean wind signal values based on all hours during 2014. Although it is shown that wind power generation, on average, slightly increases during evening and night hours, with wind signal values between 28–29 % during 21–01, the difference to the period of lowest values, between 24–25 % during 07–11, may be considered as negligible. However, hourly mean values based on all hours of 2014 does not necessarily provide appropriate

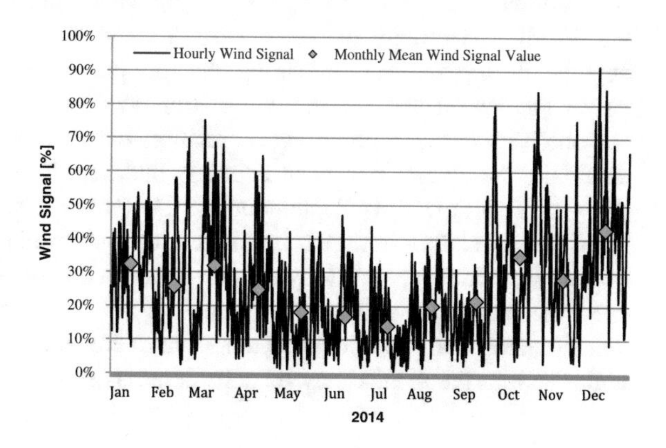

Fig. 1 Hourly dynamic wind signal and monthly mean values for the Swedish electricity market 2014

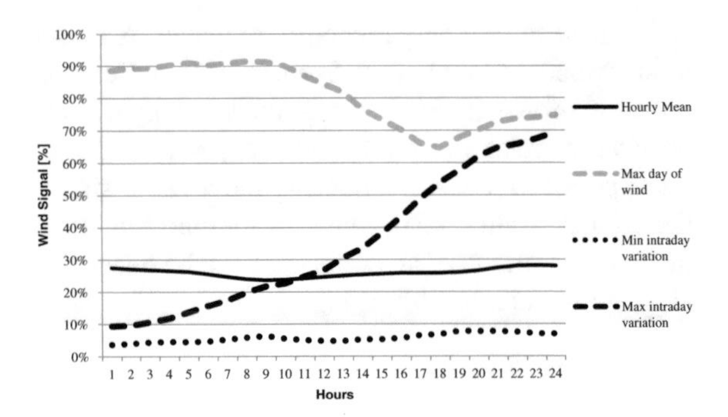

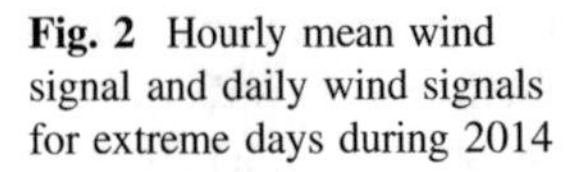
Fig. 2 Hourly mean wind signal and daily wind signals for extreme days during 2014

information on intraday variations, which is of particularly interest according to the aim of analyzing the wind signal as an potential incentive for intraday load shift. Thus, Fig. 2 also illustrates wind signal values of some extreme days in terms of hourly variation. It is shown that intraday patterns of the wind signal vary greatly between different days. From days of high volatility in wind power generation, ranging from 9–69 %, and days of relativity high and stable generation, ranging from 65–91 %. Contrary, there are days of low volatility, ranging from 4–8 %, and days of relativity low and stable generation, ranging from 1–5 %. Consequently, no clearly recurring curve shape in terms of peak or off-peak periods in wind power generation could be identified, a result highly expected given the character of wind as an unpredictable power source. However, the great variations in intraday patterns means that the potential of the signal as an incentive for load shift varies widely as well, where days of high variation provides a strong incentive to shift electricity consumption as end-use customers easily can identify wind peak and wind off-peak hours, while days of weaker variation lack a clear incentive of how to shift electricity consumption as a response to the signal.

The potential of using a wind signal as a strengthening incentive to a price-signal depends on the relation between wind power generation and electricity spot prices of the Swedish electricity market; economic and environmental benefits have to unite in order to form a strong incentive model. Figure 3 illustrates the monthly correlations between the hourly wind signal values and the electricity spot prices of the Swedish electricity market during 2014. Although the correlation is entirely negative during the year, indicating that spot prices decreases with increasing wind power generation, the monthly correlation coefficient varies widely, with no clear seasonally patterns. The strongest relation between wind power generation and electricity spot prices occurs in February and March, with a correlation coefficient $\rho = -0.51$, followed by September and December, where $\rho = -0.45$. In contrast, the weakest relation is shown in August, with a correlation coefficient $\rho = -0.14$, followed by January and May, of $\rho = -0.21$ and $\rho = -0.23$, respectively. However, the monthly t-values, highlighted in Fig. 3, all corresponds to an absolute value greater than the critical rejection value for the H_0 hypothesis of 1.96, which indicates that the results may be considered as statistical significant.

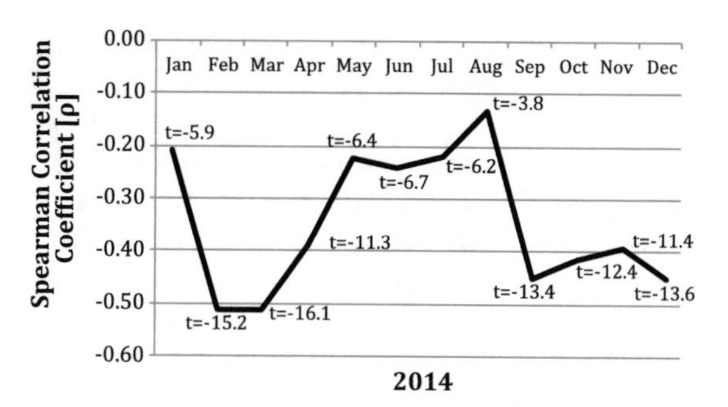

Fig. 3 Monthly correlation between wind power generation and electricity spot price of the Swedish electricity market, 2014

4 Discussion

The aim of this paper is to examine the potential of using a wind signal, reflecting the hourly variations in domestic wind power generation, as an environmental incentive for load shift in residential price-based DR programs. The rationale of such an approach is that an environmental incentive, provided as a compliment to an economic incentive, would strengthen the overall incentive model for end-use customers to adjust their daily electricity consumption patterns, shifting load from hours of high price and low wind power generation to hours of low price and high wind power generation. As a first step of the analysis, an hourly dynamic wind signal was constructed based on Swedish electricity generation for 2014, expressed as the quote of wind power generation in relation to the total installed wind power capacity in Sweden. The results show that hourly wind power generation varied significantly during the year (Fig. 1). Although no strong seasonally patterns were identified, the wind power generation slightly increased during the winter months, October to March, and was at its lowest during the summer months, April to September, corresponding to a total range in wind signal values from 1–91 %. The hourly mean value of the wind signal based on the whole year was 26 %, indicating that one fourth only of the total wind power generation capacity in Sweden 2014 was utilized on average. However, as wind power plants are located with a wide geographic spread, with variations in wind speed conditions at different times, it is arguable that a wind signal constructed on local or regional level, based on generation and total capacity for a certain geographic area, would provide a result of higher average wind power utilization.

Expectedly, given the unpredictable nature of wind, no clear recurring intraday curve pattern could be identified (Fig. 2). On an hourly average based on the whole year, wind power generation was marginally higher during night hours compared to hours during the day, but with such a small differences that it can be considered as negligible, ranging from 24–29 %. However, hourly mean values does not necessarily provide appropriate information on intraday curve shape patterns of the wind signal, which is of particularly interest according to the aim of analyzing the wind signal as an potential incentive for intraday load shift. Hence, coherent hourly values of different days were analyzed as well, showing that the daily curve shape pattern of the wind signal varied remarkable during the year, from days of relatively constant high values and large intraday variations, to days of almost no wind power generation at all and minor variations. Given this lack of a clear recurring daily curve pattern, it is reasonable to believe that the wind signal puts high request on the activeness of the end-use customers, demanding a great interest and acceptance to understand and respond to the variations in the wind signal on a daily basis. Compared to a spot price signal, where intraday curve patterns may be considered as relatively cyclic due to clear intraday patterns in demand, the unpredictability of the wind signal may imply a considerably challenge for the adoption of the signal, and to achieve response to a sufficient level. However, such a challenge may also lead to positive effects among end-use customers, such as increased awareness and knowledge about the electricity system, its environmental impact, and its relation to external weather conditions.

A fundamental requirement for the wind signal to serve as a supportive incentive to a price-signal is that both signals goes in line with each other; that load shift from a certain period during the day to another period of the day generates both economic and environmental benefits. In order to examine the potential of such a double-acting effect, the hourly relation between wind power generation and electricity spot price was analyzed for the whole year (Fig. 3). The results show that the association between wind power generation and spot price was modestly, but statistically significant, negative during the entirely year indicating that spot prices decrease with wind power generation. However, the monthly mean correlation coefficient varied markedly during the year, from $\rho = -0.14$ in August, to $\rho = -0.51$ in February and March. Given this results, with entirely negative correlation between wind power generation and spot prices, gives a positive response to the hypothesis of that providing both a wind signal and a spot price signal to end-use customers would form an enhanced and coherent incentive to shift electricity consumption from hours of high price and low wind power generation, to hours of low price and high wind power generation, beneficial in both an economical and an environmental perspective. In addition, it is reasonable to believe that two different signals, appealing to economic and environmental benefits respectively, would attract a wider range of household preferences than each signal individually; covering households that participates in DR programs of cost-saving reasons, but also households who consider the economic incentive as too weak, but who have a desire to contribute to a more climate-friendly energy system. In addition, the relevance of using a wind signal as an environmental incentive for demand response will most likely increase with the share of renewable intermittent power in the energy system, and the subsequent increased need of demand side participation. With expanded capacity of renewable generation, it may also be relevant to include additional power sources, such as solar power, in a further development of the signal.

5 Conclusions

The outcome of this study suggests that a dynamic wind signal, reflecting the hourly variations in wind power generation, could serve as an environmental incentive, supportive to the economic incentive of a spot price signal, for intraday load shift by end-use customers in residential DR programs; shifting electricity consumption from hours of high price and low wind power generation to hours of low price and high wind power generation, leading to both economical and environmental benefits. Providing such a hybrid incentive model would most likely increase the level of demand response, attracting an expanded range of residential end-use customers' motivations to respond; covering households with both economic and environmental preferences for participation. However, the lack of clear recurring daily curve pattern in wind power generation may demand high activeness and a strong commitment among electricity consumers in order to adopt and respond to the variations in the wind signal on a daily basis. Although such a challenge may be considered as a barrier for acceptance, it could also lead to increased knowledge and awareness among end-use customers related to the dynamics and the environmental impact of the electricity system. the signal.

References

Albadi, M., & El-Saadany, E. (2008). A summary of demand response in electricity markets. *Electric Power Systems Research, 78*(11), 1989–1996.

Amin, S. M., & Wollenberg, B. F. (2005). Toward a smart grid: Power delivery for the 21st century. *IEEE Power Energy Magazine, 3*(5), 34–41.

Chardon, A., Almén, O., Lewis, P.E., Stromback, J., Château B. (2008) Demand Response: A decisive breakthrough for Europe. Capgemini in collaboration with VaasaEtt and Enerdata.

Conejo, A. J., Morales, J. M., & Baringo, L. (2010). Real-time demand response model. *IEEE Transactions on Smart Grid, 1*(3), 236–242.

Darby, S. (2006). *The effectiveness of feedback on energy consumption: A review for DEFRA of the literature on metering, billing and direct displays*. Oxford: Environmental Change Institute, University of Oxford.

Farhangi, H. (2010). The path of the smart grid. *IEEE Power and Energy Magazine, 8*(1), 18–28.

Farugui, A., & Sergici, S. (2008). *The power of experimentation: New evidence on residential demand response*. Cambridge, MA: The Brattle Group.

Faruqui, A., & George, S. (2005). Quantifying customer response to dynamic pricing. *The Electricity Journal, 18*(4), 53–63.

Faruqui, A., & Sergici, S. (2010). *Household response to dynamic pricing of electricity: a survey of the empirical evidence*. Cambridge, MA: The Brattle Group.

Faruqui, A., Hledik, R., & Sergici, S. (2009). Piloting the smart grid. *The Electricity Journal, 22* (7), 55–69.

Gellings, C. W. (1985). The concept of demand-side management for electric utilities. *Proceedings of the IEEE, 73*(10), 1468–1470.

Gottwalt, S., Ketter, W., Block, C., Collins, J., & Weinhardt, C. (2011). Demand side management—a simulation of household behavior under variable prices. *Energy Policy, 39*, 8163–8174.

Grande, O.S., Sæle, H. (2005). Market based solutions for increased flexibility in electricity consumption. In *Conference proceedings of Security of supply on competitive electricity markets*. Saltsjöbaden, Stockholm, June 7–8 2005.

Ilic, M. D. (2007). From hierarchical to open access electric power systems. *Proceedings of the IEEE, 95*(5), 1060–1084.

Ipakchi, A., & Albuyeh, F. (2009). Grid of the future. *IEEE Power and Energy Magazine, 7*(2), 52–62.

Jacobson, M. Z., & Delucchi, M. A. (2011). Providing all global energy with wind, water, and solar power, Part I: Technologies, energy resources, quantities and areas of infrastructure, and materials. *Energy Policy, 39*, 1154–1169.

Kirschen, D., & Strbac, G. (2004). *Fundamentals of power system economics*. New York: Wiley.

Nilsson, A., Stoll, P., Brandt, N. (2015). Assessing the impact of real-time price visualization on residential electricity consumption, costs, and carbon emissions. *Resources, Conservation & Recycling*.

Nordpool Spot. (2015). *Historical market data*. Nordpool Spot. Retrieved from http://www.nordpoolspot.com/historical-market-data/.

Palensky, P., & Dietrich, D. (2011). Demand side management: Demand response, intelligent energy systems, and smart loads. *IEEE Industrial Informatics, 7*(3), 381–388.

Rahimi, F., & Ipakchi, A. (2010). Demand response as a market resource under the smart grid paradigm. *IEEE Transcations on Smart Grid, 1*(1), 82–88.

Siano, P. (2014). Demand response and smart grids—a survey. *Renewable and Sustainable Energy Reviews, 30*, 461–478.

Stokke, A. V., Doorman, G. L., & Ericson, T. (2010). An analysis of a demand charge electricity grid tariff in the residential sector. *Energy Efficiency, 3*, 267–282.

Stoll, P., Brandt, N., & Nordström, L. (2013). Including dynamic CO_2 intensity with demand response. *Energy Policy, 65*, 490–500.

Strbac, G. (2008). Demand side management: benefits and challenges. *Energy Policy, 36*(12), 4419–4426.

Svenska Kraftnät (SvK). (2015). *Electricity market statistics*. Svenska Kraftnät. Retrieved from http://www.svk.se/aktorsportalen/elmarknad/statistik.

Tan, Y. T., & Kirschen, D. (2007). *Classification of control for demand-side participation*. Manchester: The University of Manchester.

The International Energy Agency (IEA). (2003). *The power to choose—demand response in liberalized electricity markets*. Paris: OECD.

The Swedish Energy Agency (SEA). (2013). *The electric certificate system 2013*. The Swedish Energy Agency.

The International Energy Agency (IEA). (2014). *World energy outlook 2014*. IEA.

The Swedish Energy Agency (SEA). (2015). *Wind power statistics 2014 (in Swedish)*. The Swedish Energy Agency.

Torriti, J., Hassan, M. G., & Leach, M. (2010). Demand response experience in Europe: Policies, programmes and implementation. *Energy, 35*, 1575–1583.

U.S. Department of Energy. (2006). Benefits of demand response in electricity markets and recommendations for achieving them. In *A report to the United States Congress. Pursuant to Section 1252 of the Energy Policy Act of 2005*.

Xie, L., Young Joo, J., Ilic, M.D. (2009). Integration of intermittent resources with price-responsive loads. In *NAPS conference 2009*. North American.

Xu, Z., Togeby, M., & Ostergaard, J. (2011). Demand as frequency controlled reserve. *IEEE Transactions on Power Systems, 26*(3), 1062–1071.

Part IV
Energy System Modelling—Barriers, Challenges and Good Practice in Open Source Approaches

Wind Energy Scenarios for the Simulation of the German Power System Until 2050: The Effect of Social and Ecological Factors

Marion Christ, Martin Soethe, Melanie Degel and Clemens Wingenbach

Abstract Models of future energy systems and the development of underlying energy scenarios contribute to an answer on the question how a transformation of the energy system can be implemented. Although energy system modelling has a wide influence, the field lacks the consideration of local ecological and societal concerns, such as acceptance issues. In this paper, a methodology is developed to integrate social and ecological aspects concerning wind energy into the distribution of future wind energy capacities in Germany. Based on the calculated potential siting area (white area), an algorithm was developed to site different types of wind energy plants in the available area. Two wind expansion scenarios have been developed: one distributing future wind capacities according to technical and economic conditions (economic scenario), and the other based on the regional burden level resulting from wind energy plants (balanced scenario). The development of the burden level as a socio-ecological factor enabled the siting of wind turbines according to an equal burden level in all German districts (*Landkreise*). It was shown, that this equal distribution led to a shift of capacities from the north-west to the south-east compared to the economic approach.

Keywords Scenario development · Wind energy · Open source · Socio-ecological factors · Energy system modelling

M. Christ (✉) · M. Soethe · C. Wingenbach
Europa-Universität ZNES Flensburg, Munketoft 3b, 24937 Flensburg, Germany
e-mail: marion.christ@uni-flensburg.de

M. Soethe
e-mail: martin.soethe@uni-flensburg.de

C. Wingenbach
e-mail: clemens.wingenbach@uni-flensburg.de

M. Degel
IZT, Schopenhauerstr. 26, 14129 Berlin, Germany
e-mail: m.degel@izt.de

V. Wohlgemuth et al. (eds.), *Advances and New Trends in Environmental Informatics*, Progress in IS, DOI 10.1007/978-3-319-44711-7_14

1 Introduction

The forecasting and planning of future energy systems is a complex task. Challenges such as climate change, sustainable development and the security of energy supply necessitate a transformation of future systems. The shift to systems with larger shares of renewable energies will play a particularly significant role in answering the question of how such a transformation can be implemented (Lund and Mathiesen 2009). To handle the complexity of this task, a wide range of energy models has been developed over the last decade (Connolly et al. 2010). The knowledge gained from modelling is often used to answer questions concerning political aims, such as specific targets or investment decisions. Models produce a range of results, which are dependent on the functionality of the model and the choice of input parameters (Wiese 2014).

The results of energy models depend particularly on the definition of energy scenarios. Various approaches are used in energy scenarios to deal with possible energy requirements in the future. The scenarios help decision makers with uncertainties, as they show pathways to a preferable future (Mai et al. 2013). Scenarios help to understand the effect of input drivers on outcomes. However, they provide a range of possible futures with fixed boundaries rather than predictions or forecasts (Wilson 2000). Scenarios vary in their type and methodology. Forecasting scenarios are explorative (or descriptive), i.e. use past trends without considering if the result is desired or not, whereas backcasting (or normative) scenarios define a specific goal at the beginning and the scenario shows how this goal can be achieved and to what extent this is possible (Berkhout et al. 2002; Börjeson et al. 2006; Dieckhoff 2014). Regarding methodology, both quantitative and qualitative approaches are used. Quantitative approaches use numerical figures to describe future developments. These values are mostly obtained by modelling tools, requiring simplifications and assumptions (Varho and Tapio 2013). Qualitative approaches, using e.g. narrative stories ('storylines') to create images of the future, are mainly based on results of participatory methods such as stakeholder workshops or interviews (Dieckhoff 2014; Notten et al. 2003). Such qualitative scenarios reflect the view of the participating stakeholders and mostly describe cultural, political and social developments (Söderholm et al. 2011). However, although most energy scenarios show great technical and economic detail, they lack a consideration of social aspects (Fortes et al. 2014; Mai et al. 2013).

One approach to link qualitative stories and quantitative simulation is cross-impact balance (CIB) analysis, which validates the internal consistency of storylines (Weimer-Jehle 2006). CIB analysis helps to develop consistent scenarios, using expert knowledge to validate normative assumptions for simulations (Weimer-Jehle 2013; Weimer-Jehle et al. 2013). Other approaches build upon CIB, adjusting the method to multi-scale structures (Schweizer and Kurniawan 2016), translate storylines from one specific country into assumptions for an optimization model (Fortes et al. 2014), or link qualitative storylines with multiple, diverse quantitative models (Trutnevyte et al. 2014). The topicality of these research projects display the relevance of linking qualitative scenario development with quantitative modelling approaches. However, local societal concerns, such as acceptance issues, are still absent from the qualitative dimensions used in energy scenario development. Although global or countrywide storylines describe possible societal futures on a larger scale, local ecological or societal circumstances are neglected.

Considering the fact that there are ambitious government targets to increase the percentage of renewable energy generation in many countries, social acceptance, especially at local levels, may become a constraining factor in achieving these targets. Present research projects describe and analyse acceptance problems, particularly in the case of wind energy, which has become a subject of debate in several countries, largely due to the increasing public awareness for landscape preservation (Jobert et al. 2007; Wüstenhagen et al. 2007; Caporale and Lucia 2015; Enevoldsen and Sovacool 2016; Hammami et al. 2016; Huesca-Pérez et al. 2016; Okkonen and Lehtonen 2016). The extensive literature referring to this topic illustrates the significance of considering social acceptance in the planning process for future energy systems with high percentages of renewable energy.

In this paper, a methodology is developed to integrate social and ecological aspects concerning wind energy into the distribution of future onshore wind energy capacities. Simulation models based on backcasting scenarios, with e.g. a defined desired target for installed capacities of wind energy, use a predicted value for the installed capacity. In addition to the defined total capacity, the distribution of the future capacities inside a specific country also has to be assumed. In the German grid development plan, the methodology for the regional distribution of renewable energy plants is being steadily renewed (50Hertz Transmission GmbH et al. 2014). This paper has the objective to advance the methodology of energy scenario development by including not only technical, economic and ecological criteria, but also social criteria into the regional distribution of future wind energy plants. Energy scenarios are defined using qualitative and quantitative assumptions, suitable for the simulation of the German power system until 2050 with an open source energy simulation model. For the described scenario development in this article and the subsequent modelling, only open data and open source software have been used. The approach in this paper is one aspect of the research project 'VerNetzen', which concentrates on socio-ecological and techno-economic energy system modelling with the open source model *renpass* (Wiese 2014; Bunke et al. 2014; Degel and Christ 2015; Christ et al. 2016). The model is developed at the Europa-Universität Flensburg (Wiese 2014). A follow up version is based on the Open Energy Modelling Framework (oemof) developed by the Center for Sustainable Energy Systems (ZNES), the Reiner Lemoine Institute (RLI) in Berlin and the Otto-von-Guericke-University of Magdeburg (OVGU) (oemof—Open Energy Modelling Framework 2016a, b).

The following section presents the determined technical potential for future wind energy capacities, including the development of a database, the characterization of suitable sites and the potential siting of future wind energy plants. The development of scenarios with the distribution algorithm, and the introduction of a 'burden level' as a socio-ecological parameter are explained in the following sections. Results are presented in Sect. 3 followed by the discussion and the conclusion.

2 Methodology

Determining the technical potential for the expansion of wind energy until 2050 is important for the definition of scenarios. For this purpose a georeferenced database was developed. This database was used to determine the area where an installation of wind energy is feasible (known as the 'white area'). Furthermore, an algorithm was developed to provide an indication of the best place to site a wind energy plant according to technical and economic conditions. With the knowledge of the potential area to site future wind energy plants, scenario assumptions were set to define how much and what particular areas of the total potential were used. Determination of the specific subsets of the potential area was the result of the scenario definition. A 'burden level' was defined to include social conditions into the scenario assumption, influencing the search for the most adequate installation sites.

2.1 Georeferenced Database

The georeferenced database contains information about the German electricity grid and power plants, structural data such as landscape, settlements, nature reserves and regional data such as population, income, tourism and land use (Bundesamt für Kartographie und Geodäsie 2014; Bundesamt für Naturschutz 2015; Bundesnetzagentur 2014; Deutsche Gesellschaft für Sonnenenergie 2014; European Environment Agency 2014a, b; OpenStreetMap.org 2015; Regionaldatenbank Deutschland 2014). The use of the database system PostgreSQL with the extension PostGIS allowed the application of geometric operations to link statistical data with geographic information. This georeferenced statistical basis could be used for systematic and scientific analysis of specific regions.

2.2 Determination of the White Area

The siting potential for wind energy was determined in great detail with the aim to develop a scenario with a high spatial resolution of the distribution of wind energy plants within all German districts (*Landkreise*). Unsuitable sites such as settlements, railways or nature reserves and surrounding areas were eliminated from the land area with defined distance standards, based on the potential study carried out by the German federal environment agency (Lütkehus et al. 2013). Forested area was generally excluded in order to evaluate how wind energy expansion targets are achievable without using such highly debated areas. The calculation of the white area was performed in the PostgreSQL database using the PostGIS extension. For every German district a maximum potential area for wind energy installations was identified. The calculated white area in this paper is 27,244 km^2, corresponding to 7.6 % of Germany's total land area.

2.3 Technical Potential of Installed Wind Power

The theoretical technical potential in this paper covers the potential of wind energy plants which can be operated if the entire white area is used for the siting of modern wind turbines with a defined distance between each other. This calculated technical potential has to be understood as an estimation, based on a theoretical approach. An algorithm was developed in *python* to site wind energy plants on geometric areas. Wind speed data and the corresponding surface roughness length from the Coast-Dat2 database (Geyer 2014) were used for the calculation of potential energy outputs. Dependent on the annual average wind speed either strong wind or weak wind turbines were placed on available areas inside the defined white area. Two modern wind turbines with an installed capacity of 3.3 MW from Vestas were used (V126 and V112). The distance between the individual wind turbines was set to four times the rotor diameter (Lütkehus et al. 2013).

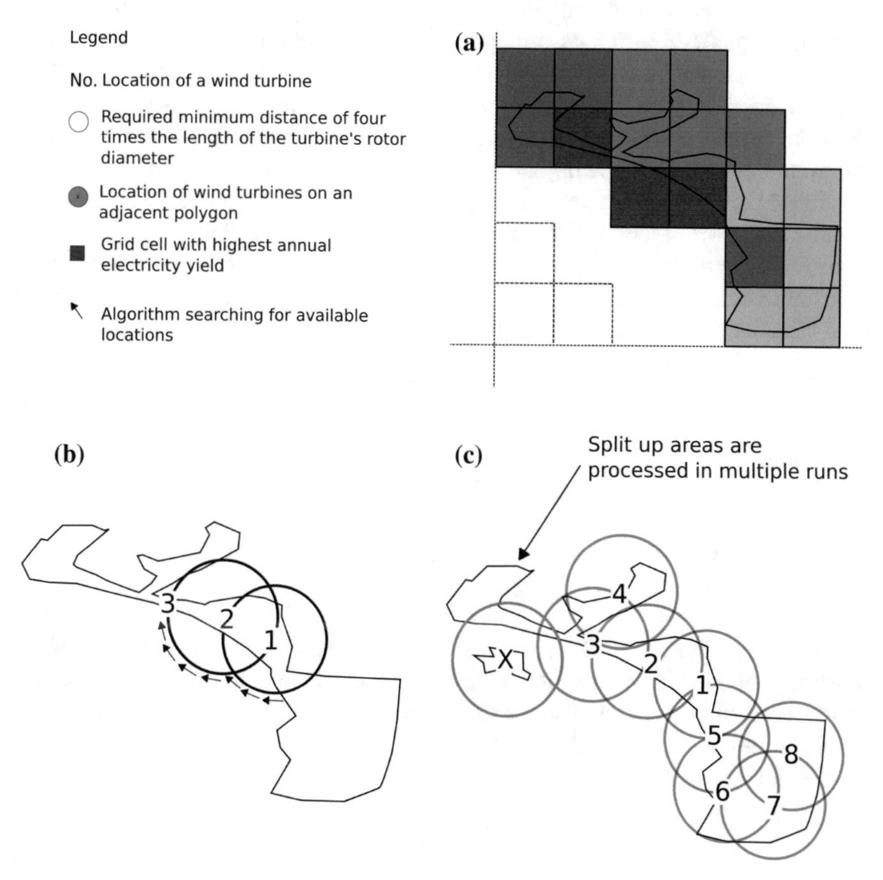

Fig. 1 Identification of locations for wind energy turbines inside the white area: **a** Exemplary white area and the corresponding potential energy output per grid cell. **b** Placement of wind turbines, starting with the most efficient location. **c** Performance of multiple runs to guarantee than the white area is fully utilized

Figure 1 shows the procedure for the placement of wind energy turbines inside the white area. The economic efficiency of a location was defined by the energy output, calculated with the nearest weather data point. The algorithm placed the wind turbines, starting with the most efficient location. Originating from the first turbine further locations were identified with an accuracy of two meters. The determined technical performance potential is 617.91 GW.

2.4 Burden Level

The burden level was used as an indicator of the societal and ecological burden of wind energy plants on a local level. Starting point for the definition of a burden level was to collect data on potential influencing parameters in general including technical, economic, ecological and political parameters (Degel et al. 2016). In addition stakeholder analysis with focus groups and expert interviews, environment and discourse analysis and the detailed analysis of numerous municipalities were carried out in cooperation with the Institute for Future Studies and Technology Evaluation (IZT) in Berlin (Bunke et al. 2014; Degel and Christ 2015). In the end two major parameters were defined, one showing the area of land occupied by wind energy plants in relation to the regional total area and the other showing the population density. These parameters provide an indication of the socio-ecological burden resulting from wind energy. The area occupied by wind energy plants $A_{occupied}$ was calculated with Eq. 1, with a maximum installed wind energy capacity derived from the summation of the total capacity of the technical potential per region.

$$A_{occupied,i} = P_i \cdot a_i, \; \forall\, i \in R \tag{1}$$

with:

$A_{occupied}$	km^2	Occupied white area
P	MW	Installed wind capacity, e.g. present, max or scenario
a	km^2/MW	Specific area used per wind capacity
R	–	Set of regions

The present burden level $b_{present}$ was calculated using Eq. 2. The burden level, as a measure of inhabitant per square kilometre, was defined as a degree of the density of the affected population. The maximum burden level b_{max} was calculated in the same way as the present burden level, whereby the occupied area is equivalent to the total white area (see Eq. 3).

$$b_{present,i} = \frac{A_{occupied,i}}{A_{region,i}} \cdot p_i, \; \forall\, i \in R \tag{2}$$

$$b_{max,i} = \frac{A_{max,i}}{A_{region,i}} \cdot p_i, \; \forall\, i \in R \tag{3}$$

with:

b	$inhabitant/km^2$	Burden level
A_{max}	km^2	Total white area
A_{region}	km^2	Region area
p	$inhabitant/km^2$	Population density

The burden level was used to determine a balanced distribution of installed wind capacity. The 'scenario - burden level' is a qualitative indicator, which influences the expansion of wind energy plants in different regions. This indicator can be applied within algorithms and offers a connection between qualitative social analysis and quantitative computing. The burden level of the population was integrated into the scenario development with an additional expansion algorithm, limiting the technical potential according to the balanced burden level b_{bal} as shown in Eq. 4. In regions where this burden level surpasses the maximum burden level, the expansion was limited by its total white area (see Eq. 5). Additionally Eq. 6 was used to include the present installed wind capacity as minimum.

$$b_i = b_{bal} = \frac{\sum P_{bal,i}}{\sum \frac{A_{region,i}}{a_i \cdot p_i}}, \; \forall\, i \in R_{bal} \tag{4}$$

$$b_i = b_{max,i}, \; \forall\, i \in \{R \mid b_{bal} > b_{max,i}\} \tag{5}$$

$$b_i = b_{present,i}, \; \forall\, i \in \{R \mid b_{bal} < b_{present,i}\} \tag{6}$$

with:

b_{bal}	$inhabitant/km^2$	Balanced burden level
P_{bal}	MW	Corresponding balanced installed wind capacity
R_{bal}	$-$	Set of regions with the balanced burden level

3 Results

Two wind expansion scenarios were developed and are described in the following. They are referred to as the 'economic scenario' and the 'balanced scenario'. For both, the installed wind capacity in 2050 was assumed to be 101.6 GW. In an energy system with a high share of fluctuating renewable energy, the sum of production needs to be higher than the demand (Bökenkamp 2014). For a stable system, a reasonable value for the ratio of renewable production to demand is 1.3 (Alonso et al. 2011). This value was applied in this study based on the scenario assumptions of the renewable energy production in 2050 used in the German pilot study of renewable energy expansion

(*Leitstudie*) (Nitsch et al. 2012), assuming that the electricity demand in 2050 is no greater than the value in 2014 (518 TWh) (ENTSO-E 2014). For a ratio of renewable production to demand of 1.3, the renewable capacities of the *Leitstudie* had to be increased. For the simulation and scenario development in the *VerNetzen* project, the wind onshore capacity was increased by a factor of two, wind offshore by a factor of 1.5 and solar energy by a factor of 2.

For the economic scenario, the yearly full-load hours of each region in Germany were used as steering factor for the distribution of the capacity of 101.6 GW. The total capacity was expanded linearly between 2014 and 2050, taking into account the already installed capacity in 2014. Regions with relatively high yearly full load hours were assigned a higher share of installed wind capacity per year. The amount of capacity added to each region depended on the technical potential and economic limits. The algorithm just used installation sites where an economic operation was feasible according to the yearly full-load hours. The burden level had no influence on the calculation of the economic scenario.

For the balanced scenario, the burden level was equally balanced in each German district. Two alternative approaches have been used to calculate the balanced scenario: in the first approach, the 'balanced greenfield' approach, the wind capacities currently installed were not considered. The assumed 101.6 GW were distributed according to an equal burden level in all German districts. In the second approach, the 'balanced' approach, the existing wind energy capacities of 2014 were taken into account, so that just the difference between the assumed 101.6 GW and the currently installed capacity was equally distributed. Regions which currently have a high installed wind capacity can therefore exceed the equal burden level in 2050, limited by their present burden level, which is already higher. This approach is based on the assumption, that regions which currently have a high share of wind energy will still hold a high share until 2050 due to experience and repowering projects.

Figure 2 shows the German districts (excluding urban districts) with their corresponding population density and the ratio of used area for wind energy plants to the total district area. The lines show the present burden level in 2014 and the burden level in the different scenarios, economic, balanced greenfield and balanced. The highest burden level in 2050 was determined in the economic scenario in Alzey-Worms with a regional burden level of almost 12 inhabitants per square kilometre and an installed wind capacity of 539.24 MW. In Alzey-Worms, 27.6 % of the total white area was used in 2050, in comparison to 18.21 % in 2014. In contrast to Alzey-Worms, in Nordfriesland a burden level of just 3 inhabitants per square kilometre was derived in the economic scenario by using a similar share of the white area (27.7 %). The equal burden level in the greenfield approach amounts to 1.34 inhabitants per square kilometre. With the consideration of the present installed capacity in 2014, the equal burden levels decreased by 6.7 % to 1.25 inhabitants per square kilometre. Districts with low shares of white area and therefore with a low siting potential for wind energy plants stay below the calculated equal burden.

Figure 3 illustrates the installed capacity for each German district in 2050 for two different scenarios, the economic and the balanced scenario. In the economic scenario the highest installed capacity in 2050 was determined in Nordfriesland with

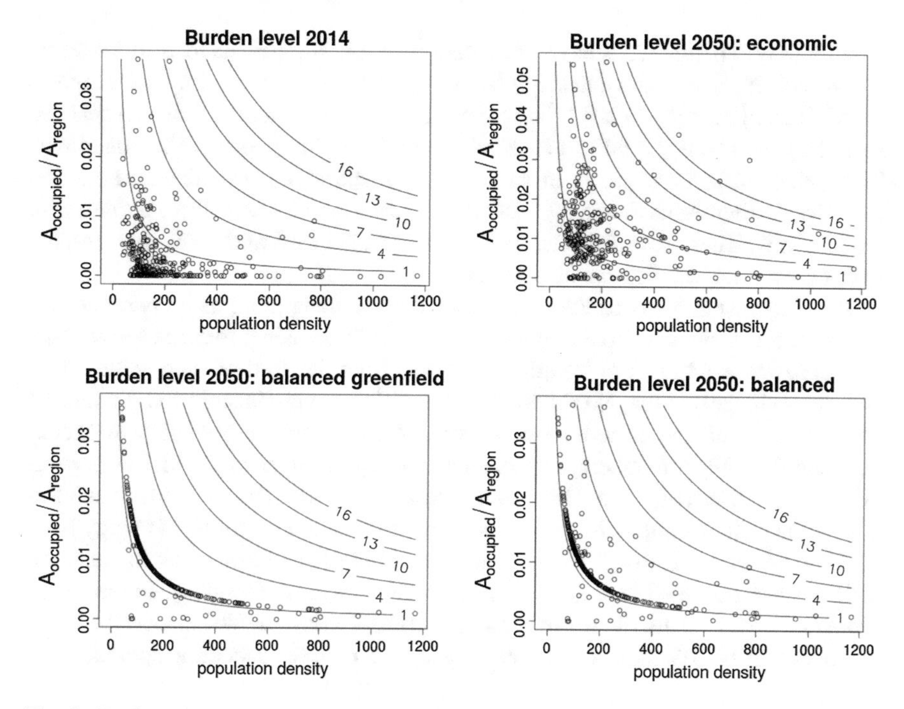

Fig. 2 Burden levels in 2014 and 2050 for all German districts excluding cities: present burden level 2014, burden level 2050: economic scenario, balanced greenfield scenario (equally distributed) and balanced scenario (equally distributed based upon installed wind capacity in 2014)

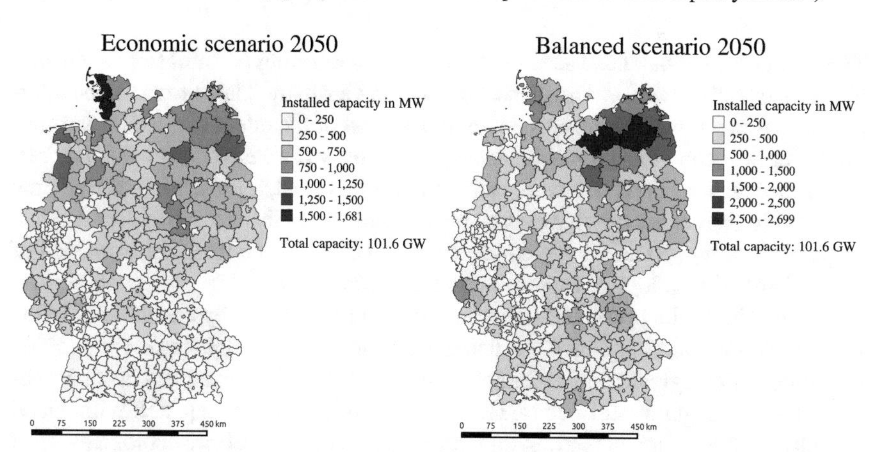

Fig. 3 Installed wind capacity in 2050 for each German district in MW: economic scenario and balanced scenario. 101.6 GW of total installed capacity

1,680.6 MW. This is 373.8 MW more than in 2014. Nordfriesland is followed by Dithmarschen, Uckermark and Prignitz. These regions accounted also for the highest installed wind capacity in 2014. The technical-economic potential limit was reached in 140 cases, mainly in cities and southern regions with poorer wind conditions. The highest installed capacity in the balanced scenario was calculated in the district Mecklenburgische Seenplatte with 2,698.6 MW and Ludwigslust-Parchim with 2,626.9 MW. These two districts, with areas of 5,501 km^2 and 4,767 km^2 respectively, are the largest districts in Germany.

As a key result, the capacities were distributed among more districts in the balanced scenario compared to the economic scenario. The main difference between the two scenarios is the shift of installed capacities from the northern and western federal states Niedersachsen, Nordrhein-Westfalen, Schleswig-Holstein and Rheinland-Pfalz to the south and eastern states Bayern, Mecklenburg-Vorpommern, Brandenburg and Baden-Württemberg. Especially large regions in the east with low population densities take a high share of installed capacities. The federal state Mecklenburg-Vorpommern for example, increased its share of the land area for wind energy from 0.9 % in the economic to 2.2 % in the balanced approach, and would thereby have the highest share occupied land area of all the federal states in Germany. In the south, lower capacities are distributed over more districts. In Bayern, the balanced scenario led to around 10 GW more wind energy capacities than the economic approach.

4 Discussion

The economic and balanced scenarios provide two differing possibilities for the distribution of future wind energy capacities within Germany. They are not to be interpreted as predictions or forecasts but an illustration of the effect of specific assumptions. For the economic approach no social circumstances were considered. The balanced approach, on the other hand, included a burden level, representing the regional socio-ecological effect of wind energy installations. In both scenario approaches the target of installing 101.6 GW wind capacity in 2050 could be met. The exclusion of forested area for ecological reasons had no detrimental effect.

The results of the presented scenarios have to be analysed by considering information about local acceptance, participation concepts and legal frameworks. Even by reducing the regional burden limits by means of a shift of capacities, local acceptance problems might still occur. The burden level cannot be equated with social acceptance, it is rather an instrument to display the local social and ecological effect of wind energy. The expansion of wind energy within Germany is a general challenge, which might be dealt with in an easier way by considering local acceptance problems at an early stage. The developed scenarios can help to define regional political targets or local installation or burden limits which could be widely accepted in those specific regions.

5 Conclusion

The presented approach is a first step in the inclusion of local societal and ecological factors in energy system modelling. The developed scenarios enable the simulation of the German power system in 2050 to show effects of a more socio-ecological wind energy distribution in comparison to present economic scenarios. The economic and the balanced scenarios illustrate the possible spread of future wind energy siting, focusing on economic or on socio-ecological aspects, respectively.

The developed methodology can be interpreted as a basis for the connection of social science and scientific computing. Results of regional social analysis concerning acceptance issues of renewable energy installations can be included into the scenario development by setting local burden limits or by defining specific targets. Energy system models can be advanced by considering local social or ecological aspects in the scenario setting. Political recommendations, based on simulation results are in this way not only dependent on technical and economic conditions, but also influenced by ecological and social aspects.

The use of open data and open source leads to transparency and reproducibility. The scenarios can be used, on the one hand, to improve results and resulting recommendations of energy system models, and on the other hand, to build a bridge to participation processes at a local level.

References

50Hertz Transmission GmbH, Amprion GmbH, TenneT TSO GmbH, TransnetBW GmbH (2014) Regionalisierungsmethodik Erneuerbarer Energien für das Szenario B 2024 im Netzentwicklungsplan Strom 2014, 2. Entwurf. Grundlage für das Szenario B 2024. http://www.netzentwicklungsplan.de/sites/default/files/media/documents/Erl%C3%A4uterung_Regionalisierung_NEP_2014.pdf. Cited 23. Feb 2015

Alonso, O., Galbete, S., & Sotés, M. (2011). Sizing and management of energy storage for a 100% renewable supply in large electric systems. In R. Carbone (Ed.) *Energy Storage in the emerging era of smart grids, chap. 15* (pp. 321–348). Croatia: InTech.

Berkhout, F., Hertin, J., & Jordan, A. (2002). Socio-economic futures in climate change impact assessment: using scenarios as learning machines. *Global Environmental Change*, *12*(2), 83–95. doi:10.1016/S0959-3780(02)00006-7.

Bökenkamp, G. (2014). The role of norwegian hydro storage in future renewable electricity supply systems in germany: Analysis with a simulation model. Ph.D. thesis, Europa-Universität Flensburg. http://www.zhb-flensburg.de/dissert/boekenkamp/Gesine_B%C3%B6kenkamp_Dissertation/dissertation%20boekenkamp.pdf. Cited 15. Mar 2016

Börjeson, L., Höjer, M., Dreborg, K. H., Ekvall, T., & Finnveden, G. (2006). Scenario types and techniques: Towards a user's guide. *Futures*, *38*(7), 723–739. doi:10.1016/j.futures.2005.12.002.

Bundesamt für Kartographie und Geodäsie (2014). Digitales Landschaftsmodell 1 : 250 000 (AAA—Modellierung) DLM 250 (AAA). http://www.geodatenzentrum.de/docpdf/dlm250.pdf. Cited 15. Feb 2016

Bundesamt für Naturschutz (2015). WFS Layer der Schutzgebiete von Deutschland. http://geodienste.bfn.de/wfs/schutzgebiete. Cited 15. Feb 2016

Bundesnetzagentur (2014). Kraftwerksliste der Bundesnetzagentur—Stand: 29.10.2014. http://www.bundesnetzagentur.de/DE/Sachgebiete/ElektrizitaetundGas/Unternehmen_Institutionen/Versorgungssicherheit/Erzeugungskapazitaeten/Kraftwerksliste/kraftwerksliste-node.html. Cited 02. June 2015

Bunke, W.D., Christ, M., & Degel, M. (2015). VerNetzen: Sozial-ökologische und technisch-ökonomische Modellierung von Entwicklungspfaden der Energiewende. Tagungsband: Bundesnetzagentur Wissenschaftsdialog 2014. Technologie, Landschaft und Kommunikation, Wirtschaft pp. 107–125. http://www.netzausbau.de/SharedDocs/Downloads/DE/Publikationen/Tagungsband_14.pdf?__blob=publicationFile. Cited 09. Mar 2016

Caporale, D., & Lucia, C. D. (2015). Social acceptance of on-shore wind energy in apulia region (southern italy). *Renewable and Sustainable Energy Reviews*, *52*, 1378–1390. doi:10.1016/j.rser.2015.07.183.

Christ, M., Soethe, M. & Degel, M. (2016) Windausbauszenarien unter Berücksichtigung gesellschaftlicher und ökologischer Rahmenbedingungen für die Simulation des deutschen Energiesystems bis 2050. Tagungsband: EnInnov2016. 14. Symposium Energieinnovation. Energie für unser Europa (110–111). http://www.tugraz.at/fileadmin/user_upload/Events/Eninnov2016/files/lf/Session_B2/LF_Christ_WIND.pdf. Cited 09 Mar 2016

Connolly, D., Lund, H., Mathiesen, B., & Leahy, M. (2010). A review of computer tools for analysing the integration of renewable energy into various energy systems. *Applied Energy*, *87*, 1059–1082. doi:10.1016/j.apenergy.2009.09.026.

Degel, M. & Christ, M. (2015). VerNetzen: Sozial-ökologische und techisch-ökonomische Modellirung von Entwicklungspfaden der Energiewende. Tagungsband: FONA Statuskonferenz des BMBF zur Fördermaßnahme Umwelt- und gesellschaftsverträgliche Transformation des Energiesystems (197–202). https://www.fona.de/mediathek/pdf/Tagungsband_Statuskonferenz_Transformation_des_Energiesystems_2015.pdf. Cited 22. Mar 2016

Degel, M., Christ, M., Bunke, W.D., Mester, K., Soethe, M., Grünert, J. & Becker, L. (2016). Projektabschlussbericht—VerNetzen: Sozial-ökologische und technisch-ökonomische Modellierung von Entwicklungspfaden der Energiewende. In preparation.

Deutsche Gesellschaft für Sonnenenergie e.V. (2014). EEG Anlagenregister Nov. 2014. http://www.energymap.info/download/eeg_anlagenregister_2014.11.utf8.csv.zip. Cited 15. Jan 2015

Dieckhoff, C.e.a. (2014). Zur Interpretation von Energieszenarien (Schriftenreihe Energiesysteme der Zukunft). acatech—Deutsche Akademie der Technikwissenschaften e. V. http://www.leopoldina.org/uploads/tx_leopublication/2014_ESYS_Energieszenarien.pdf. Cited 02. Feb 2014

Enevoldsen, P., & Sovacool, B. K. (2016). Examining the social acceptance of wind energy: Practical guidelines for onshore wind project development in france. *Renewable and Sustainable Energy Reviews*, *53*, 178–184. doi:10.1016/j.rser.2015.08.041.

ENTSO-E (2014). Production, consumption, exchange package. https://www.entsoe.eu/db-query/country-packages/production-consumption-exchange-package. Cited 15. Mar 2016

European Environment Agency (2014a). Corine Land Cover 2006 seamless vector data V-3. http://www.eea.europa.eu/data-and-maps/data/natura-5#tab-gis-data. Cited 24. Feb 2015

European Environment Agency (2014b). Natura 2000 data—the European network of protected sites. http://www.eea.europa.eu/data-and-maps/data/natura-5#tab-gis-data,abgerufenam10.November2014. Cited 24. Feb 2015

Fortes, P., Alvarenga, A., Seixas, J. & Rodrigues, S. (2014). Long-term energy scenarios: Bridging the gap between socio-economic storylines and energy modeling. *Technological Forecasting & Social Change*, *91*. doi:10.1016/j.techfore.2014.02.006

Geyer, B. (2014). High-resolution atmospheric reconstruction for europe. 1948–2012. *Earth System Science Data*, *6*, 147–164. doi:10.5194/essd-6-147-2014

Hammami, S.M., Chtourou, S. & Triki, A. (2016). Identifying the determinants of community acceptance of renewable energy technologies: The case study of a wind energy project from tunisia. *Renewable and Sustainable Energy Reviews*, *54*, 151–160. http://dx.doi.org/10.1016/j.rser.2015.09.037

Huesca-Pérez, M. E., Sheinbaum-Pardo, C., & Köppel, J. (2016). Social implications of siting wind energy in a disadvantaged region—the case of the isthmus of tehuantepec, mexico. *Renewable and Sustainable Energy Reviews*, *58*, 952–965. doi:10.1016/j.rser.2015.12.310.

Jobert, A., Laborgne, P., & Mimler, S. (2007). Local acceptance of wind energy: Factors of success identified in french and german case studies. *Energy Policy*, *35*, 2751–2760. doi:10.1016/j.enpol.2006.12.005.

Lund, H., & Mathiesen, B. (2009). Energy system analysis of 100systems—the case of denmark in years 2030 and 2050. *Energy*, *34*, 524–531. doi:10.1016/j.energy.2008.04.003.

Lütkehus, I., Salecker, H. & Adlunger, K. (2013). Potenzial der Windenergie an Land. https://www.umweltbundesamt.de/sites/default/files/medien/378/publikationen/potenzial_der_windenergie.pdf. Cited 10. Jan 2016

Mai, T., Logan, J., Blair, N., Sullivan, P. & Bazilian, M. (2013) Re-assume—a decision maker's guide to evaluating energy scenarios, modeling, and assumptions. http://iea-retd.org/wp-content/uploads/2013/07/RE-ASSUME_IEA-RETD_2013.pdf. Cited 06. May 2014

Nitsch, J., et. al. (2012). Langfristszenarien und Strategien für den Ausbau der erneuerbaren Energien in Deutschland bei Berücksichtigung der Entwicklung in Europa und global. http://www.dlr.de/dlr/Portaldata/1/Resources/bilder/portal/portal_2012_1/leitstudie2011_bf.pdf. Cited 11. Aug 2015

oemof—Open Energy Modelling Framework: Documentation (2016a). Overview —the idea of an open framework. http://oemof.readthedocs.io/en/stable/overview.html. Cited 05. Jun. 2016

oemof—Open Energy Modelling Framework (2016b). Github repository. https://github.com/oemof. Cited 05. Jun. 2016

Okkonen, L., & Lehtonen, O. (2016). Socio-economic impacts of community wind power projects in northern scotland. *Renewable Energy*, *85*, 826–833. doi:10.1016/j.renene.2015.07.047.

OpenStreetMap.org (2015). Geofabik aufbereiter OpenStreetMap Datensatz. http://download.geofabrik.de/europe/germany-latest.osm.pbf. Cited 18. Mar 2015

Regionaldatenbank Deutschland (2014) Datensätzen der statistischen Ämter des Bundes und der Länder. https://www.regionalstatistik.de/genesis/online/logon. Cited 23. Feb 2015

Schweizer, V.J. & Kurniawan, J.H. (2016). Systematically linking qualitative elements of scenarios across levels, scales, and sectors. *Environmental Modelling & Software*. doi:10.1016/j.envsoft.2015.12.014. In Press

Söderholm, P., Hildingsson, R., Johansson, B., Khan, J., & Wilhelmsson, F. (2011). Governing the transition to low-carbon futures: A critical survey of energy scenarios for 2050. *Futures*, *43*(10), 1105–1116. doi:10.1016/j.futures.2011.07.009.

Trutnevyte, E., Barton, J., O'Grady, Á., Ogunkunle, D., Pudjianto, D., & Robertson, E. (2014). Linking a storyline with multiple models: A cross-scale study of the uk power system transition. *Technological Forecasting and Social Change*, *89*, 26–42. doi:10.1016/j.techfore.2014.08.018.

van Notten, P. W., Rotmans, J., van Asselt, M. B., & Rothman, D. S. (2003). An updated scenario typology. *Futures*, *35*(5), 423–443. doi:10.1016/S0016-3287(02)00090-3.

Varho, V., & Tapio, P. (2013). Combining the qualitative and quantitative with the Q2 scenario technique—The case of transport and climate. *Technological Forecasting and Social Change*, *80*(4), 611–630. doi:10.1016/j.techfore.2012.09.004.

Weimer-Jehle, W. (2013). ScenarioWizard 4.1 Constructing Consistent Scenarios Using Cross-Impact Balance Analysis. ZIRIUS. http://www.cross-impact.de/Ressourcen/ScenarioWizardManual_en.pdf. Cited 26. Feb 2016

Weimer-Jehle, W., Prehofer, S., & Vögele, S. (2013). Kontextszenarien. Ein Konzept zur Behandlung von Kontextunsicherheit und Kontextkomplexität bei der Entwicklung von Energieszenarien. *Technikfolgenabschätzung—Theorie und. Praxis*, *2*, 27–36.

Weimer-Jehle, W. (2006). Cross-impact balances: A system-theoretical approach to cross-impact analysis. *Technological Forecasting and Social Change*, *73*(4), 334–361. doi:10.1016/j.techfore.2005.06.005.

Wiese, F. (2014). Renpass–Renewable Energy Pathways Simulation System—Open source as an approach to meet challenges in energy modeling. Ph.D. thesis, Europa-Universität Flensburg. http://renpass.eu. Cited 10. Jan 2016

Wilson, I. (2000). From scenario thinking to strategic action. *Technological Forecasting and Social Change*, *65*(1), 23–29. doi:10.1016/S0040-1625(99)00122-5.

Wüstenhagen, R., Wolsink, M., & Bürer, M. J. (2007). Social acceptance of renewable energy innovation: An introduction to the concept. *Energy Policy*, *35*, 2683–2691. doi:10.1016/j.enpol.2006.12.001.

AC Power Flow Simulations within an Open Data Model of a High Voltage Grid

Ulf Philipp Müller, Ilka Cussmann, Clemens Wingenbach and Jochen Wendiggensen

Abstract The development of transparent and reproducible electricity grid models poses a scientific challenge, mainly due to a lack of data transparency in grid economy. This paper presents a modelling method based on Open Data and Open Source. It is focused on a model of the high voltage grid of the German federal state Schleswig-Holstein. The developed model enables to perform static AC power flow simulations. Topological OpenStreetMap information is combined with literature-based assumptions of typical line data. Generation capacities were assigned to the grid buses using freely available registers. The demand was allocated to these buses using the correlating distribution of population. The output of the AC power flow simulations respectively bus voltages and line currents for different worst-case scenarios are presented, analysed and discussed. The results show the potential of scientific electricity grid modelling using Open Data and Open Source. Realistic grid congestion within the high voltage grid of Schleswig-Holstein was identified, providing a basis for further discussion. For a comprehensive validation of the simulated results relevant data need to be opened for the public.

Keywords Electricity grid modelling · Electricity grid transparency · Open Data · Open Source

U.P. Müller (✉) · I. Cussmann · J. Wendiggensen
Centre for Sustainable Energy Systems, Flensburg University of Applied Sciences,
Kanzleistr. 91-93, 24943 Flensburg, Germany
e-mail: ulf.p.mueller@fh-flensburg.de

I. Cussmann
e-mail: ilka.cussmann@fh-flensburg.de

J. Wendiggensen
e-mail: jochen.wendiggensen@fh-flensburg.de

C. Wingenbach
Centre for Sustainable Energy Systems, Europa-Universität Flensburg,
Auf dem Campus 1, 24943 Flensburg, Germany
e-mail: clemens.wingenbach@uni-flensburg.de

V. Wohlgemuth et al. (eds.), *Advances and New Trends in Environmental Informatics*, Progress in IS, DOI 10.1007/978-3-319-44711-7_15

1 Introduction

The transformation of the German electricity system toward higher shares of renewable energy (RE) production has been influencing the electricity grid immensely. By the end of 2015 an installed power of RE sources of 94 GW had been realized (Bundesnetzagentur 2015). Therefore the installed power already exceeds the German peak load (ca. 83 GW) (Graichen et al. 2016). The majority (90 %) of the RE capacity is installed in the distribution grid (Büchner et al. 2014). This development implies a shift from unidirectional to bidirectional power flows leading to congestion and critical voltage variations. Specifically the 110 kV grid, which was historically intended for electricity distribution, is increasingly used as a transmission grid (van Leeuwen et al. 2014). The comparably high shares of RE of Germany's most northern federal state Schleswig-Holstein (see Sect. 2.2), mainly attributable to excellent wind energy resources, leads to the fact that this state is now facing serious congestion problems in the 110 kV grid. In 2014 already 8 % (1 TWh) of the total RE production had to be switched off due to congestion management (Ministerium für Energiewende, Landwirtschaft, Umwelt und ländliche Räume Schleswig-Holstein 2015). This leads to high economic costs as the deficits in RE production need to be compensated according to the Renewable Energy Sources Act (Bundesministerium für Wirtschaft und Energie 2014). At the same time other power plants need to absorb these losses in production which can increase external costs.

It is widely consented to eradicate the necessity of such congestion management measures. In this context grid operators have to optimize, reinforce and expand their grids (50Hertz Transmission et al. 2012b; 2014b). Grid expansion measures, especially the construction of overhead lines, have been confronted by problems of public acceptance. At the same time information about models and data with respect to grid planning is highly restricted and proprietary. This inhibits the reproducibility of simulations and therefore hampers the study of necessary grid expansion measures. This consequently weakens public acceptance (Ciupuliga and Cuppen 2013; Roland Berger, Strategy Consultants 2015; Wiese 2015; Wiese et al. 2014). Recently much research has been undertaken to address this problem, including a large number of research projects such as open_eGo (Flensburg University of Applied Sciences et al. 2015–2018), Vernetzen (Europa-Universität Flensburg 2013–2016), SciGRID (NEXT ENERGY—EWE Research Centre for Energy Technology 2014–2017), osmTGmod (osmTGmod 2016), GridKit (GridKit 2016), openMod.sh (Europa-Universität Flensburg 2014–2016), OPSD (Europa-Universität Flensburg et al. 2015–2017), GENESYS (RWTH Aachen 2012–2016) and oemof (oemof 2016).

In this work a methodology is presented to model power grids based on Open Data and Open Source software. The oemof.powerflow app and complex data processing methods were developed (Sect. 2). Within this section it is focused on the description of the Open Data model. The presented methodology and the following results (Sect. 3) demonstrate that it is possible to simulate realistic power flows in German high voltage (HV) grids in a totally reproducible and transparent approach. Particularly the status quo grid congestions and voltage variations of the 110 kV grid of Schleswig-Holstein are simulated and analysed at high geographic resolution for static worst-case scenarios. Finally recently realised and planned measures of grid optimization, reinforcement and expansion are compared to the simulated results of the current situation.

2 Methodology

2.1 The Oemof.powerflow App

The presented results were achieved by developing and applying the Open Source power flow simulation software *oemof.powerflow* (oemof.powerflow 2016), which is an application within the Open Energy Modelling Framework (*oemof*) (oemof 2016) and uses functionalities from the python package *PYPOWER* (Pypower 2015). *oemof* applies a generic concept based on a bipartite directed graph to describe energy systems. This graph can then be used for different modelling approaches (e.g. unit-commitment, power flow). Currently *oemof.powerflow* is extended to be able to use functionalities from the more up-to-date *PyPSA*-package (Python for Power System Analysis) (PyPSA 2016).

The results of the power flow simulations focus on two parameters: the capacity usage rates of the power lines s and the slow voltage variations at the grid buses u.

These parameters are defined in Eqs. 1 and 2. The active power P and reactive power Q are simulated variables which display the absolute usage of the power lines. *rate_a* reflects the nominal long-term apparent power rating of the power lines. Consequently s is a relative measure for the capacity usage of each power line l in the set of all power lines L (Eq. 1). s may not exceed a usage rate of 50 % concerning the worst-case (one power line consisting of 2 parallel circuits) in order to satisfy the (n-1) criterion. In the best-case power lines can be operated up to $s = 100\,\%$.

$$s_l = \frac{\sqrt{P_l^2 + Q_l^2}}{rate\_a_l}, \ \forall\, l \in L \tag{1}$$

U is simulated and represents an absolute measure for the voltage at each bus. The *base_kv* reflects the nominal voltage of the grid (here: $base\_kv = 110\,kV$). The relative value of u is used in order to determine the slow voltage variation at each corresponding bus b in the set of all buses B (Eq. 2). u should stay within the displayed boundaries with respect to the relevant grid code (Verband der Netzbetreiber 2004).

$$u_b = \frac{U_b}{base\_kv}, \forall\, b \in B \qquad (2)$$
$$s.t.\ 0.9 \le u_b \le 1.12, \forall\, b \in B$$

2.2 Approach for the HV Grid Model of Schleswig-Holstein

In order to simulate AC power flows a complex set of grid data is needed. In the presented model the data format refers to the conventions of *PYPOWER* (Pypower 2015) and *MATPOWER* (Zimmermann and Murillo-Sanchez 2015), which are similar to the *IEEE Common Data Format* (IEEE Working Group on a Common Format for Exchange of Solved Load Flow Data 1973). Accordingly information about the *buses*, *branches* and the *demand* and *generation* at the buses have to be obtained.

Due to the lack of available data the developed model relies on OpenStreetMap (OSM) (OpenStreetMap Foundation 2015a, b), literature-based assumptions, power plant registers and statistics. The topology of the grid (buses and branches) is derived from OSM data. Within the scope of this paper the deriving method of the OSM data towards a calculable topology consists of a heuristic approach using the branch information as OSM ways. Similar procedures have been automatized by *SciGRID* (Medjroubi and Matke 2015; NEXT ENERGY—EWE Research Centre for Energy Technology 2014–2017) and *osmTGmod* (osmTGmod 2016, Scharf 2015). In contrast to *SciGRID*, which focuses on a deterministic approach, the developed heuristic approach is comparable with the one in (Scharf 2015) considering the way elements within the OSM's conceptual data model (Ramm and Topf 2010). Due to the lack of existing routing algorithm, which are standard for transport purposes, approaches like the ones mentioned have been recently developed.

The methodology of the model is described in detail within the following three subsections *Grid*, *Generation* and *Demand*.

Grid

The modelled topology of the 110 kV grid of Schleswig-Holstein is shown in Fig. 1. The 110 kV substations which were modelled as PQ buses are derived from OSM by employing the power value *substation*. These buses receive generation and demand characteristics. In order to filter only the 110 kV equipment the OSM key *voltage* was used. The key *frequency* enables to exclude the one-phase railway circuits (16.7 Hz instead of 50 Hz). In special cases also the key *cables* helped to determine railway specific circuits (divisible by 2 instead of 3).

Transformers which connect the 110 kV grid to the lower and higher voltage levels were not included. The lower voltage grids including the entire demand and generation are abstracted directly and without losses to the 110 kV grid nodes. The interconnections to the 220 and 380 kV grids were modelled as ideal slack buses (voltage is set as input value, $u = 1$) (Milano 2010). This implies a perfect performance of the overlaying grid. The connecting substations were identified by analysing the OSM key *voltage* of the incoming power lines. The identified substations were matched

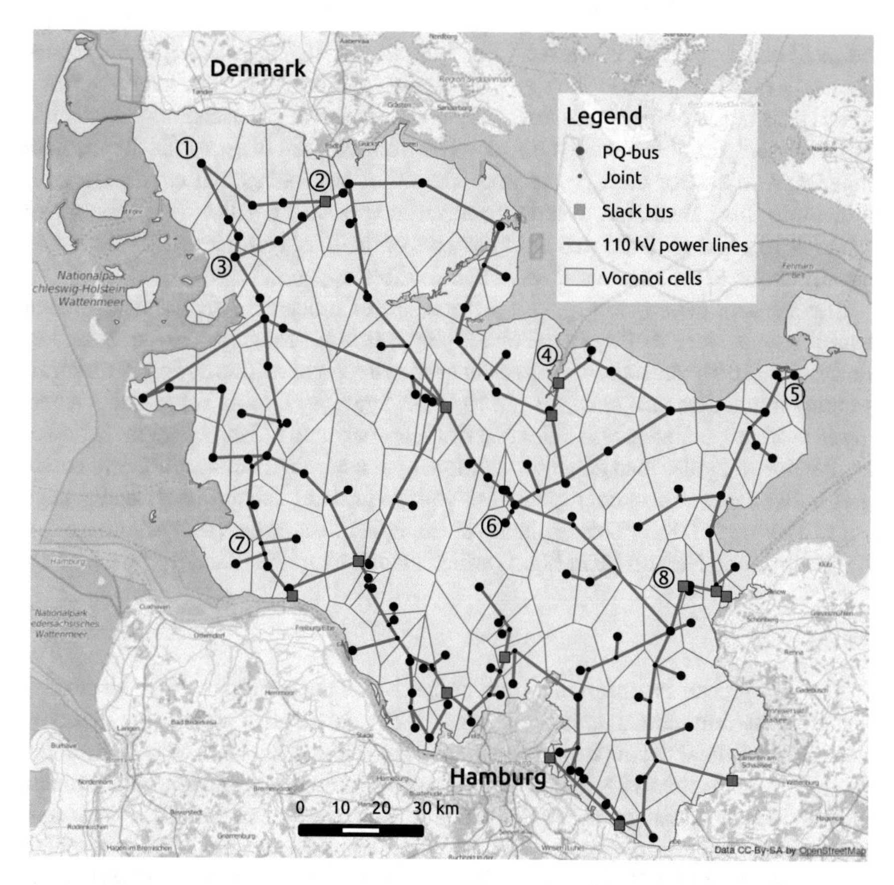

Fig. 1 Model of the 110 kV grid in Schleswig-Holstein based on OpenStreetMap (OpenStreetMap Foundation 2015a)

with the interconnecting substations on the ENTSO-E map (ENTSO-E Grid Map 2014) resulting in 14 slack buses in total. Their locations are visualised in Fig. 1.

The joints represent most commonly T-intersections between three different power lines. Since in OSM the power circuits are ambiguously defined (OpenStreetMap Foundation 2015b) it was assumed that all particular circuits meet in the modelled joints. Therefore the joints are inherent to the model and do not explicitly exist as mapped nodes in OSM. This depicts a necessary simplifying abstraction due to the specific OSM mapping convention (apart from the general lack of official information). It is expected that for simulations on the normal operating mode (n-0 case) this simplification is less relevant than in the context of n-1 contingency criterion testing (which is excluded within this work).

The power lines were modelled with all necessary parameters combining OSM information with typical parameters from literature. The typical parameters were derived from (Hofmann and Oswald 2010) and (Agricola et al. 2012) representing the most common 110 kV overhead line type (Al/St 265/35). Underground cables were not modelled due to lack of publicly available information. The length-specific typical parameters of resistance, reactance and capacitance or susceptance were multiplied by the particular line length (derived from osm) in order to obtain absolute impedance values for each power line. The OSM keys *wires* and *cables* are integrated into this calculation in order to reach more realistic values (Medjroubi and Matke 2015). The total thermal limit currents of the lines being length-independent also differ due to the number of *wires* and *cables* mapped in OSM.

The Voronoi cells (see Fig. 1) are the result of a nearest neighbour algorithm assuming that every facility in the underlying medium voltage grid is connected to the nearest 110 kV substation. As a result each cell represents one region in which all generation and demand ($base\_kV \leq 110\,kV$) is aggregated to its substation. Generally in more populated areas (cities and the suburban area of Hamburg) the cells are smaller due to higher load density. Additionally the nominal voltage of a grid correlates with its size. The existing medium voltage grids in Schleswig-Holstein range from 10 kV up to 60 kV. This explains for example the large size of the Voronoi cell of the substation Niebüll (① in Fig. 1) which includes e.g. the island Sylt via a 60 kV grid.

Generation

To gain comprehensive information about the generation facilities within the considered area different official registers were used. The power plant register published by the German federal network agency contains data on power plants with an installed capacity of more than 10 MW and additionally clustered information on RE plants with an installed capacity of less than 10 MW (Bundesnetzagentur 2015). This register does not include any georeferencing. Therefore power plants with an installed capacity of over 10 MW were georeferenced manually. As the generalistic power plant register does not provide sufficient and detailed information on RE plants the renewable power plant register was used as an addition (Deutsche Gesellschaft für Sonnenenergie e. V. 2014). The listed RE power plants are georeferenced by their postal code. A higher spatial resolution is not essential for a power flow simulation in the 110 kV grid which focuses mainly on the macrostructure of the electricity grid. As a further approach to achieve a comprehensive knowledge about relevant generation facilities a list of combined heat and power (CHP) plants by the Federal Office for Economic Affairs and Export Control was used to identify power plants which were not listed in the registers mentioned above (Federal Office for Economic Affairs and Export Control 2014).

The applied power plant registers were retrieved in 2014 and resulted in an overall installed capacity of 5960 MW for power plants with a nominal voltage of 110 kV or less. 84 % of the installed capacity are accounted for volatile energy sources such as wind and photovoltaics. In a next step the power plants were allocated to the centre of the closest 110 kV-substation using a nearest neighbour algorithm. The results are visualised in Fig. 2a. The regions in which power plants are allocated to a specific substations are displayed as Voronoi cells with a specific colour according to the annual electricity consumption within this area. Hot spots with a high

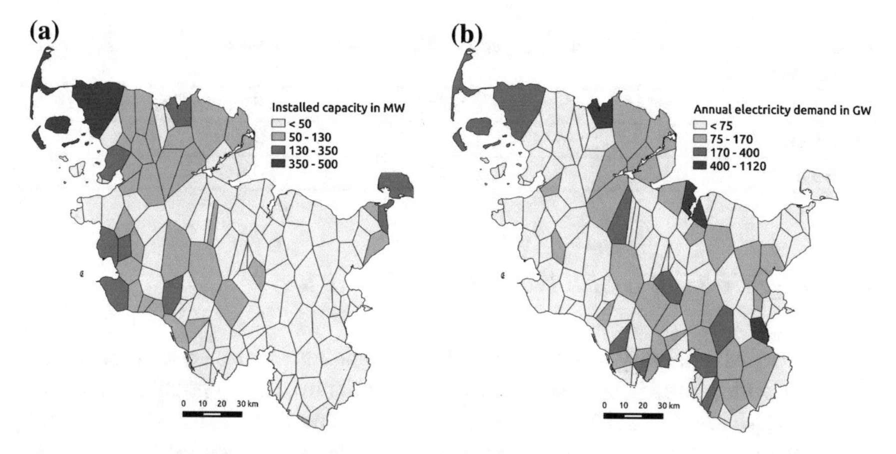

Fig. 2 Distribution and allocation of installed generation capacity in 2014 (**a**) and electricity demand in 2010 (**b**) to the closest 110 kV substation illustrated as Voronoi cells

installed capacity are located at the west coast and in the east near Göhl (⑤ in Fig. 1), mainly due to a high installed capacity of wind and photovoltaic power plants in these regions.

Demand

The lack of information on the spatial and temporal distribution of electricity demand is a major problem when collecting realistic data for power flow simulations.

The statistical office of Schleswig-Holstein publishes an energy balance which provides data on the annual electricity demand. For 2010 the energy balance reports a total electricity consumption of 11.6 TWh (Statistisches Amt für Hamburg und Schleswig-Holstein 2012). As the spatial distribution of electricity consumption is essential to perform a power flow calculation the total consumption was allocated to the 110 kV substation according to the local population. Data on the population per municipality is provided by the Federal Bureau of Statistics (Statistische Ämter des Bundes und der Länder 2012). The calculated electricity consumption per municipality was allocated to the nearest 110 kV substation and visualised in Fig. 2. The figure depicts a concentration of demand in the southern part of Schleswig-Holstein and around regional centres like Kiel (④ in Fig. 1), Neumünster ⑥ or Flensburg ②. Large Voronoi cells such as Niebüll ① show a high demand due to their areal extent. Sparsely populated regions with a low electricity consumption are located at the west coast and in the most eastern part of Schleswig-Holstein. Both regions show a high installed capacity of RE at the same time. The data bases for demand and generation are lacking consistency due to different observed time periods. As there are no information on the electricity demand in 2014 available, a constant electricity consumption between 2010 and 2014 is assumed. This refers to equivalent assumptions in recognized surveys on the German distribution and transmission grid (50Hertz Transmission et al. 2014b, Agricola et al. 2012).

Table 1 Simultaneity factors for the generation with respect to the worst-case scenarios (based on Agricola et al. 2012)

	Wind	Solar	Biomass/CHP/run-of-river
High demand	0	0	0
High feed-in	1	0.85	0.8

2.3 *Definition of Static Worst-Case Scenarios*

The static grid reliability can be assessed by simulating worst-case scenarios. This method is widely used in science and industry in order to identify possible violations on s and u (see Sect. 2.1) (Agricola et al. 2012; Büchner et al. 2014; Mahmud et al. 2014; Probst et al. 2013). Two extreme situations have to be defined to cover the range of possible worst cases within normal operating mode.

The high demand scenario represents the highest possible stress situation concerning low u values at the buses as well as high s values of the power lines. It is characterized by the highest annual demand and simultaneously no feed-in within the 110 kV grid and lower.

The other worst case refers to a high feed-in scenario representing the highest possible stress situation concerning high u values and critical s values. Here the maximal simultaneous feed-in and minimal demand is assumed.

The simultaneity of generation and demand is a statistical measure which differs depending on the characteristics of the grid. In this case the simultaneity factors for the generation are chosen for the German HV grid due to (Agricola et al. 2012) and (Patzack et al. 2016) and can be observed in Table 1. The simultaneity factors for the demand was derived from the ENTSO-E demand time series for Germany of the year 2011 (ENTSO-E 2015). The maximal and minimal value were factorized by the total yearly consumption. Consequently the minimal factorized value belongs to the high feed-in scenario whereas the maximal factor represents the simultaneity for the high demand scenario.

3 Results

The generated results of the static worst case scenarios as they are described in Sect. 2.3 were evaluated based on the criteria discussed in Sect. 2.1. The results of the high demand scenario show a unproblematic grid situation without any bottlenecks or violations of tolerable voltage variations in the 110 kV grid. The load case characterizes a situation which matches the existing structures of the electricity grid in Schleswig-Holstein. In contrast the feed-in case reveals multiple bottlenecks in the 110 kV grid. These results are visualised in Fig. 3a, which depicts various power lines with a capacity usage rate of over 50 %. These tense grid situations occur mainly between regions with a high installed electrical capacity and their closest slack buses where electricity can be transported to the overlayed grid level. This can be observed

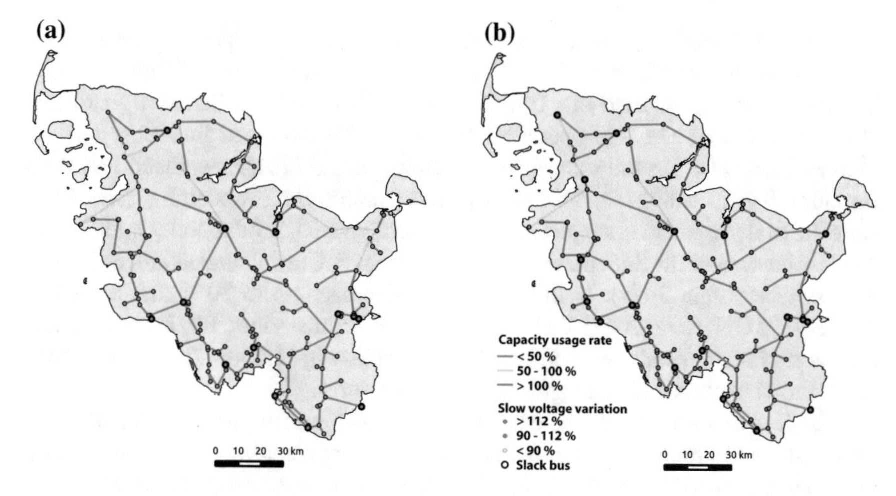

Fig. 3 Results of high feed-in case (**a**) and high feed-in case with additional slack buses (**b**)

at the westcoast as well as in Ostholstein (⑤ in Fig. 1), where regions with a high share of RE are not directly connected to the 380 kV grid. The highest capacity usage rate of 294 % arises in Dithmarschen ⑦. Similarly the west-east connector between Niebüll ① and Flensburg ② was highly overloaded ($s = 235\,\%$). These high capacity usage rates are theoretical values which in reality are mitigated by congestion management such as temporary reduction of RE feed-in.

Concerning the slow voltage variations at the buses, no problems were detected. The substation Göhl in Ostholstein ⑤ shows the highest simulated voltage level of 105 % which displays a tolerable voltage variation according to (Verband der Netzbetreiber 2004).

4 Discussion

The lack of reference data and models disables a holistic validation of the results. Nevertheless the results were compared with realised and planned grid optimization, reinforcement and expansion actions. The grid operator of the 110 kV grid in Schleswig-Holstein has realized a number of such measures. Three prominent ones are the overhead line monitoring, capacity usage monitoring and the construction of a 110 kV overhead line from Breklum (③ in Fig. 1) to Flensburg ② (Ministerium für Energiewende, Landwirtschaft, Umwelt und ländliche Räume Schleswig-Holstein 2015). The lastly mentioned overhead line was put into operation in 2011 (Ministerium für Energiewende, Landwirtschaft, Umwelt und ländliche Räume Schleswig-Holstein 2015). Our model displays high usage rates on this particular line (see Fig. 3a), which most likely confirms its relevance and hints towards adequate modelling.

Within the overhead line monitoring, lines are utilised depending on the surrounding weather conditions (E.ON Netz 2007; Goldschmidt 2014; Ministerium für Energiewende, Landwirtschaft, Umwelt und ländliche Räume Schleswig-Holstein 2015). It was first put into operation on the 110 kV overhead line from Niebüll ① to Flensburg ②. On the same line as well as on the 110 kV overhead line from Breklum ③ to Flensburg ②, the capacity usage monitoring is operating since 2012. Here the (n-1) criterion is subjected to the controllable RE production (Goldschmidt 2014; Ministerium für Energiewende, Landwirtschaft, Umwelt und ländliche Räume Schleswig-Holstein 2015). Both measures enable each up to 50 % additional line capacity (E.ON Netz 2007; Goldschmidt 2014; Ministerium für Energiewende, Landwirtschaft, Umwelt und ländliche Räume Schleswig-Holstein 2015). The results (see Sect. 3) explain and can legitimate these measures.

Additionally measures on the extra-HV level can lower the stress of the HV grid. This is shown in Fig. 3b. Within the grid development planning of the transmission grid operators two important activities have been commissioned (50Hertz Transmission et al. 2012a, 2013, 2014a, TenneT TSO 2014, 2015a, b). In Fig. 3b four additional slack buses at the west coast model a 380 kV overlay line which is currently under construction. The simulation shows that this leads to a massive relief on all of the west-east 110 kV interconnectors. The simulated results appear to be in line with the grid planning of the grid operators. The second mentioned activity refers to the planned east coast 380 kV-overlay in the region of Ostholstein which is modelled as an extra slack bus in Göhl ⑤ (TenneT TSO 2015a). This results into a relief of the 110 kV interconnectors towards Kiel ④ and Lübeck ⑧ (see Fig. 3b). Nevertheless in some hours of the year a few lines will be still congested if no additional measures are realised.

Further model validation remains to be researched on. For instance in OSM underground cables are seldomly mapped. This is especially a problem in urban areas. The influence in the rather rural state of Schleswig-Holstein is considered to be small. The results show that the Voronoi method displays a straightforward and adequate method for the assignment of generation and demand to the HV grid buses. Alternatively restrictive data of grid operators would be needed or other more complex methods are to be developed by e.g. studying the correlation of municipality shapes with the structure of electrical grids. Additionally the influence of assumptions on the input parameters has to be kept in mind. Eminently the influence of statistically assumed simultaneity factors is significant (Patzack et al. 2016) and demands further sensitivity analysis. Load profile simulations throughout one or more years can improve this matter substantially.

5 Conclusion

A general approach to use exclusively Open Source and Open Data in order to model real electrical grids was developed, particularly adjusted and implemented concerning the HV grid of the German federal state Schleswig-Holstein. The *oemof.powerflow* app was designed and linked to the developed grid data model enabling the performance of AC load flow simulations. The results reflect recent congestion problems of power lines within the 110 kV grid of Schleswig-Holstein. Especially the power lines connecting the wind power penetrated west coast areas with the 380 kV transformer stations at the central axis are affected. Furthermore the simulations manifest that the congestion problems can be mostly overcome by the measures realised and planned by the transmission grid operators. These results demonstrate the feasibility of transparent and reproducible AC grid modelling as well as the necessity of grid expansion measures or alternative flexibility measures.

Further research will focus on holistic approaches integrating the developed power flow simulation methods into economic optimization methods including flexibility options such as redispatch and storage solutions. Currently the data model is extended to the entire German grid including all voltage levels. The application of the developed approach on grids in other countries is principally possible but strongly depends on data availability. Facilitating further validation and improvement of the results' accuracy a broad, consistent and publicly available data base is highly eligible.

References

Pypower 5.0.1 (2015). https://github.com/rwl/PYPOWER

GridKit is an power grid extraction toolkit (2016). https://github.com/bdw/GridKit

oemof-Open Energy Modelling Framework—A modular open source framework to model energy supply systems (2016). https://github.com/oemof

osmTGmod (2016). Open source German transmission grid model based on OpenStreetMap. https://github.com/wupperinst/osmTGmod

Powerflow simulations based on oemof (2016). https://github.com/openego/oemof.powerflow

PyPSA (2016). Python for Power System Analysis. https://github.com/FRESNA/PyPSA

50Hertz Transmission, Amprion, TenneT TSO, Transnet BW (2012a). Anhang—Netzentwicklungsplan Strom 2012—2. berarbeiteter Entwurf der bertragungsnetzbetreiber. Technical Report, 50Hertz Transmission and Amprion and TenneT TSO and Transnet BW

50Hertz Transmission, Amprion, TenneT TSO, Transnet BW (2012b) Netzentwicklungsplan Strom 2012—2. berarbeiteter Entwurf der bertragungsnetzbetreiber. http://www.netzentwicklungsplan.de/content/netzentwicklungsplan-2012-2-entwurf

50Hertz Transmission, Amprion, TenneT TSO, Transnet BW (2013). Anhang—Netzentwicklungsplan Strom 2013—2. Entwurf der Übertragungsnetzbetreiber. Technical Report, 50Hertz Transmission and Amprion and TenneT TSO and Transnet BW

50Hertz Transmission, Amprion, TenneT TSO, Transnet BW (2014a) Anhang—Netzentwicklungsplan Strom 2014—2. Entwurf der bertragungsnetzbetreiber. Technical Report, 50Hertz Transmission and Amprion and TenneT TSO and Transnet BW

50Hertz Transmission, Amprion, TenneT TSO, Transnet BW (2014b) Netzentwicklungsplan Strom 2014: Zweiter Entwurf der Übertragungsnetzbetreiber. http://www.netzentwicklungsplan.de/_NEP_file_transfer/NEP_2014_1_Entwurf_Teil1.pdf

Agricola, A.C., Höflich, B., Richard, P., Völker, J., Rehtanz, C., Greve, M., Gwisdorf, B., Kays, J., Noll, T., Schwippe, J., Seack, A., Teuwsen, J., Brunekreeft, G., Meyer, R., & Liebert, V. (2012). Ausbau- und Innovationsbedarf der Stromverteilnetze in Deutschland bis

2030 (kurz: dena-Verteilnetzstudie): Endbericht. http://www.dena.de/fileadmin/user_upload/Projekte/Energiesysteme/Dokumente/denaVNS_Abschlussbericht.pdf

Büchner, J., Katzfey, J., Floercken, O.D., Moser, A., Schuster, H., Dierkes, S., van Leeuwen, T., Verheggen, L., Uslar, M., & van Amelsvoort, M. Verteilernetzstudie für Deutschland (Verteilernetzstudie): Forschungsprojekt Nr. 44/12: Abschlussbereicht. http://www.bmwi.de/BMWi/Redaktion/PDF/Publikationen/Studien/verteilernetzstudie,property=pdf,bereich=bmwi2012,sprache=de,rwb=true.pdf

Bundesministerium für Wirtschaft und Energie: Gesetz für den Ausbau erneuerbarer Energien (Erneuerbare-Energien-Gesetz - EEG 2014): EEG 2014 (21.07.2014). http://www.gesetze-im-internet.de/bundesrecht/eeg_2014/gesamt.pdf

Bundesnetzagentur (2015). List of power plants. http://www.bundesnetzagentur.de/cln_1911/DE/Sachgebiete/ElektrizitaetundGas/Unternehmen_Institutionen/Versorgungssicherheit/Erzeugungskapazitaeten/Kraftwerksliste/kraftwerksliste-node.html

Ciupuliga, A., & Cuppen, E. (2013). The role of dialogue in fostering acceptance of transmission lines: the case of a France–Spain interconnection project. Energy Policy

Deutsche Gesellschaft für Sonnenenergie e.V. (2014). Die EEG-Anlagen der Region "Bundesrepublik Deutschland". http://www.energymap.info/energieregionen/DE/105.html

ENTSO-E (2014). ENTSO-E Grid Map 2014. https://www.entsoe.eu/Documents/Publications/maps/2014_Map_ENTSO-E-4.000.000.zip

ENTSO-E (2015). Consumption data. https://www.entsoe.eu/data/data-portal/consumption/Pages/default.aspx

E.ON Netz (2007). Freileitungs-Monitoring. http://apps.eon.com/documents/ene_flyer-freil-monito_0907_ger.pdf

Europa-Universität Flensburg: VerNetzen (2013–2016) Sozial-ökologische, technische und ökonomische Modellierung von Entwicklungspfaden der Energiewende. https://www.uni-flensburg.de/eum/forschung/laufende-projekte/vernetzen/

Europa-Universität Flensburg (2014–2016). Open Source Energie-Modell Schleswig-Holstein (openMod.sh). https://www.uni-flensburg.de/eum/forschung/laufende-projekte/openmodsh/

Europa-Universitüt Flensburg, DIW Berlin, Technical University of Berlin, and Neon Neue Energieökonomik (2015–2017). OPSD: Open Power System Data; A free and open data platform for power system modelling. http://open-power-system-data.org/

Federal Office for Economic Affairs and Export Control (2014). Beim BAFA nach dem Kraft-Wärme-Kopplungsgesetz zugelassenen KWK-Anlagen

Flensburg University of Applied Sciences, Europa-Universität Flensburg, NEXT ENERGY—EWE Research Centre for Energy Technology, Carl-von-Ossietzky Universität Magdeburg, Reiner Lemoine Institut (2015–2018). open_eGo: open electricity Grid optimization: Development of a holistic grid planning tool as an integral part of an open energy modelling platform aiming at the determination of an optimal grid and storage expansion in Germany. https://www.uni-flensburg.de/eum/forschung/laufende-projekte/open-ego/

Goldschmidt, T. (2014). Schleswig-Holstein neu denken—Wärme und Wind in Schleswig-Holstein. In: *Fachtagung BUND Klimaschutz in der Metropole*

Graichen, P., Kleiner, M., & Podewils, C. (2016). *Die Energiewende im Stromsektor: Stand der Dinge 2015*. Technocal Report, Agora Energiewende

Hofmann, L., & Oswald, B. (2010). *Gutachten zum Vergleich Erdkabel—Freileitung im 110-kV-Hochspannungsbereich*. Technical Report, Leibniz Universität Hannover

van Leeuwen, T., Dierkes, S., Verheggen, L., Schuster, H., Köhne, F., & Moser, A. (2014). Ermittlung von Transitflüssen im Hochspannungsnetz durch mehrere Verknüpfungspunkte mit dem Übertragungsnetz. In: *13. Symposium Energieinnovation, 12.-14.2.2014*, Graz/Austria

Mahmud, M., Hossain, M., & Pota, H. (2014). Voltage variation on distribution networks with distributed generation: worst case scenario. *IEEE Systems Journal*, *8*(4), 1096–1103.

Medjroubi, W., & Matke, C. (2015). SciGRID Open Source Transmission Network Model - USER GUIDE V 0.2. Technical Report, NEXT ENERGY EWE-Forschungszentrum für Energietechnologie e. V. http://www.scigrid.de/releases_archive/SciGRID_Userguide_V0.2.pdf

Milano, D. F. (2010). *Power System Modelling and Scripting*. London Limited: Springer-Verlag.
Ministerium für Energiewende, Landwirtschaft, Umwelt und ländliche Räume Schleswig-Holstein (2015). Abregelung von Strom aus Erneuerbaren Energien und daraus resultierende Entschädigungsansprüche in den Jahren 2010 bis 2014. Technical Report
NEXT ENERGY—EWE Research Centre for Energy Technology: SciGRID (2014–2017). Open Source Reference Model of European Transmission Networks for Scientific Analysis. http://www.scigrid.de/
OpenStreetMap Foundation (2015a). OpenStreetMap. www.openstreetmap.org
OpenStreetMap Foundation (2015b) OpenStreetMap Wiki. https://wiki.openstreetmap.org
Patzack, S., Erle, N., Vennegeerts, H., & Moser, A. (2016). Einfluss von auslegungsrelevanten Netznutzungsfällen auf die Netzdimensionierung. In: *14. Symposium Energieinnovation in Graz/Austria*
Probst, A., Tenbohlen, S., Seel, M., & Braun, M. (2013). Probabilistic grid planning with considerartion of dispersed generationand electric vehicles. In: CIRED 2013, Stockholm
Ramm, F., & Topf, J. (2010). OpenStreetMap: Die freie Weltkarte nutzen und mitgestalten. Lehmanns Media
Roland Berger, Strategy Consultants (2015). Study regarding grid infrastructure development: European strategy for raising public acceptance. Technical Report
RWTH Aachen (2012–2016). GENESYS-Genetic Optimization of a European Energy Supply System. http://www.genesys.rwth-aachen.de/
Scharf, M. (2015). *Entwicklung eines Modells des deutschen Übertragungsnetzes auf Basis der offenen Geodatenbank OpenStreetMap*. Bachelor-thesis: University of Applied Sciences Flensburg.
Statistische Ämter des Bundes und der Länder (2012). Gemeindeverzeichnis Gebietsstand: 31.12.2011 (Jahr). https://www.destatis.de/DE/ZahlenFakten/LaenderRegionen/Regionales/Gemeindeverzeichnis/Administrativ/Archiv/Administrativ.html
Statistisches Amt für Hamburg und Schleswig-Holstein (2012). Energiebilanz Schleswig-Holstein 2010. https://www.statistik-nord.de/fileadmin/Dokumente/Sonderver
TenneT TSO (2014). Planungsbericht Westküstenleitung—TenneT im Dialog. Technical Report
TenneT TSO (2015a) Ostküstenleitung 380-kV-Netzausbau Kreis Segeberg—Raum Lübeck—Raum Göhl—Siems. http://www.tennet.eu/de/fileadmin/downloads/Netz-Projekte/Onshore/Ostkuestenleitung/Broschuere-Ostkuestenleitung_V12-November_2015_final.pdf
TenneT TSO (2015b). Brunsbüttel bis Süderdonn—Baustart der 380-kV-Westküstenleitung. http://www.tennet.eu/de/fileadmin/downloads/Netz-Projekte/Onshore/WiCo/Westkuestenleitung_Baufactsheet.pdf
Verband der Netzbetreiber—VDN—e.V. beim VDEW (2004). EEG-Erzeugungsanlagen am Hoch- und Höchstspannungsnetz. https://www.vde.com/de/fnn/dokumente/documents/rl_eeg_hh_vdn2004-08.pdf
Wiese, F. (2015). Renpass—renewable energy pathways simulation system: Open source as an approach to meet challenges in energy modeling. Dissertation, Europa-Universität Flensburg
Wiese, F., Bökenkamp, G., Wingenbach, C., & Hohmeyer, O. (2014). An open source energy system simulation model as an instrument for public participation in the development of strategies for a sustainable future. *Wiley Interdisciplinary Reviews: Energy and Environment*, *3*(5), 490–504.
Working Group on a Common Format for Exchange of Solved Load Flow Data (1973). Common Format For Exchange of Solved Load Flow Data. IEEE Transactions on Power Apparatus and Systems
Zimmermann, R., & Murillo-Sanchez, C. (2015). MATPOWER 5.1, User's Manual. Technical Report, Power Systems Engineering Research Center. http://www.pserc.cornell.edu//matpower/manual.pdf

Part V
Sustainable Mobility

Empirical Study of Using Renewable Energies in Innovative Car-Sharing Business Model "in Tandem" at the University of Hildesheim

Mohsan Jameel, Olexander Filevych and Helmut Lessing

Abstract The renewable energy is becoming an integral part for the electric vehicle charging station to reduce greenhouse gases. The combination of this clean technology with appropriate business models reinforce the positive impact on the environment. In this paper we present empirical study of photovoltaic charging system used in a new car-sharing business model "in Tandem" implemented at the University of Hildesheim under the project name e2work. The data analyzed from photovoltaic (PV) over the period of one year, exhibits an interesting insight into the renewable energies supply of electric vehicles under preset conditions of the business model and thus helps to dimension an appropriate size and storage capacity of a photovoltaic system considering different weather conditions and mobility demands. It ensures the quick transferability and easy implementation of both the business model and the appropriate infrastructure to any place.

Keywords Electro-mobility · Photovoltic charging station · Car-sharing · Electric-vechicles

1 Introduction

Facing multiple challenges of climate change, new technological solutions embedded in economically viable business cases are required. With the rise of the shared-used economy, the model of mobility use shifts from vehicle ownership to vehicle

M. Jameel (✉)
Information Systems and Machine Learning Lab, University of Hildesheim,
31141 Hildesheim, Germany
e-mail: mohsan.jameel@ismll.uni-hildesheim.de

O. Filevych · H. Lessing
Institute of Business and Operations Research, University of Hildesheim,
31141 Hildesheim, Germany
e-mail: filevy@uni-hildesheim.de

H. Lessing
e-mail: lessing@cs.uni-hildesheim.de

V. Wohlgemuth et al. (eds.), *Advances and New Trends in Environmental Informatics*, Progress in IS, DOI 10.1007/978-3-319-44711-7_16

sharing offering at least one key strategy to minimize the steadily increasing environmental pollution (Heinrichs and Grunenberg 2012; Linsenmeier 2011). In Germany, the massive growth of the number of car-sharing companies due to the growing demand for ecological friendly and flexible mobility services in the last five years shows clearly the size and dynamic of this mobility market (Linsenmeier 2011). The mobility on demand has among other things not only the potential to reduce overall noise pollution and motorized traffic in metropolitan cities, but at the same time it offers to new approaching technologies an attractive setting to enter. However, the most operators of car-sharing fleet still prefer vehicles with internal combustion engine (ICE) over the vehicles with alternative powertrains like battery electric vehicle (BEV). The high purchasing costs, technical restrictions like limited range, long recharging time, insufficient recharging infrastructure as well as acceptance barriers might be the main reasons for slow integration of pure electric vehicle in existing car-sharing business models (Carroll and Walsh 2010; Dudenhöffer 2013; Franke et al. 2012; Hidrue et al. 2011). Companies using only electric vehicles in its car-sharing fleets are rare phenomenon. Moreover, in most cases metropolitan areas with high population density offers the only economical environment for such enterprises to survive. Their operational area is therefore limited and, except for a few cases, often does not include rural areas at all.

This paper presents a new business case, which incorporates both the use of BEVs and an intelligent charging infrastructure directly powered by renewable energies. In this research, the use of renewable resources for charging purposes is considered important to increase the environmental benefits of BEVs during their operating lifetime. Thus, the development and implementation of a photovoltaic(PV) system with intelligent charging management has been an integral element of a new two-way car-sharing business model "in Tandem" developed and implemented at the University of Hildesheim. The empirical data collected from a PV system over the period of an entire year displays the interplay between the appropriate dimensioning of generation capacity of the PV system, the right size of additional storage capacity and potentially needed direct connection on the grid all in dependence of weather conditions and the charging behavior of the users. The perfect match of all these elements in accordance to the specific setup of the business model ensure the quick transferability and easy implementation of both the business case and its appropriate infrastructure to any place.

This paper is organized as follows, Sect. 2 provides an overview of the innovative electric car-sharing business model "in Tandem", Sect. 3 discuss renewable energy in context of charging BEVs, Sect. 5 presents an empirical study of electric car-sharing business model using renewable energy, implemented at University of Hildesheim and Sect. 4 concludes the research and provides insight into the possible future work.

2 Innovative Electric Car-Sharing Business Model in "Tandem"

The business case "in Tandem", which has been developed at the University of Hildesheim is based on a cellular structure. The innovative element, a "cell", represents the organizational and economic activity between the employer and a selected private user (tandem partner). A dynamic cellular network can be created by connecting multiple cells via Information and Communication Technology(ICT) based processes. This dynamic cellular network can bring shift in the base car sharing to free-floating mechanism. In comparison to other car-sharing models, both the simple cell structure, clear assignment of each BEV to the tandem partner and its allocation of financial risks equally between two parties, increase simultaneously vehicle utilization and reduces financial burdens on both sides. The business model includes the possibility to allocate the financial burdens like leasing rate, vehicle insurance, cost of maintenance in different percentages independent of the actual usage of BEVs and only depending on financial circumstances of the parties involved in each cell.

The Fig. 1 presents the typical cell structure consisting of tandem partner and business employer, which require car for business trips. It also presents a simplified schematic overview of a renewable energy based charging concept as exemplified

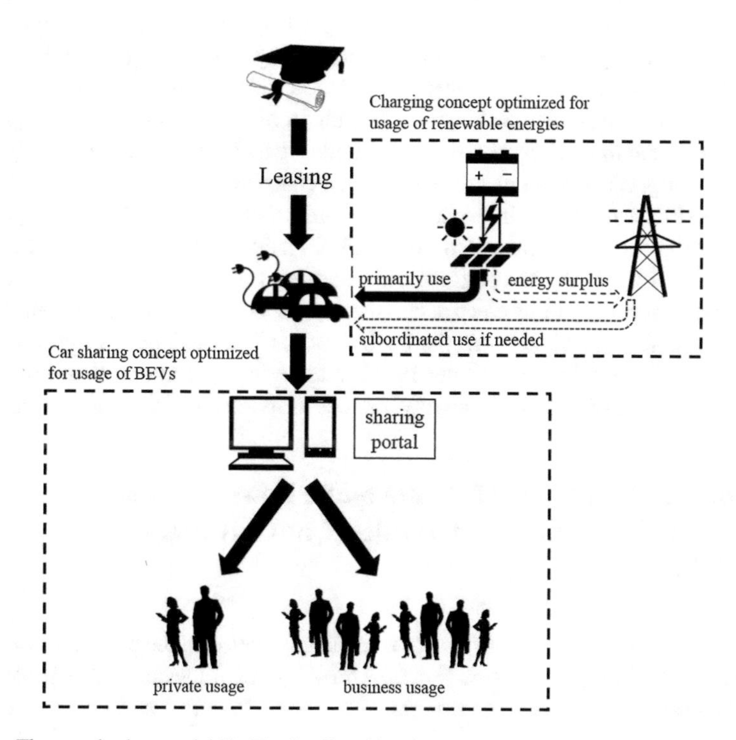

Fig. 1 The car-sharing model "in Tandem" and its charging concept developed at the University of Hildesheim

at the University of Hildesheim. The car fleet leased or purchased in a cell is used both for business and private use. During the working hours i.e. 9:00–16:00 hrs, the car stands on predefined parking lot with specific charging infrastructure and it can be reserve for business trips. After working hours and on weekends, the car is free for private use. The intelligent reservation system based on mathematical calculation provides trouble-free coordination between the private and business use.

3 Renewable Energy Source for Charging Electric Vehicles

The tail-pipe emissions caused by motorized private sector, is one of the major concerns especially in metropolitan cities with high pollution index. Thus, automotive manufacturers are pushing the BEVs as an environmental friendly alternative to the ICE specifically to reduce locally emitted pollution of their car fleet required to meet the EU consumption targets set by 2020. However, the lack of attractive purchase conditions as well as some technology restrictions still hinder the widespread of BEVs. Furthermore, depending on the energy mix, BEVs are not systematically environmental friendly and in addition often put some strain on normal distribution power by disrupting power load management (Goli and Shireen 2015).

These issues have therefore attracted research interest in powering the BEVs using primarily renewable energy sources. According to literature, the most suitable source of renewable energy especially in build-up areas are photovoltaic panels. For example, Neumann et al. (2012) analyzed the potential of generating renewable energy with PV installation at the parking lot. Tulpule et al. (2013) studied the impact of the economic parameters and feasibility of building a charging station directly powered by PV suggests that an installation of such charging infrastructure at workplace is easier to implement than in private households. Considering the average parking time at workplace of 5 h presented 2009 in US National Household Travel Survey (Santos et al. 2011), we investigated the benefits of a PV-powered battery charging station in support of our electric car-sharing business model "in Tandem". This technical architecture along with the business case "in Tandem" can prevent high expenditures for purchasing an oversized PV charging station and therefore speed up the widespread of renewable energy sources in motorized transport sector.

4 Empirical Analysis of Photovoltaic Powered Electric Car-Sharing Model "in Tandem" at University of Hildesheim

The business model "in Tandem" finds its local application in the context of the eAutarke Zukunft subproject "e2work"[1] at the University of Hildesheim. The project has five VW brand pure electric driven vehicles (four VW e-up! and one VW eGolf)

[1] http://e2work.de.

allocated to three different locations. Two charging sites has one wall socket (230 V, 10 A) each, for charging one BEV. The third charging site has two charging points. One charging point has four wall sockets and the second consists of a PV system with two charging outlets (230 V, 16 A, single-phase). Generally, two strategies can be used in order to cover the user mobility demands for business trips required by the business model. The first assumes a quick charging station (≥22 kW) supplying a lower number of BEVs to ensure a full charged vehicle at any time of the business use. The second foresees a slow charging infrastructure (≤3.7 kW) supplying a greater number of vehicles. For the technical charging concept of the tandem model the first option seems unreasonable due to high purchasing costs of the quick charging infrastructure, requirements concerning energy supply, legal specifications and in particular the long installation process. Simple, easy manageable, transferable and quick implementable charging infrastructure is best suited for the technical concept of the business model "in Tandem", although it should always be adopted to the specific mobility demand of each cell if needed.

4.1 *Photovoltaic Power Charging Station at the University of Hildesheim*

The PV powered charging station setup at one of the charging points to study the dynamic of power demand and renewable energy for electric car-sharing model. Keeping in view the utility of the BEV charging station, we choose parking lot at the main campus. The benefits of using existing parking lot are availability of the sunlight, no new construction site and minimum distance from the point of use. The charging station has capacity to support charging for two BEVs and optionally, three e-bikes. The total power of photovoltaic cells is 3.84 kWp and uses Sunny Boy 3000TL solar inverter. The Internal buffer consists of 24 cells lead-acid battery to store surplus energy for future use. The system has two 3.7 kW Wall-b home charging points for BEVs and extra provision of three Wall Sockets for e-bikes. The PV power source has enough capacity to meet the charging requirements of a single vehicle. In this study we charge two cars under this charging station. To support charging of two vehicles without delays and to meet the power requirements when solar power is not available, additional power is drawn from the power distribution grid. The power distribution grid is not only use to meet the power requirements but surplus power can be given back to the grid. The Campbell Scientific CR800 data logger was used to measure different system parameters such as Photovoltaic generation (after the inverter), power supply from/to the power grid, battery charge/discharge cycles (DC), consumption of charging points and additionally battery temperature and solar radiations. The technical details are summarized in Table 1. A centralize data server periodically retrieve data from the datalogger, stores it in the mysql database, which is visualized and analyzed on the web portal. The information portal acts as a monitoring tool, i.e. if any component stops working, it can easily be identified on the web portal, instead of periodic visits to the installation sites.

Table 1 Technical details of solar powered charging station at University of Hildesheim

Photovoltaic power:	3.84 kWp (16 × 240 Wp)
Buffer:	Lead-acid battery (24 cells, each 266 Ah C_{10} and 227 Ah C_5) with nominal capacity 10 or 12 kWh
Charging points:	2 Wallb-e Home (max. 3.7 kW AC), 3 Wall socket (max. 2.3 kW AC)
Data logger:	Campbell Scientific CR800 with NL201 network interface
Solar inverter:	Sunny Boy 3000TL
Battery inverter:	SMA Sunny Iceland 6.0
Lighting:	2 × 10 W LED with light sensor

4.2 Empirical Data

The PV system was commissioned in September 2014 and all the activities related to the charging station were logged from October 2014 till September 2015, covering a period of one year. Throughout this period two BEVs (i.e. one VW e-up! and one VW eGolf) were allocated to this charging station. The data was logged every 15 min and stored in the database. The data is visualized on the web portal in form of the graphs representing information in hourly, daily and monthly intervals. The hourly information is the fine-grain representation. It provides capability to monitor all the activities at the solar charging station. The Fig. 2 represents information in hourly interval of the four randomly selected days (8-Dec-14, 17-Mar-15, 17-Jun-15 and 3-Sept-15), each from different quarter of the year. The PV in the graph represents amount of solar energy generated, BEV represents actual charging demand, Grid represents the amount of energy acquired from (positive values) or given to (negative values) the grid and Battery represents the charging (negative values) and discharging (positive values) behavior of the internal buffer. It is observed that most of the charging activity happen during day time i.e. from 7:00 h in the morning till 17:00 h in the evening.

The system dynamics can also be interpreted from the Fig. 2. The dynamics of the system can be divided into two scenarios. In the first scenario, when a BEV arrives at the charging station it starts acquiring power from the PV and internal buffer. If the demand is greater than the combined power of the two sources additional power is drawn from the power distribution grid, otherwise the demand is managed from the internal sources. In second scenario, a BEV is fully charged or in absence of power demand, the power generated by PV source is used to fill in the internal buffer first and than surplus power is given back to the grid. These two scenarios can be seen in the Fig. 2, which also validates the correct functioning of the system.

The Figs. 3 and 4 presents data in aggregated form of the daily and monthly consumption of the system respectively. The Fig. 3 shows the accumulated daily production and consumption of energy by the system and is laid out in four quarters of the observation period. In the first quarter, due to winter months most of the power demands were met from the power grid. The power supply from the PV starts to get

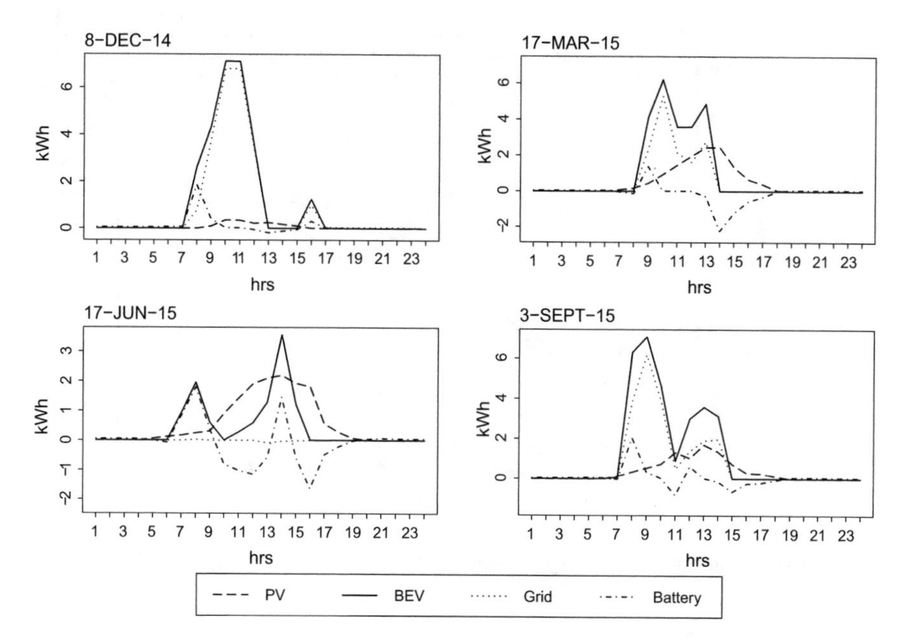

Fig. 2 The graph represents information in hourly interval of the four randomly selected days (8-Dec-14, 17-Mar-15, 17-Jun-15 and 3-Sept-15), each from different quarter of the year

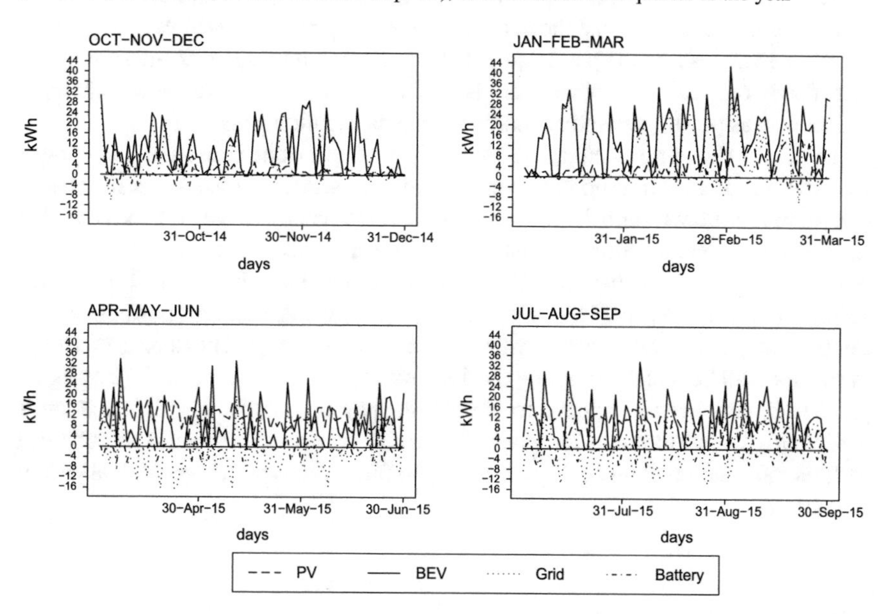

Fig. 3 The daily graph over one year from October, 01 2014 to October, 31 2015

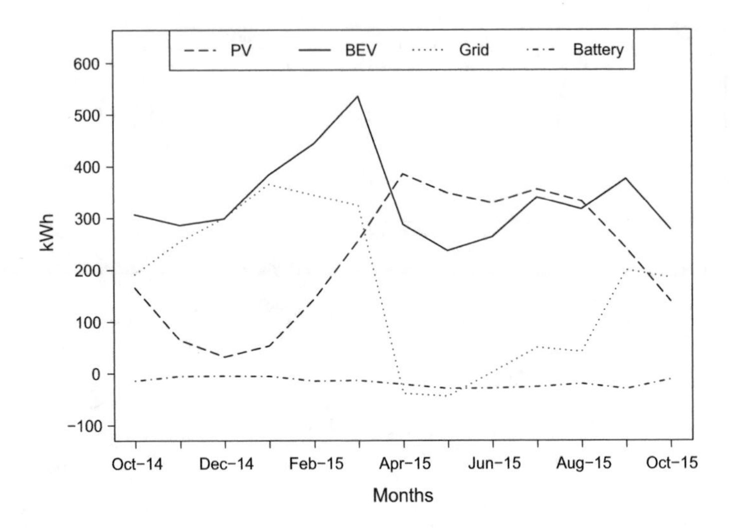

Fig. 4 The monthly graph over one year from October, 01 2014 to October, 31 2015

better in the second quarter as more sunlight become available. The third and fourth quarters show less and less dependency on the power grid. It can also been seen that in these quarters not just the dependency on the grid is reduced but also surplus power is given back to the power grid. The daily graph represents an interesting fact that mostly on weekends, when there is no BEV demand, the solar power charging station acts as power generation source for the distribution grid.

The Fig. 4 shows the accumulated monthly production and consumption of power over the observation period. This information is particularly useful as it helps in determining the per month demand and production capacity of the system. When PV source is available, in summer months, despite being underpowered for the demand of two BEVs, it stills completely cover the demand and also been able to give back surplus power to the grid. This validates that the power demand for business model can be managed in economical way through renewable energy. The intelligent charging station will help in determining the dynamic load of the system and an intelligent prediction model can be applied to forecast the power demand. This information is particularly useful in countries where the electricity price is dynamic and one can use this information to buy cheap electricity to further reduce the operational cost of the system.

4.3 Analysis

The empirical data acquired over period of one year, gives useful insight into the requirements of the business model for a specific implementation at University of

Hildesheim. It is an intelligent system that adjust itself according to the operating condition. The Eq. 1 defines the power dynamics of the system.

$$P_{BEV} + P^{-}_{grid} + P^{+}_{Bat} + \mathscr{L} = P_{PV} + P^{+}_{grid} + P^{-}_{Bat}\,, \tag{1}$$

The P_{BEV} defines the power demand and P_{PV} represents power generation through solar panel. The P^{+}_{grid} and P^{-}_{grid} represent power acquired and given to the grid respectively. The P^{+}_{Bat} and P^{-}_{Bat} define the charging and discharging of the internal buffer respectively. The $\mathscr{L}$ is the system losses due to inefficiency of lead-acid battery (typically 50–70 % efficiency can be achieved [23]) and others (i.e. light bulb) and it can be calculated using the Eq. 1. The system operates under dynamic demand conditions and as stated earlier the P_{PV} is not standalone power generation source. In order to gauge the environmental impact and understand how much demand has been managed from environmental friendly P_{PV}, we calculated the efficiency in percentile ($\mathscr{E}_{PV}$ and $\mathscr{E}_{grid}$) using Eqs. 2 and 3 respectively.

Table 2 Monthly data from 1st October 2014 till 31st September 2015

Month	P_{PV} (kWh)	P^{+}_{grid} (kWh)	P^{-}_{grid} (kWh)	P_{BEV} (kWh)	P^{-}_{Bat} (kWh)	P^{+}_{Bat} (kWh)	$\mathscr{L}$ (kWh)	$\mathscr{E}_{PV}$ (%)	$\mathscr{E}_{grid}$ (%)
Oct 2014	165	245	54	306	60	74	36	41 %⇒	59
Nov 2014	64	278	24	286	24	30	26	18 %⇓	82
Dec 2014	31	307	7	298	21	26	28	9 %⇓	91
Jan 2015	53	378	13	383	27	32	29	12 %⇓	88
Feb 2015	143	405	61	443	42	56	29	26 %⇓	74
Mar 2015	256	467	143	535	52	66	31	36 %⇒	64
Apr 2015	385	183	222	287	95	117	37	70 %⇑	30
May 2015	348	148	192	236	84	113	38	74 %⇑	26
Jun 2015	328	168	166	263	106	134	38	70 %⇑	30
Jul 2015	355	249	199	339	110	136	39	61 %⇑	39
Aug 2015	332	224	183	316	91	112	37	61 %⇑	39
Sep 2015	241	299	99	375	88	118	35	47 %⇒	53
Total	2707	3359	1369	4075	805	1019	443	44 %⇒	56

$$\mathscr{E}_{PV} = \frac{P_{PV}}{P_{BEV} + P^{-}_{grid} + P^{+}_{Bat} - P^{-}_{Bat} + \mathscr{L}}, \tag{2}$$

$$\mathscr{E}_{grid} = \frac{P^{+}_{grid}}{P_{BEV} + P^{-}_{grid} + P^{+}_{Bat} - P^{-}_{Bat} + \mathscr{L}}, \tag{3}$$

The Table 2 presents efficiency of P_{PV}, P_{grid} and system loss, which were calculated from empirical data by applying Eqs. 1, 2 and 3. The table is coded with arrows representing the environmental impact of the system. The efficiency of $\mathscr{E}_{PV}$ greater than 60 % coded with $\Uparrow$, shows high utilization of PV source. The $\mathscr{E}_{PV}$ above 30 % and less than 60 % coded with $\Rightarrow$, shows an intermediate use of PV. The $\mathscr{E}_{PV}$ less than 30 % coded with $\Downarrow$, shows a high dependence on the power grid. It is observed that the PV power generated by the system peaks at about 70 % in the summer months, where we have sufficient amount of sunlight available. The PV power generation was at lowest (below 10 %) in December month. The overall power generation by the PV source for the whole year is 44 %, whereas 56 % was drawn from the grid.

5 Conclusion and Future Work

In this paper we presented data from a PV-powered charging station operating in a context of an innovative electric car-sharing business model. Our data shows, that even PV with less power output can sustainably meet the energy demand of BEVs requirement by the business model, when smart charging concept is applied. In summer month, the system is able to meet more than 60 % of the energy demand with photovoltaics. But in winter months, when sunshine is insufficient, photovoltaic system is able to cover less than 20 % of the energy demand. Over the year, almost 44 % of energy demand can be covered through renewable energies.

By further developing modeling mechanisms for optimal sizing of the photovoltaic systems and real user behavior, the coverage rate of renewable energies can be increased. This will be the main focus of further research work. Additionally, the further optimizing of the business model using only pure electric vehicles will be undertaken, so an easy transferability and quick implementation of the whole concept outside the framework of a university can be permanently established. This research work gives insight into the interplay between different components of our charging concept under the special setup of the business case in tandem provides the decision maker with information needed to promote affordable electric mobility powered by clean energies.

References

Carroll, S., & Walsh, C. (2010). The smart move trial: Description and initial results. Centre of Excellence for Low Carbon and Fuel Cell. Technologies report.

Dudenhöffer, K. (2013). Why electric vehicles failed. *Journal of Management Control, 24*(2), 95–124.

Franke, T., Neumann, I., Bühler, F., Cocron, P., & Krems, J. F. (2012). Experiencing range in an electric vehicle: Understanding psychological barriers. *Applied Psychology, 61*(3), 368–391.

Goli, P., & Shireen, W. (2015). Control and management of PV integrated charging facilities for PEVs. In *Plug In Electric Vehicles in Smart Grids* (pp. 23–53). Springer.

Heinrichs, H., & Grunenberg, H. (2012). Sharing economy: Auf dem weg in eine neue konsumkultur?

Hidrue, M. K., Parsons, G. R., Kempton, W., & Gardner, M. P. (2011). Willingness to pay for electric vehicles and their attributes. *Resource and Energy Economics, 33*(3), 686–705.

Linsenmeier, C., & Heidemarie, I. (2011). Mein Haus, mein Auto, mein Boot war gestern. Teilen statt besitzen! Was sagen die Deutschen zum Sharing-Trend.

Neumann, H. M., Schär, D., & Baumgartner, F. (2012). The potential of photovoltaic carports to cover the energy demand of road passenger transport. *Progress in Photovoltaics: Research and Applications, 20*(6), 639–649.

Santos, A., McGuckin, N., Nakamoto, H. Y., Gray, D., & Liss, S. (2011). Summary of travel trends: 2009 national household travel survey. Technical report.

Tulpule, P. J., Marano, V., Yurkovich, S., & Rizzoni, G. (2013). Economic and environmental impacts of a PV powered workplace parking garage charging station. *Applied Energy, 108*, 323–332.

Trends in Mobility: A Competitive Based Approach for Virtual Mobility Providers to Participate in Transportation Markets

Alexander Sandau, Jorge Marx Gómez and Benjamin Wagner vom Berg

Abstract Mobility is necessary to participate on day-to-day life, but comes along with negative implications on our environment. A change in our mobility behavior caused by the growing digitalization and arise of new mobility services is essential to reduce the environmental impact of our mobility. To achieve such goal, new services with a strengthened competition has to be established. In addition, sharing of workforce and operating supplies between companies are addressed to adopt new business models. B2B sharing, including occasional workers, has huge potential to generate new services and business models, especially in the service sector. The described approach of a mobility broker provides such a tool for sharing between businesses, but also between businesses and occasional working persons in the field of mobility services.

Keywords Mobility broker • Mobility services • Digitalization in mobility • Share economy • B2B sharing

1 Introduction

Mobility has strong negative impacts on our environment, but is also indispensable to participate on day-to-day life. A detailed view on mobility shows, that emissions are one of the worst outcomes of our mobility behavior. The annual energy consumption of transport in the European Union raised by 38 % between 1990 and 2007. Therefore, road transport accounts for the largest share of energy con-

A. Sandau (✉) · J. Marx Gómez · B. Wagner vom Berg
Department of Business Information Systems, University of Oldenburg, Oldenburg, Germany
e-mail: alexander.sandau@uni-oldenburg.de

J. Marx Gómez
e-mail: jorge.marx.gomez@uni-oldenburg.de

B. Wagner vom Berg
e-mail: benjamin.wagnervomberg@uni-oldenburg.de

V. Wohlgemuth et al. (eds.), *Advances and New Trends in Environmental Informatics*, Progress in IS, DOI 10.1007/978-3-319-44711-7_17

sumption with 74 % of the total EU energy demand in transportation in 2013 (European Environment Agency 2015).

One possibility to counteract this adverse development is changing the modal split. Today, our mobility behavior is mainly shaped by motorized individual traffic (private car) which is the worst option (usual passenger occupancy rates) in an environmental perspective. In 2012, private cars serve 83.3 % of the inland passenger transportation in the EU-28 and this trend is increasing steadily (eurostat 2015). On a long term, a shift in our behavior is the only way to reach the tipping point of this development.

By all negative effects, it is very important to show up new alternatives to our established behavior and create sustainable business models. This can be achieved due to the creation of new mobility services, which are mainly enabled by digitalization and ICT penetration. New ICT based services are going to be established worldwide in many areas, for example Smart Home Technologies, Internet of things or quantified self. In addition, a stronger service orientation is focused, like food delivery (HelloFresh Deutschland AG & Co. KG. 2016), mobility services (Chalmers University of Technology 2016), health monitoring (FitBit Inc. 2016) or leasing of clothes (MudJeans 2016) are in common use (c.f. Product Service System, Tietze et al. 2011). Over the last years huge changes happened in the mobility sector. To mention are some of the developments like car sharing and ride hailing, electric vehicles or partly autonomous driving.

These developments are mainly fueled by the rapid development of information and communication technology that opens up possibilities that are initial to entrepreneurial activities to create new innovative products and services (c.f. Tietze et al. 2011).

The next chapter describes mobility trends and influential actors in the private and public mobility sector. Sect. 3 introduces an ICT based broker to enhance the competition and resource sharing, so private and public companies are enabled to share their human resources and operating supplies, like vehicles.

2 Mobility Trends and New Mobility Providers

The private and public mobility sectors are strongly influenced by digitalization. Digitalization offers big advantages for rural and urban mobility markets, for new and established mobility providers and, of course customers and citizens within bounds of a suitable regulation. New players on the mobility market lead to a more intense competition on passenger transportation, which leads to lower prices and an extensive supply. The following phenomenological analysis (c.f. Berger and Luckmann 1966) introduces important players and important trends on the private and public mobility sector.

2.1 Private Mobility Sector

In terms of private mobility, autonomous (self-driving) cars are in strong development across conventional car manufactures, but also in tech-companies like apple and google (Spickermann et al. 2013). Tesla Motors Inc. (2016) is one competitor in this market for self-driving cars, which introduced a lane-assistant and rudimentary autonomous behavior of vehicles through a software update. From the beginning, the cars are equipped with several sensors, as camera, radar, ultrasound, GPS and further. The delivery of additional functions as a software update over the air is a new approach to provide additional functions to customers in the automotive market. This new function (partly autonomous driving) is provided as payable feature, which will cost 3300€ for activation and 2700€ when buying a fabric new model. Long-established market competitors are not addressing such new forms of distributing new functions and features. They still are only focusing on selling cars and maintenance services (c.f. Tietze et al. 2011). In addition, they established services like leasing and financing of new vehicles, to increase the market of pre-owned cars and to lift their number of sales. Furthermore, they made small efforts in establishing car-sharing and car-pooling services like car2go by Daimler or drive now by BMW.

The term of new and innovative mobility service provider is often linked with Uber (Uber Technologies Inc. 2016), an American multinational mobile ride hail (service on demand) company founded 2009 in San Francisco. In an economic perspective, the company serves as an intermediate on a bilateral mobility market. In this market, the network effect is very important for each side. A huge customer base leads to greater workload and potentially greater revenue, which leads to more registered drivers. As more driver are registered on the platform, more customer are using the services and this leads to shorter waiting time and a bigger scope of the services (Brühn and Götz 2014).

Such network effects are not only limited to Uber, also a deregulated taxi business, rental car business or similar business models can benefit. One of the biggest advantages is the penetrating use of ICT, which leads to decreasing mediating costs as more trips are mediated. Uber uses ICT to handle the whole business relation of drivers and customer pre, during and after contracting. Starting with the mediation of customer, pricing, price transparency, payment and assessment on an ICT base (Brühn and Götz 2014).

One of the most known service from Uber is uberPOP. The service distributes passengers to private drivers (Drive Partners). The investigation of the cost structure of uberPOP shows, that there are mostly no fixed costs, because the drivers using their own cars not only for business purposes, so they are handled as occasional driver. Due to this, drivers don't have to pay sales tax (only below 17.500 euro turnover annually in Germany). So the costs structure for the driver contains of variable cost per kilometer (fuel etc.), opportunity costs (in Germany 8.5 €/hr) and brokerage (24.2 % of the end price) (Haucap 2015).

In Europe Blablacar (Comuto SA 2016) dominates the ride sharing market with a community of more than 20 million users, with a strong network effect (as described for Uber). The platform acts as intermediate and connects drivers and passengers (peer to peer). In difference to Uber, the driver doesn't act as service on demand provider, but rather to optimize their vehicle occupancy. Blablacar is offering additional services like breakdown service (accommodation for passengers or onward transportation) and an increased accident insurance on his driver included in the intermediate fee. According to Blablacar the average car occupancy is around 2.8 people (vs. 1.6 in normal average). The company established through their service a new mobility behavior, like Airbnb (AirBnB Inc. 2016) did for accommodation. (Longhi et al. 2016).

Beside new mobility companies, existing car manufactures addressing the changing consumption behavior with new mobility services, like car sharing. Some of them are already on the market and acting more or less successful (car2go Daimler AG 2016; DriveNow BMW AG 2016; quicar Volkswagen AG 2016). Many OEMs are behind the pace and trying to catch up with the trend. They invest huge amounts in companies like Lyft, Uber, Gett and Didi.

In contrast to the OEMs, the business model of automotive retailers is strongly focused on a regional net product. The main revenue streams are the disposal of vehicles and associated services. However, the car manufactures cutting provisions and margins of retailers, which leads to existence-threatening situations and decreasing employee rates. Some experts predict decreasing private disposal numbers for Western Europe in mid to long term (Wagner vom Berg et al. 2016).

Therefore car dealers and retailers facing huge challenges to guarantee their survival. They are trying to address new business fields like mobility consulting or establishment of new mobility services (e.g. car sharing). Also in cooperation with regional power suppliers, they are operating regional charging stations for electric vehicles (e.g. electric car sharing). Through their local focus, authorized car dealers and free car retailers are predestinated to establish regional mobility services. One strong benefit of car retailers and car dealers are widely distributed retailer networks with many local resources and a strong regional customer base. (Wagner vom Berg et al. 2016)

Beside the car sharing services, the retailer are also offering conventional rental car services, which will be merged with car sharing fleets, when existing software solutions are available in the market. The German car rental market has ca. 8.500 companies with around 36.000 vehicles (Bundesministerium für Verkehr, Bau und Stadtentwicklung 2012).

Along this new business fields, the aligned companies have to be innovative too. They have to address and support this new products and services, for example insurance companies. Especially for the sharing economy new insurance products, resp. packages have to be developed. Other aligned business for example are JustPark (JustPark 2016) that offers free parking spaces for rent from private persons. This business can be adopted on urban located companies offering their spaces, or the companies paying parking spaces next to their location for their employees.

A huge shift is noticeable in the private mobility sector, conventional automotive companies trying to transform to mobility providers.

2.2 Public Mobility Sector

Besides traditional automotive companies, conventional public transport companies like (Deutsche Bahn AG 2016) also addressing new mobility services as car sharing. The railroad company established a service called Flinkster (Deutsche Bahn AG 2016) that provides a car sharing offer for the so called last mile of a trip. In general, the car sharing service is station based, but in a few German cities Flinkster is offering a free floating approach. To afford such a service Flinkster is cooperating with car2go provided by Daimler.

An important new trend in public transport service is a stronger demand orientation with sensible responsive of customer needs. This is currently addressed by research projects on international and national level.

One approach for a demand tailored public transport services is initiated through the availability of (partly) autonomous vehicles. One of these projects is the (CityMobil2 2016) co-founded by the EU, where a pilot platform for automated road transport system is implemented for test purposes in several European cities. The automated transport vehicles operating without a driver in a collective mode, where they play a useful role in the transport mix. They can supply transport services (individual or collective) in areas of low or dispersed demand complementing the main public transport network (Alessandrini et al. 2015).

New forms of demand oriented public transport are also addressed by the research project (Reallabor Schorndorf DLR e. V. 2016) of the German Aerospace Center (DLR). The project addresses a digital supported demand oriented bus concept without fixed public transport stations. The bus traffic is scheduled on the individual needs and demands of the people. Through this approach, the vehicles are used more efficient and reducing the negative impacts in an environmental and social manner. In collaboration with citizens individual district based buses are implemented, which supplements the offer in low demand times. Due the lack of stations and strict time schedules, new communication challenges between the passenger, bus driver and command and control station are appearing. Nevertheless, the increasing digitalization enables the citizens to submit their demand requests via mobile applications or by phone in advance. The ICT system organizes the bus schedule and informs the passengers, where and when the bus picks them up. Beside rush hour, smaller vehicles replacing the busses to operate more economical and ecological.

Besides multi-person transportation, the taxi business can round up the regional mobility portfolio. According to a representative survey of IFAK in 2014, one third of the German population requested at least one taxi trip, out of these the median is around five usages in half a year (BZP—Deutscher Taxi- und Mietwagenverband e. V. 2014). The distribution of Taxis is mainly proceeded via phone calls in taxi offices (74 %), followed by taxi ranks (14 %) and roadside (10 %). Only 0.4 % of the taxi requests are submitted via internet or mobile applications. This shows the huge potential in new distributing forms of taxis. A report from the BMVBS in 2012 reveals that approx. 22.000 taxi companies with 54.000 taxis are operating in Germany (Bundesministerium für Verkehr, Bau und Stadtentwicklung 2012). Since 1990 the passenger volume and distance remains constant, but the revenue for taxi drivers increased by 45 % in 2012 compared to 1990, due to raising kilometer fees.

The comparison of established mobility services like taxi and innovative mobility services like uberPOP give interesting insights. According to a report by Haucap the prices for an average taxi ride of 6.5 km in 15 min in German metropolises is between 15 and 17 euros. In comparison, uberPOP accounts between 8 and 11€ for the same ride (Haucap 2015). This is a good example, how new and innovative approaches and the growing digitalization leads to more competition in established markets and new business fields for companies. On a long term, such developments may lead to a broader offer of mobility services, a reduction of harmful emission due a more efficient occupancy rate and of course lower costs.

As shown in these two subsections, OEMs are under huge pressure to transform their creation of value to service providing and coincidently developing autonomous vehicles. At the same time, the mobility market asking for customer centered mobility services, which fit the needs of the citizens. The following section introduces a mobility broker for the transition time, in which the citizens get a tool to use their locally resources efficient for self-organized mobility services.

3 Broker

The prior introduced trends, services and products reveal a big shift in the mobility market initiated by a growing digitalization. In this section, a competition module is introduced to strengthen the competition in the public mobility sector. The main objective is to show up an easy tool to share resources more efficient and to give more stakeholder a market access.

3.1 Strengthen the Competition in Transport Services by Enhancing Access

Especially in rural areas, the public transport system is weak developed. Regular trips are rarely provided and accompany with long waiting times. In addition, the combination of several vehicle modes (intermodal) during a trip are often accompany with waiting times on interchange points. A stronger competition in rural areas may lead to a stronger portfolio of mobility services. This can be achieved through new mobility services from rural partners and stakeholders. After establishing such competitors, some areas may still be undersupplied, for such regions a combination of payable mobility and voluntary mobility services is the key element.

The following competition module serves as an enabler to establish new competitors in the market of transport services. These competitors can participate in the private or public mobility sector. A fundamental facet is the broker that gives the possibility to create virtual mobility associations/providers. A virtual mobility provider is not a fixed company; it is an incorporation of a provider of operating supplies (vehicle) and a service operator (driver). So in this case vehicle owner and licensed driver are brought together to apply for public and private transport orders, even on a voluntary base. For vehicle owner, or fleet manager, which are trying to increase the workload of a fleet or vehicle, licensed drivers can be acquired by such a system. Together a higher net product can be achieved on public or private transport routes. Possible drivers are freelancers, self-employees or volunteers with country specific licenses (if necessary). In addition, this broker can be used to handle the sharing of resources and contracting between companies to establish new products and services.

Local municipalities up to the federal government can place public route orders in such a system. Private demands can be orders from private persons like ordering a taxi or an Uber car.

The creation of virtual mobility providers is not essential necessary, existing transport associations or disruptive transport companies can apply for such route orders in the same way, as they fulfill the legal requirements.

3.2 Architecture of the Competition Module

The following figure shows the competition process of the broker with negotiation, contracting and clearing (Fig. 1).

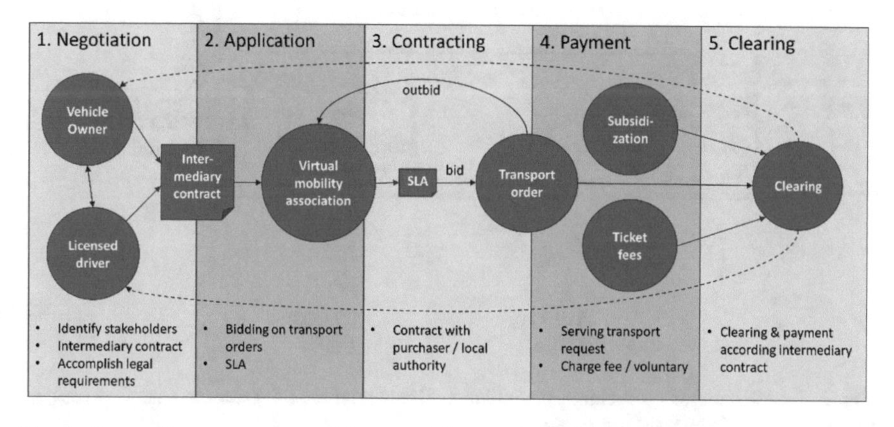

Fig. 1 Competition process to participate on private and public transport orders

The proposed competition module includes the broker for establishing virtual mobility associations. At its heart, the broker component contains the negotiation of vehicle owner and licensed driver (cf. Sect. 3.1). Inside the component a contract is negotiated, which includes, beside the contracting partners, the operator of such a broker respectively platform. Each party negotiate a fixed or dynamic quota of the received income and the operator charge a fix usage fee, except volunteer services. The technical representation of such contracts can be achieved by using *Smart Contracts* (cf. Omohundro 2014) (Fig. 2).

Additional to the contracting inside the broker component, the parties have to guarantee that they accomplished the legal requirements to apply on public or private mobility orders. One restriction in Germany for private mobility services is the requirement of a passenger transportation license.

Besides the broker, there is a module for mobility associations to register for e.g. auctions on public or private transport route orders. The so-called (virtual) mobility associations are a construct to have an unified legal association which meets all requirements. In addition, it is possible, that private and public mobility companies are sharing their resources. For example, if a public mobility company has a lack of drivers, a private bus company or similar may commit drivers to the public company. The juristic mapping of this construct may be represented as a temporal employee agency, or a rental service, in which the participating companies have to allow such service mapping in house.

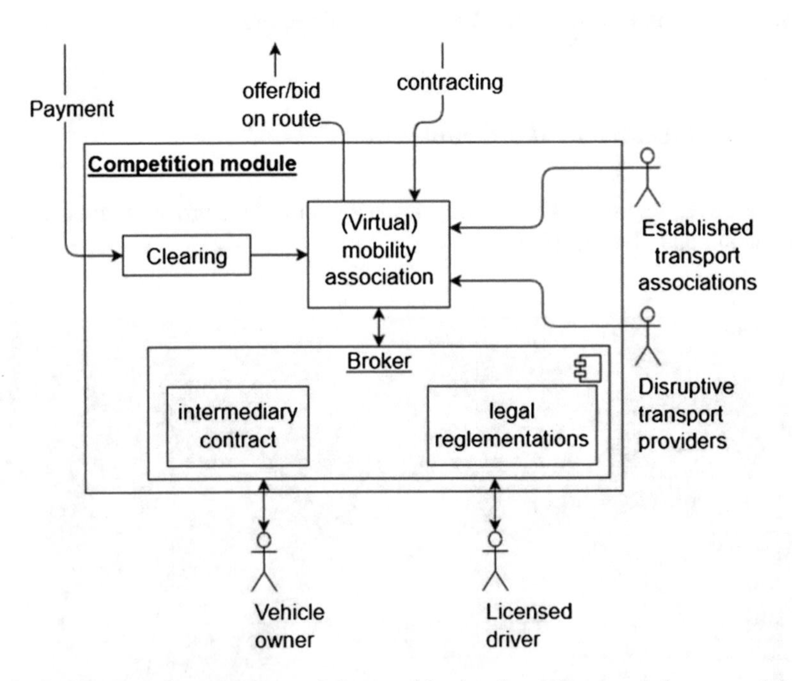

Fig. 2 Architecture of competition module to enable virtual mobility associations to participate on public and private transport orders

The (virtual) mobility associations are bidding on transport orders. This transport orders can be private or public, particular or regular and voluntary or compensated. As an association gets the acceptance, a contract between the customer and the association based on the bid and associated service agreements is signed.

The clearing component collects and handles all payment events of the customers for private and public trips and the subsidization of public trips. The payments of private trips are processed by the clearing component and according to the previously negotiated contract distributed. On handling payments of public transport orders, the clearing is more complex by consideration of subsidization by the federal government.

The proposed architecture is in implementation as web platform and is based on an open source framework for multi-vendor marketplaces. The adoption of an established marketplace is very suitable to map customers and vendors, but also to create demands and offers. The used open source framework is enhanced by the specific requirements, e.g. juristic conditions in the registration process and the clearing processes.

4 Conclusion

The public mobility sector is influenced by several major trends, e.g. cost pressure and securing the public mobility of the population with transport services. The introduced trends and developments in the transport market showing the potential to transform mobility offers to a more sustainable manner. The proposed broker contributes to the problematic situation in rural regions, which a lack of public mobility services as well as in cities with huge transportation amount. Lowering the barriers to participate on these regional markets can have serious implications on actors, and as well on citizens. In Sect. 2.1 introduced example of Uber shows that ICT supported mobility service leading to new employment relationships. For young or older people new forms of casual working are established to improve their income or pension. This evolves to a bigger mobility portfolio with more efficient transport services, even in rural areas. In addition, local companies have a new business field to participate, with the benefits of a more efficient payload on their resources with additional income. Also, the broker enable companies to develop new products and by sharing of resources and knowledge.

The proposed work will be prototypical implemented as part of the research project NEMo. This an acronym (in German) for "Sustainable satisfaction of mobility demands in rural regions". As a result of demographical change it is becomes more and more difficult for rural municipalities to offer a basic set of public transport services. On the other side the mobility demand in rural regions is increasing because of the agglomeration of health care and shopping infrastructure in nearby cities and the proceeding urbanization.

Based on this problem statement the research project NEMo is developing sustainable and innovative mobility offers and supporting business models out of the social communities, especially in rural regions. Information and communication technology plays an important role as enabler for participation to achieve an integrated mobility portfolio. NEMo is an inter- and transdisciplinary project. Besides the scientific participants (ICT, business information systems, social entrepreneurship, service management, law), many partners from industry (e.g. mobility providers) and public (e.g. municipalities) are involved in the project. This provides the optimal basis for a practical oriented evaluation in rural regions in cooperation with the partners of the project.

Acknowledgments This work is part of the project "NEMo—Sustainable satisfaction of mobility demands in rural regions". The project is funded by the Ministry for Science and Culture of Lower Saxony and the Volkswagen Foundation (VolkswagenStiftung) through the "Niedersächsisches Vorab" grant program (grant number VWZN3122).

References

AirBnB Inc. (2016). *Hot it works—Airbnb.* https://www.airbnb.de/help/getting-started/how-it-works.

Alessandrini, A., et al. (2015). Automated vehicles and the rethinking of mobility and cities. *Transportation Research Procedia, Issue, 5*, 145–160.

Berger, P. L., & Luckmann, T. (1966). *The social construction of reality: A treatise in the sociology of knowledge*. s.l.: Anchor.

BMW AG. (2016). *How DriveNow works.* https://de.drive-now.com/en/#!/howto.

Brühn, T., & Götz, G. (2014). Die Modelle Uber und Airbnb: Unlauterer Wettbewerb oder eine neue Form der Sharing Economy? *ifo Schnelldienst 21/2014*, November 13 (pp. 3–27).

Bundesministerium für Verkehr, Bau und Stadtentwicklung. (2012). *Bericht über die Sondererhebung zum Taxen -und Mietwagenverkehr.*, s.l.: s.n.

BZP—Deutscher Taxi- und Mietwagenverband e.V. (2014). Kundenzufriedenheit mit Taxi-. Berlin: Deutscher Taxi- und Mietwagenverband e.V.

Chalmers University of Technology. (2016). *New urban mobility service receives international innovation award.* https://www.chalmers.se/en/areas-of-advance/Transport/news/Pages/International-innovation-award-awarded-to-a-new-urban-mobility-service.aspx.

CityMobil2. (2016). *CityMobil2—Cities demonstrate automated road passenger transport.* www.citymobil2.eu.

Comuto SA. (2016). *Blablacar.de.* https://www.blablacar.de/wie-es-funktioniert.

Daimler AG. (2016). *Car2Go.* www.car2go.com.

Deutsche Bahn AG. (2016). www.deutschebahn.com.

Deutsche Bahn AG. (2016). *Flinkster.* www.flinkster.de.

DLR e. V. (2016). *Busfahren on demand.* Retrieved May 18, 2016 from http://www.dlr.de/dlr/desktopdefault.aspx/tabid-10122/333_read-16600#/.

European Environment Agency. (2015). *Final energy consumption by mode of transport.* Copenhagen: European Environment Agency.

Eurostat. (2015). *Energy, transport and environment indicators*. Luxembourg: Publications Office of the European Union.

FitBit Inc. (2016). *Official webpage of Fitbit.* www.fitbit.com.

Haucap, J. (2015). *Chancen der Digitalisierung auf Märkten für urbane Mobilität: Das Beispiel Uber.* No: DICE Ordnungspolitische Perspektiven. 73.

HelloFresh Deutschland AG & Co. KG. (2016). https://www.hellofresh.de/so-funktionierts/.

JustPark. (2016). *Pre-Book Parking to Save.* www.justpark.com.

Longhi, C., Mariani, M. M., & Rochhia, S. (2016). Sharing and tourism: The rise of new markets in transport. *Documents de travail GREDEG—working paper series,* issue 01.

MudJeans. (2016). *Lease a Jeans.* http://www.mudjeans.eu/lease-philosophy/.

Omohundro, S. (2014). Cryptocurrencies, cryptocurrencies, smart contracts, and artificial intelligence. *AI MATTERS, 1*(2), 19–21.

Spickermann, A., Grienitz, V., & von der Gracht, H. A. (2013). Heading towards a multimodal city of the future? Multi-stakeholder scenarios for urban mobility. *Technological Forecasting & Social Change.*

Tesla Motors Inc. (2016). www.teslamotors.com.

Tietze, F., Schiederig, T., & Herstatt, C. (2011). Firms' transition towards green product-service-system innovators. In *Working papers/technologie- und innovationsmanagement, Technische Universität Hamburg-Harburg,* issue 62.

Uber Technologies Inc. (2016). www.uber.com.

Volkswagen AG. (2016). *Germany: Quicar—Sahre a Volkswagen.* Retrieved May 20, 2016 from http://thinkblue.volkswagen.de/en_US/blue_projects/germany_car_sharing.

Wagner vom Berg, B., et al. (2016). *Mobility 2020—IKT-gestützte Transformation von Autohäusern zum regionalen Anbieter nachhaltiger Mobilität.* s.n.: Ilmenau.

Part VI
Life Cycle Assessment

Regionalized LCI Modeling: A Framework for the Integration of Spatial Data in Life Cycle Assessment

Juergen Reinhard, Rainer Zah and Lorenz M. Hilty

Abstract Life Cycle Assessment (LCA), the most prominent technique for the assessment of environmental impacts of products, typically operates on the basis of average meteorological and ecological conditions of whole countries or large regions. This limits the representativeness and accuracy of LCA, particularly in the field of agriculture. The production processes associated with agricultural commodities are characterized by high spatial sensitivity as both inputs (e.g. mineral and organic fertilizers) and the accompanying release of emissions into soil, air and water (e.g. nitrate, dinitrogen monoxide, or phosphate emissions) are largely determined by micro-spatial environmental parameters (precipitation, soil properties, slope, etc.) and therefore highly context dependent. This spatial variability is vastly ignored under the "unit world" assumption inherent to LCA. In this paper, we present a new calculation framework for regionalized life cycle inventory modeling that aims to overcome this inherent limitation. The framework allows an automated, site-specific generation and assessment of regionalized unit process datasets. We demonstrate the framework in a case study on rapeseed cultivation in Germany. The results from the research are (i) a framework for generating regionalized data structures, and (ii) a first examination of the significance of further use cases.

Keywords Regionalization · Site-specific LCI modeling · LCA · Raster data

J. Reinhard (✉) · L.M. Hilty
University of Zurich, Binzmühlenstrasse 14, 8050 Zürich, Switzerland
e-mail: juergen.reinhard@uzh.ch

R. Zah
Quantis, Reitergasse 11, 8004 Zürich, Switzerland

V. Wohlgemuth et al. (eds.), *Advances and New Trends in Environmental Informatics*, Progress in IS, DOI 10.1007/978-3-319-44711-7_18

1 Introduction

In order to address the challenges associated with climate change and other environmental threats, environmental considerations need to be integrated in many types of decisions (Finnveden et al. 2009). This highlights the importance of methods and tools for measuring and comparing the environmental impacts of human activities for the provision of goods and services. One of the most prominent methods in this regard is Life Cycle Assessment (LCA). LCA is a technique for the comprehensive quantitative assessment of the environmental impacts of products in a life-cycle perspective (Finnveden et al. 2009). It focuses on the compilation and environmental evaluation of all inputs and outputs associated with a product throughout its life cycle with the goal to pinpoint ecological weaknesses, compare possible alternatives, evaluate the main impact factors, design new products, measure the environmental relevance of a material or product and establish recommendations for actions (Guinée et al. 2002).

The environmental impact of a product is caused by the exchange of energy and matter between its technical (product) system and its surrounding environment (Carlson et al. 1998). The technical system is represented as a linear model composed of a network of nodes called unit process datasets (UPDs) (Carlson et al. 1998). Each UPD includes data on (i) the *intermediate exchanges*, i.e., the input of energy and material flows and the output of products and waste flows, and (ii) the *exchanges with environment*, i.e., the inputs of natural resources and outputs of emissions (Carlson et al. 1998). The UPDs are linearly linked via their intermediate exchanges. The calculation of the product system for the reference flow (product) of interest reveals the throughput of all exchanges with environment, the so-called Life Cycle Inventory (LCI). Life Cycle Impact Assessment (LCIA)—the succeeding step of the LCA procedure—then takes the inventory data on these exchanges as an input to determine the impacts on the surrounding environment.

It is common practice to consider average meteorological and ecological conditions of a whole country and geographical region for the compilation of the intermediate exchanges and exchanges with environment listed in the UPDs. That is, LCA in practice is mostly structured around the use of either *site-generic*[1] or *site-dependent*[2] unit process datasets (UPD), whereas *site-specific*[3] data is used very rarely. The high cost of primary data collection only allows for rudimentary site-specific UPD generation. That is, LCA usually must focus on average values at the expense of specificity (Mutel and Hellweg 2009).

[1]Site-generic values represent an average over large geographic regions, such as continents or the globe (Mutel et al. 2012).

[2]Site-dependent values follow country or state boundaries (Mutel et al. 2012).

[3]Site-specific values are usually used only for individual locations, such as a particular plot, factory or landfill (Mutel et al. 2012).

While this approach is suitable for determining important key drivers of the environmental impacts, it limits the range of questions which can be properly addressed with the LCA technique. For example, UPDs representing agricultural cultivation comprise exchange flows characterized by a high sensitivity to the natural variability of the surrounding environment. The type and amount of resources used (e.g. water, land), the intermediate flows required (e.g. the application of mineral and organic fertilizer or the use of machinery) and the accompanying release of emissions into soil, air and water (e.g. nitrate, di-nitrogen monoxide, phosphate emissions) are determined by micro-spatial environmental parameters (precipitation, soil properties, slope, etc.) and therefore highly context dependent. That is, for such processes even small changes in local, bio-geographical conditions can alter the type and magnitude of the included exchange flows and hence also their environmental impacts (Geyer et al. 2010a).

Regionalized LCI modeling is motivated by the "recognition that industrial production characteristic vary throughout space" (Mutel et al. 2012). We believe that the development of a computerized technique for spatially explicit (regionalized) LCI modeling in LCA would decrease the costs of collecting and computing more and better information about agricultural production characteristics. The potential of regionalized *LCI* modeling has already been acknowledged (Mutel et al. 2012; Mutel and Hellweg 2009; Seto et al. 2012) although most of the recent literature on regional aspects in LCA focuses on regionalized LCIA modeling—recognizing that the location of a source and the conditions of its surroundings influence the environmental impact (Hauschild 2006). To the best of our knowledge, regionalized LCI modeling is to date only applied within six case studies (Dresen and Jandewerth 2012; Geyer et al. 2010a, b; Reinhard et al. 2011; Scherer and Pfister 2015; Zah et al. 2012). All of these studies succeed in the parameterization of the spatial properties of selected exchanges. However, none of the studies analyzed provides a *general framework* for regionalized LCI modeling in LCA.

In this paper, we present a prototype calculation framework[4] that allows the automated, site-specific (regionalized) generation and assessment of cradle-to-gate agricultural UPDs. We transform publicly available spatial (raster) data (Table 1) into comprehensive UPDs using default data from version 3.2 of the ecoinvent database and the emission models from the World Food Life Cycle Database (WFLDB) Guidelines (Nemecek et al. 2015). We present a case study of rapeseed production in Germany to illustrate the framework. Using a resolution of 30 arc seconds (~1 × 1 km), we generate and assess roughly 580'000 regionalized UPDs for rapeseed cultivation in Germany. We conclude with a discussion of limitations and further use cases of the presented framework.

[4]The framework is currently still under development and therefore not publicly available.

Table 1 The compiled repository of spatial raster datasets

(No.) Institution	Content	Resolution	Spatial and temporal extent	Source
(1) Global Agro-Ecological Zones (GAEZ)	Precipitation	5 * 5 min	World, mean, 1961 - 1990	(IIASSA/FAO 2012)
(2) Global Agro-Ecological Zones (GAEZ)	Length of crop growing cycle, rapeseed	5 * 5 min	World, mean, 1961 - 1990	(IIASSA/FAO 2012)
(3) Publication	Irrigation requirement	Country	World, 2000	(Pfister et al. 2011)
(4) ISRIC, Soil grids	Soil property data (pH, coarse fragments, organic carbon content, bulk density, clay content)	30 * 30 arc second	World, 2014	(Hengl et al. 2014)
(5) Earth Stat	Fertilizer application (N, P & K) for major crops	5 * 5 min	World, 2000	(Mueller et al. 2012)
(6) Earth Stat	Harvested area and yield for 175 Crops	5 * 5 min	World, 2000	(Monfreda et al. 2008)
(7) Earth data	P content of soils	0.5 * 0.5°	World, 2014	(YANG 2014)
(8) Earth data	Amount of nitrogen manure produced and present on the landscape	0.5 * 0.5°	World, 2001	(Potter et al. 2011a; Potter et al. 2010)
(9) Earth data	Amount of phosphorous in manure produced and present on the landscape	0.5 * 0.5°	World, 2001	(Potter et al. 2010), (Potter et al. 2011b)
(10) European Soil Data Centre (ESDAC)	Soil erodibility factor	30 * 30 arc second	Europe, 2014	(Panagos et al. 2014)
(11) European Soil Data Centre (ESDAC)	Length slope factor	30 * 30 arc second	Europe, 2015	(Panagos et al. 2015)
(12) Natural Earth	Country shape files		World, 2015	(Natural Earth 2016)

2 Method

2.1 Framework for Regionalized LCI Modeling

We compiled a repository of publicly available raster data (see Table 1) indicating harvested area, yield, fertilizer application rates, irrigation requirement of all major crops as well as data on precipitation, soil properties and terrain. Almost all of the

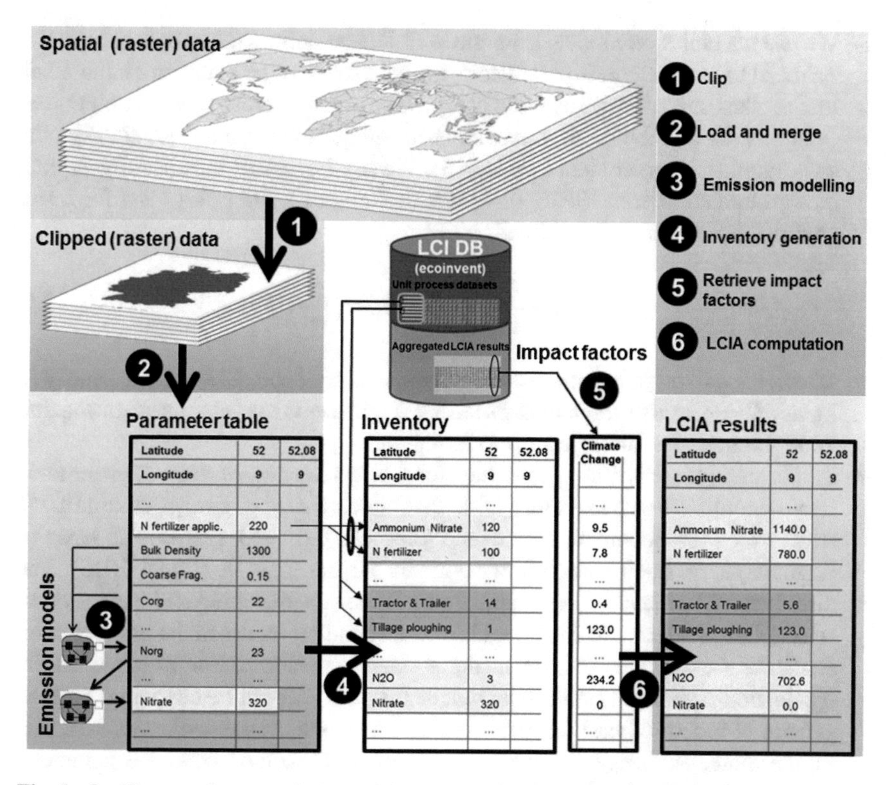

Fig. 1 Our framework to transform spatial raster data into UPDs. A detailed explanation is given in the text. *Source* authors

data has a global scale (see Table 1). This ensures that the framework is applicable in all regions of the world.

We process this repository into UPDs with the following procedure, implemented in python 3.4 (see Fig. 1):

1. We clip the raster data to the spatial extend (e.g. country borders, province borders, etc.) of interest. We use GDAL (Geospatial Data Abstraction Library) and shape files from natural earth for this task.
2. We load the crop and country specific raster data into data frames using Pandas, a Python Data Analysis Library. We merge[5] these data frames into one table. Each column of this parameter table represents, for a given latitude-longitude combination, the corresponding grid cell values of all clipped raster datasets (see Fig. 1, parameter table). The parameter table serves as the basis for the further processing of the data.

[5]To date, the merging of the data is the most time consuming step (see discussion section).

3. We use the emission models from the WFLDB guidelines (Nemecek et al. 2015, Table 3) to calculate new parameters, i.e., add rows to the parameter table. Each model takes several rows as input and produces one or more rows as output. For example, the calculation of nitrate emissions requires the parameter *OC*, i.e., the soil organic carbon content (see step 3). We use Eq. 1 and the spatially explicit background data from ISRIC soil grids to compute *SOC* in kg C/ha for a soil depth of 0.3 m for each grid cell.

$$\mathrm{SOC} = \frac{C_{org}}{1000} * 3000 * BD * (1 - CF) \tag{1}$$

   where: C_{org} = organic carbon content in soil in g/kg soil, BD = Bulk density in kg/m^3, CF = coarse fragments in %, 3000 represents the amount of m^3/ha and 1000 the transformation from g to kg/kg soil.
4. In this step (see 4 in Fig. 1), we transform selected rows of the parameter table into a comprehensive inventory table that match the common agricultural UPD structure of the ecoinvent database. We access the UPD data on the basis of brightway2, an open source framework for advanced LCA (Mutel 2015). The direct access to the ecoinvent database facilitates the matching and expansion of selected parameter table entries with a corresponding intermediate product or an elementary exchanges present in the ecoinvent database. For example, the N application rates from Earth Stat (Mueller et al. 2012) can be converted into the generic N-based mineral fertilizer products available in the ecoinvent database. Furthermore, exchange flows which cannot be computed from the parameter table due to the lack of regionalized data or models (e.g. transport tractor and trailer, mechanical field work) can be extrapolated on the basis of default data obtained from existing ecoinvent UPDs. Each column of the final inventory table then represents a regionalized UPD for a particular grid cell in ecoinvent nomenclature. That is, step 4 of the framework already yields regionalized cradle-to-gate agricultural UPDs.
5. We generate a vector of impact factors for each exchange flow in the inventory table. We compute the aggregated LCIA impacts for each intermediate exchange (e.g. fertilizer, mechanical field work) and retrieve the characterization factor for each elementary exchange (e.g. N2O emission). We can generate this vector for any LCIA indicator implemented in the ecoinvent database.[6] Again, we use the brightway2 framework for this task.
6. Finally, we assess the environmental impacts of the regionalized inventory. We multiply the LCIA impact factors for a particular row with the corresponding exchange flow value retrieved from the inventory table. The resulting LCIA table (see Fig. 1, step 6) contains for each grid cell the LCIA impacts associated with each exchange flow value in the inventory table. The column sums of this

[6]That is, the LCIA vector can be generated for 692 LCIA indicators.

table then represent the total environmental impacts of each grid cell for a particular LCIA indicator.

We illustrate the application of this framework using a case study of rapeseed cultivation in Germany.

2.2 Case Study of Rapeseed Cultivation

We generate and assess UPDs for a resolution of ~ 1 × 1 km, i.e., in the resolution of the soil property raster data.[7] We compute UPDs for each grid cell where rapeseed cultivation takes place. Gaps in raster data are excluded from the analysis. The exchange flows in each regionalized UPD refer to the cultivation of one hectare of rapeseed in a cradle-to-gate perspective.[8]

We first generate the parameter table by clipping and merging all data in Table 1 to the geographical extent of Germany (steps 1 and 2). In step 3, we compute Ammonia, Nitrous oxide, Dinitrogen monoxide, Nitrate and Phosphorus emissions according to the WFLDB guidelines (Nemecek et al. 2015, Table 3).

We next generate the inventory table based on selected columns in the parameter table and default data from the German rapeseed cultivation dataset (step 4). Table 2 shows, for selected inventory flows, how we integrate such default data to generate an inventory table that match the common agricultural UPD structure of the ecoinvent database.

After elimination of data gaps,[9] the final inventory table contains roughly about 588'000 columns, i.e. UPDs. We apply three LCIA midpoint indicators to assess the environmental impacts for each UPD in the inventory table (steps 5 and 6): climate change (CC, IPCC2013 GWP100a), marine eutrophication (MEP, ReCiPe Midpoint (H)) and freshwater eutrophication (FEP, ReCiPe Mipoint (H)).

[7]This means that due to the difference in resolution between the crop specific raster datasets (see no. 5 and no. 4 in Table 1) and the soil property raster datasets (see no. 4, 10 and 11 in Table 1) the 100 rapeseed UPDs generated for a ~10 × 10 km grid cell will always have the same yield and fertilizer input. That is, within a ~10 × 10 km grid cell, the only varying parameters are the soil properties and the therefrom computed emissions. Although this computation strategy comes at higher computational costs—for one parameter, we have to extract the values of roughly one million grid cells—it pays off by enabling us to assess the influence of changing soil properties on emissions in a ceteris paribus examination.

[8]That is, all upstream interventions are included. The further usage of the rapeseed (e.g. as feed or biofuel) is not considered.

[9]Roughly 6 % of the UPDs (or 40'000 UPDs) were excluded due to data gaps. The extrapolation of missing data was not in the scope of this article.

Table 2 Integration of default values from the ecoinvent database

Parameter table	Inventory table	Computation	Comment/Source
N application rate (NAR)	Ammonium nitrate, as N	= NAR * 0.55	Fertilizer type and proportion is taken from the ecoinvent datasets "rapeseed production, DE". Factors (2.3 and 1.2) transform from P to P_2O_5 and K to K_2O, respectively.
	Nitrogen fertilizer, as N	= NAR *0.45	
P application rate (PAR)	Phosphate fertilizer, as P_2O_5	= PAR * 2.3	
K application rate (KAR)	Potassium chloride, as K_2O	= KAR * 1.2	
Tractor and Trailer; Tillage; Ploughing and harrowing; Sawing; Fertilizing by broadcaster; Combine harvesting; Pesticide, unspecified.		= Value in generic ecoinvent dataset	Taken by default from the ecoinvent dataset "rapeseed production, DE"

Table 3 Comparison with generic inventory data from different LCI databases

Exchange flow	Unit	Nem	AB	AF	EI3.2	Our result	
						Ø	CV
Nitrogen fertilizer	kg N/ha		162	200	113	207	0.22
Phosphate fertilizer	kg P_2O_5/ha		34	46	54	66	0.32
Potassium chloride	kg K_2O/ha		25	82	45	73	0.19
Nitrate (NO_3)	kg NO_3-N/ha	30–140	32	71	12	110	0.73
N_2O	kg N_2O-N/ha	0.5-2.5*	2.75	3.12	2.72	3.4	0.31
Phosphorus, river	kg/ha		0.53	n.a.	0.12	0.77	1.47

Nem = literature review of (Nemecek et al. 2014); AB = Agribalyse, rapeseed production in France; AF = AgriFootprint, rapeseed production in Germany; EI3.2 = ecoinvent version 3.2, rapeseed production in Germany; CV = coefficient of variation
*Asterisks indicate that values are not rapeseed specific

3 Results

Figure 2 shows the spatially explicit environmental impacts per hectare rapeseed cultivated for CC, FEP and MEP. The spatial distributions of the environmental impacts differ depending on the LCIA indicators applied. While rapeseed cultivation in Lower Saxony and Schleswig-Holstein causes the highest CC impacts, both states show very low impacts for FEP.

CC impacts are dominated by N_2O emission (resulting from N-based fertilizer application) and the energy intensive production of N-based mineral fertilizer; both

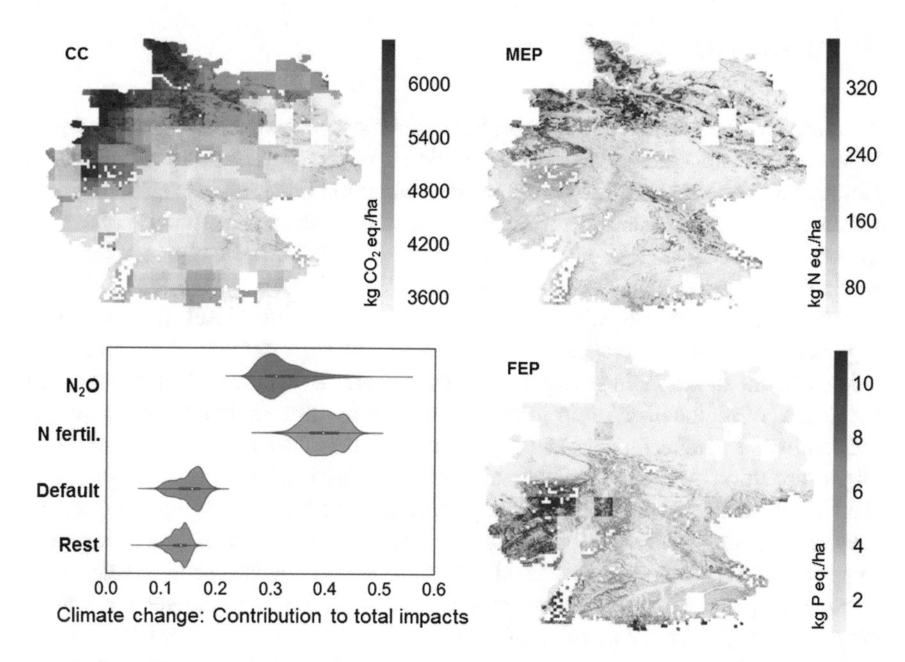

Fig. 2 Heat map showing the spatially explicit environmental impacts (in kg impact per hectare) of roughly 588'000 UPDs for the LCIA indicators CC, MEP and FEP (see Table 3). *White areas* represent blind spots caused by data gaps. The *violin plot* (*left bottom*) shows the relative contribution of four types of exchanges to the total climate change impact: the two exchanges contributing the most (N_2O and N-based fertilizers), all exchanges modeled with default data (Default) and all remaining exchanges (Rest). The vertical extent of the "violins" shows the frequency of data points

typically cause more than 70 % of the impacts (see violin plot). Consequently, the spatial distribution of CC impacts correlates largely with the application intensity of N-based mineral fertilizer. The impacts according to MEP and FEB are dominated by the contributions of nitrate (90 %) and phosphorus emissions (50 %), respectively. Nitrate emissions are dependent on many factors. Therefore, the MEP heat map indicates an impact pattern much more diverse than the pattern observed for CC impacts. Phosphorus emissions are largely dependent on soil erosion, which is particularly large for the federal state of Rhineland-Palatinate.

Table 3 contrasts the mean of important inventory flows with corresponding values in existing rapeseed UPDs and literature data. In general, our results appear reasonable; though often higher than the values from comparable inventories.

The coefficient of variation (CV) indicates a high spatial sensitivity of Phosphorus and Nitrate emissions, i.e., the size of these flows varies greatly throughout space. This confirms that the spatially explicit computation of these flows is important to obtain accurate cultivation process datasets.

4 Discussion

The case study presented above is probably the most comprehensive cradle-to-gate LCA on climate change and eutrophication impact of rapeseed cultivation in Germany. However, it still relies on many assumptions such as the use of standard values for crop tillage and management practice to compute phosphorous flows or the static modeling of field operations on the basis of default data obtained from the original rapeseed cultivation datasets in the ecoinvent database.

While the use of default values in the case study decreases the degree of regionalization, the general possibility to revert to such defaults should be considered as one of the main strength of the framework. It facilitates the fine-tuning of the degree of regionalization to the spatial sensitivity of exchange flows and to the availability of spatial models and input data. On the other hand, computations in "high resolution" (30 arc seconds and lower) for large geographic areas such like the US or Brazil can currently not be performed on a modern laptop computer as they are to time intensive.[10] However, computations in such resolutions are important for the appropriate consideration of certain spatial characteristics such like the length slope factors and erodibility. We are currently implementing a more efficient approach for the merging of the raster data into one parameter table (step 2) —the most time intensive step—which is expected to facilitate computations in "high resolutions" for large geographies also on a modern laptop computer. In addition, the relevance of the spatial repository should be updated. For example, the spatial data on harvested rapeseed area (see Table 1, no. 6) referring to the year 2000 does not consider the recent 25 % increase to 1.5 million hectare (FAOSTAT 2016). Finally, the framework doesn't support UPD export in common LCA data formats such like the ILCD or the ecospold2 format.

Despite these limitations, the overall framework harbors great potential for improving LCI modeling, as the following use cases may show. First, the framework might be relevant in every project setting where the assessment of the environmental impacts of spatially sensitive processes are key, i.e., when food, feed, fibre and (bio-)fuel production are under study. In this context, it offers an improved basis for decision making as it allows to explicitly consider and compute variations in micro-spatial conditions, which are typically ignored in the common "unit world" paradigm of LCA. These variations may be relevant, for example, when a large-scale bioenergy producer wants to compare the environmental performance of alternative bioenergy-cropping systems in an explicit spatial setting.

Second, the framework can be used as a tool to improve LCI databases by adding more accurate, average agricultural UPDs. To date, UPDs in such databases are mostly site-generic. The consistent, spatially explicit bottom-up computation of UPDs allows for the generation of more accurate averages for relatively homogeneous regions. That is, the spatial scale (i.e. the region covered) of an average UPD

[10]For Brazil, we have to merge about roughly 25 × 20 million data entries. With the current merging approach, this would require more than a month on a modern laptop computer.

does not necessarily have to follow political (country or state) boundaries but could be determined by a spatial cluster analysis which considers location and the spatial variation in exchange flow range. As a welcome side-effect, the framework facilitates a better evaluation of spatial uncertainty of exchange flows[11] as it allows the exploration of the spatial sensitivity of each regionalized exchange flow for any spatial extent. Improving both the representativeness of and spatial uncertainty information in agricultural UPDs is an improvement at the very foundation of LCA, as it makes the entire background system involved in every LCA more reliable.

Third, our framework offers a basis for the integration of more sophisticated emission models that were developed outside the LCA context. Emission models applied in agricultural UPDs are often adjusted to the "averaging" nature of LCA, i.e., they lack important compartments due to their generic spatial and temporal orientation. For example, the computation of nitrate ignores losses with waterborne and windborne sediment. However, as shown in the result section, the emissions in agricultural UPDs often contribute significantly to the total environmental impact. Our framework offers a modular and expandable test bed for the application advanced models. This could improve the representativeness and accuracy of the emissions recorded in agricultural UPDs.

Fourth, the framework could be used as a background computation engine in web-based LCA tools (often called "footprinters"). For example, our framework could improve the usability and accuracy of the RSB tool (Reinhard et al. 2011). The RSB tool is a web-based tool for the carbon footprint computation of various user-defined biofuel pathways. The user has to enter a great amount of context-specific information (soil type, land use, precipitation, etc.) into a questionnaire. Our framework could reduce these data entries and improve the accuracy of inventory computations.

Finally, our framework delivers regionalized UPDs, which is the fundament also for regionalized LCIA.

5 Conclusion

We presented a framework for the automated generation of site-specific agricultural unit process datasets (UPDs). Our approach is to transform publicly available, spatial raster data into site-specific UPDs on the basis of the WFLDB Guidelines. Although the framework has been demonstrated by means of a case study on rapeseed cultivation in Germany, it can be applied to any major crop or country on the globe.

Acknowledgments We thank our colleague Mireille Faist-Emmenegger for her support with WFLDB emission models and the three anonymous reviewers for their helpful comments.

[11]To date, exchange flow specific variability in LCI databases (used for Monte-Carlo analysis) is largely based on rough estimates.

References

Carlson, R., Tillman, A.-M., Steen, B., & Löfgren, G. (1998). LCI data modelling and a database design. *International Journal of Life Cycle Assessment, 3*, 106–113.

Dresen, B., & Jandewerth, M. (2012). Integration of spatial analyses into LCA—calculating GHG emissions with geoinformation systems. *International Journal of Life Cycle Assessment, 17*, 1094–1103.

FAOSTAT: http://faostat.fao.org/site/567/DesktopDefault.aspx?PageID=567#ancor.

Finnveden, G., Hauschild, M. Z., Ekvall, T., Guinée, J., Heijungs, R., Hellweg, S., et al. (2009). Recent developments in life cycle assessment. *Journal of Environmental Management, 91*, 1–21.

Geyer, R., Lindner, J. P., Stoms, D. M., Davis, F. W., & Wittstock, B. (2010a). Coupling GIS and LCA for biodiversity assessments of land use. *International Journal of Life Cycle Assessment, 15*, 692–703.

Geyer, R., Stoms, D., Lindner, J., Davis, F., & Wittstock, B. (2010b). Coupling GIS and LCA for biodiversity assessments of land use. *International Journal of Life Cycle Assessment, 15*, 454–467.

Guinée, J. B., Gorrée, M., Heijungs, R., Huppes, G., Kleijn, R., de Koning, A., van Oers, L., Sleeswijk, A. W., Suh, S., de Haes, H. A. U., de Bruijn, H., van Duin, R., & Huijbregts, M. A. J. (2002). *Handbook on life cycle assessment*. Operational Guide to the ISO Standards.

Hauschild, M. (2006). spatial differentiation in life cycle impact assessment: A decade of method development to increase the environmental realism of LCIA. *International Journal of Life Cycle Assessment, 11*, 11–13.

Hengl, T., de Jesus, J. M., MacMillan, R. A., Batjes, N. H., Heuvelink, G. B. M., Ribeiro, E., et al. (2014). SoilGrids1 km—Global soil information based on automated mapping. *PLoS One, 9*, e105992.

IIASSA/FAO: Global Agroecological Zones (GAEZ v3.0). IIASA, Laxenburg/FAO, Rome (2012).

Monfreda, C., Ramankutty, N., & Foley, J. A. (2008). Farming the planet: 2. Geographic distribution of crop areas, yields, physiological types, and net primary production in the year 2000. *Global Biogeochemical Cycles, 22*.

Mueller, N. D., Gerber, J. S., Johnston, M., Ray, D. K., Ramankutty, N., & Foley, J. A. (2012). Closing yield gaps through nutrient and water management. *Nature, 490*, 254–257.

Mutel, C.: A new open source framework for advanced life cycle assessment calculations. https://brightwaylca.org/.

Mutel, C. L., & Hellweg, S. (2009). Regionalized life cycle assessment: Computational methodology and application to inventory databases. *Environmental Science and Technology, 43*, 5797–5803.

Mutel, C. L., Pfister, S., & Hellweg, S. (2012). GIS-based regionalized life cycle assessment: how big is small enough? Methodology and case study of electricity generation. *Environmental Science and Technology, 46*, 1096–1103.

Natural Earth: Admin 0—Country data. www.naturalearthdata.com/http//www.naturalearthdata.com/download/110m/cultural/ne_110m_admin_0_countries_lakes.zip.

Nemecek, T., Bengoa, X., Lansche, J., Mouron, P., Riedener, E., Rossi, V., & Humbert, S. (2015). Methodological guidelines for the life cycle inventory of agricultural products. World Food LCA Database (WFLDB), Version 3.0. http://www.quantis-intl.com/microsites/wfldb/files/WFLDB_MethodologicalGuidelines_v3.0.pdf.

Nemecek, T., Schnetzer, J., & Reinhard, J. (2014). Updated and harmonised greenhouse gas emissions for crop inventories. *International Journal of Life Cycle Assessment*, 1–18.

Panagos, P., Borelli, P., & Meusburger, K. (2015). A New European slope length and steepness factor (LS-Factor) for modeling soil erosion by water. *Geosciences, 5*, 117–126.

Panagos, P., Meusburger, K., Ballabio, C., Borrelli, P., & Alewell, C. (2014). Soil erodibility in Europe: A high-resolution dataset based on LUCAS. *Science of the Total Environment, 479–480*, 189–200.
Pfister, S., Bayer, P., Koehler, A., & Hellweg, S. (2011). Environmental impacts of water use in global crop production: Hotspots and trade-offs with land use. *Environmental Science and Technology, 45*, 5761–5768.
Potter, P., Ramankutty, N., Bennett, E. M., & Donner, S. D. (2010). Characterizing the spatial patterns of global fertilizer application and manure production. *Earth Interactions, 14*, 1–22.
Potter, P., Ramankutty, N., Bennett, E. M., & Donner, S. D. (2011a). Global Fertilizer and Manure, version 1: Nitrogen in Manure production. http://dx.doi.org/10.7927/H4KH0K81.
Potter, P., Ramankutty, N., Bennett, E. M., Donner, S. D. (2011b). Global Fertilizer and Manure, version 1: Phosphorus in Manure production. http://dx.doi.org/10.7927/H49Z92TD.
Reinhard, J., Emmenegger, M. F., Widok, A. H., & Wohlgemuth, V. (2011). RSB tool: A LCA tool for the assessment of biofuels sustainability. In *Proceedings of the Winter Simulation Conference, Winter Simulation Conference*, Phoenix, Arizona (pp. 1048–1059).
Scherer, L., & Pfister, S. (2015). Modelling spatially explicit impacts from phosphorus emissions in agriculture. *International Journal of Life Cycle Assessment, 20*, 785–795.
Seto, K. C., Reenberg, A., Boone, C. G., Fragkias, M., Haase, D., Langanke, T., Marcotullio, P., Munroe, D. K., Olah, B., & Simon, D. (2012). Urban land teleconnections and sustainability. *Proceedings of the National Academy of Sciences.*
Yang, X., Post, W. M., Thornton, P. E., & Jain, A. K. (2014). Global gridded soil phosphorus distribution maps at 0.5-degree resolution. http://dx.doi.org/10.3334/ORNLDAAC/1223.
Zah, R., Gmuender, S., Gauch, M. (2012). *Evaluación del ciclo de vida de la cadena de producción de biocombustibles en Colombia*. Banco Interamericano de Desarrollo (BID).

Open Calculator for Environmental and Social Footprints of Rail Infrastructures

Francisco Barrientos, Gregorio Sainz, Alberto Moral, Manuel Parra, José M. Benítez, Jorge Rodríguez, Carlos Martínez, Francisco Campo and Rubén Carnerero

Abstract In EU27 0.2 % of global emissions correspond to rail transport. Infrastructure supposes 28 % of these emissions, half of them caused during construction. This shows the high environmental impact of these activities. Life cycle assessment (LCA) techniques combined with intelligent data analysis improves sustainability of railway infrastructure construction processes as a whole, considering environmental, economic and social aspects. The goal is the development of methodologies and tools to optimize decision making process, reducing carbon and water footprints of

---

F. Barrientos (✉) · A. Moral
CARTIF, Parque Tecnológico de Boecillo 205, 47151 Valladolid, Spain
e-mail: frabar@cartif.es

A. Moral
e-mail: albmor@cartif.es

G. Sainz
Systems Engineering and Control Department, School of Industrial Engineering, University of Valladolid, 47011 Valladolid, Spain
e-mail: gresai@eii.uva.es

M. Parra · J.M. Benítez
University of Granada, DECSAI,CITIC-UGR, IMUDS, ETSI Informática y Telecomunicación, 18071 Granada, Spain
e-mail: manuelparra@ugr.es

J.M. Benítez
e-mail: J.M.Benitez@decsai.ugr.es

J. Rodríguez · C. Martínez
VIAS Y CONSTRUCCIONES S.A., c/ Orense 11, 28020 Madrid, Spain
e-mail: jorge.rodriguez@vias.es

C. Martínez
e-mail: carlos.martinez@vias.es

F. Campo · R. Carnerero
INGURUMENAREN KIDEAK INGENIERIA, c/ Fandería 2, Oficina 102, 48901 Barakaldo, Bizcaia, Spain
e-mail: f.campo@ik-ingenieria.com

R. Carnerero
e-mail: r.carnerero@ik-ingenieria.com

V. Wohlgemuth et al. (eds.), *Advances and New Trends in Environmental Informatics*, Progress in IS, DOI 10.1007/978-3-319-44711-7_19

railway infrastructure construction projects from their earliest stages, i.e. design and planning processes. Environmental and social impact of most relevant tasks have been reviewed and analyzed in order to set the impact of railway networks construction processes. Multi-objective optimization has been performed to find a trade-off solution for project units scheduling. A tool has been developed from this information compilation, providing selected specific footprint values and environmental indicators as open data to the community.

Keywords LCA · Computational intelligence · Carbon footprint · Water footprint · Rail infrastructure

1 Introduction

Life Cycle Assessment (LCA) techniques, combined with intelligent analysis of data coming from rail infrastructure works, is helping to reduce their carbon and water footprints. The goal is to improve the rail infrastructure construction processes with regard to their environmental impact, mainly in those aspects related to climate change like carbon and water footprints and other environmental indicators. This is a rising trend, as illustrated by environmental product declarations (EPD) of railway infrastructures Botniabanan (2010).

No tools have been found for this task, except for a concrete and cement EPD tool EPD International (2014). General purpose tools like SimaPRO Pré Consultants (2016) or GaBi LCA Software Thinkstep (2016), are suitable for this type of calculations, but require huge efforts in characterizing each item for every project unit. However, a specific purpose tool, like the descrived in this paper, saves time and reduces error probability in characterizing an infrastructure construction project. It should be noted that railway infrastructures like tunnels and bridges are equivalent to road infrastructures, except for superstructure layer and roadway surface, so this tool can be used for both projects type.

At a fast pace, water footprint is getting relevant, due to scarcity of this natural resource. This paper presents a pioneer study on this matter, because no other reference about water footprint of rail infrastructures has been found on the specialized literature. The study here described is aligned with the European Water Initiative EUWI (2002), fostering a correct water resources management. Also social impact estimation of rail infrastructures is a novel approach.

A threefold (environmental, economic and social) in-depth analysis has been performed on railway infrastructure project units. A multi-objective optimization has been performed in order to find a trade-off solution for project units scheduling.

This paper is organized as follows: Sect. 2 describes the problem related to environmental and social impact of rail infrastructures' construction process. Section 3 explains the proposed methodology. The analysis of some preliminary results is shown in Sect. 4. Finally, main conclusions and further work are outlined in Sect. 5.

Table 1 Kilometers of high speed lines in Europe

Country	In operation	Under construction	Planned	Total country
France	1,896	210	2,616	4,722
Germany	1,285	378	670	2,333
Italy	923	0	395	1,318
Poland	0	0	712	712
Portugal	0	0	1,006	1,006
Spain	2,056	1,767	1,702	5,525
Sweden	0	0	750	750
Switzerland	35	72	0	107
United Kingdom	113	0	204	317
Total	**6,308**	**2,427**	**8,055**	**16,790**

2 Environmental and Social Impact of Railway Infrastructures

Rail transport causes 0.2 % of global emissions in EU27, and about 28 % of total emissions associated with rail transport are due to the infrastructure Skinner et al. (2010). Almost half of these emissions are caused by the construction of the infrastructure, showing the high environmental impact of this activity. This environmental unfriendly process has a considerable room for improvements. Most of those emissions are due to materials production and transport, being this last issue a key point for environmental impact reduction. Table 1 illustrates the growing number of km of high speed lines in EU UIC (2011).

Construction of 1 km of railway supposes around 1.04 CO_2 tons emissions, assuming most of the energy used for its construction comes from fuel oil ($1\,\text{kWh} = 0.2674\,\text{kg}\ CO_2\ eq$ Carbon 2011). It is estimated that the amount of energy needed to build railway infrastructure, including tunnels and bridges, is $45,000$ GJ/km track Simonsen (2010). Both, direct and indirect emissions, show the high environmental impact of these activities. Thus, implementation of LCA techniques in railway construction can contribute to reduce those emissions Campo et al. (2014).

3 Methodology

The goal of this study is the development of a methodology to optimize decision making process, reducing carbon and water footprints of railway infrastructure construction. Other environmental indicators are also taken into account, see Sect. 3.1, as well as economic and social indicators. Thus, the environmental, economic and social impact of rail infrastructure construction process has been reviewed and analyzed, as illustrated by Fig. 1.

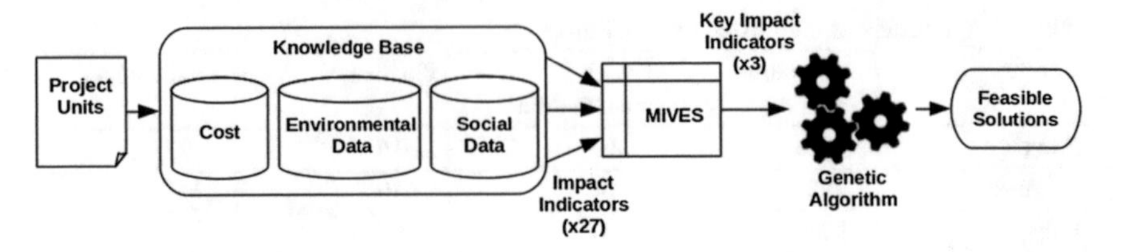

Fig. 1 Proposed methodology schema

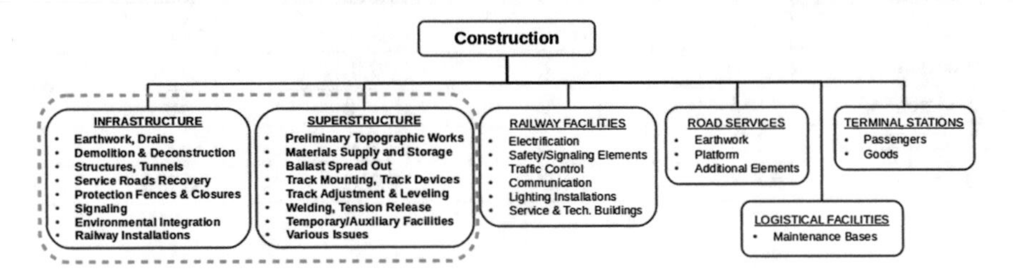

Fig. 2 Phases of railway infrastructure construction

3.1 Project Units Compilation

This study performed a comprehensive compilation of railway networks construction materials, solutions and processes ADIF (2008), and calculation of sustainable indicators to analyze environmental, economic and social impacts of construction processes Barrientos et al. (2016). Not every phase in the construction process is taken into account. This paper focuses on Infrastructure (subgrade and substructure) and Superstructure layers, as illustrated by Fig. 2 (*dashed line*), because they produce the highest environmental impacts.

The first step was to establish what is railway infrastructure, taking into account that a sustainability assessment tool should consider the entire subgrade and substructure, and those items of superstructure performed more frequently, according to construction companies' experience. Catenary and signaling project units were excluded of this study.

To define the scope of the study, the second step has been rail structure life cycle definition, which include the following stages: raw material extraction, transportation of raw materials, manufacturing, distribution, construction, use, maintenance and storage, and end of life.

Based on this definition, the scope of the study was defined as 'cradle to gate'. This assumption is supported by the work currently done on Environmental Product Declarations (EPD) EPD International (2015), where it was found that in the product category rules, maintenance and end of life phases appeared as optional because of the limited information that exist, hindering the development of a reliable assessment of these stages.

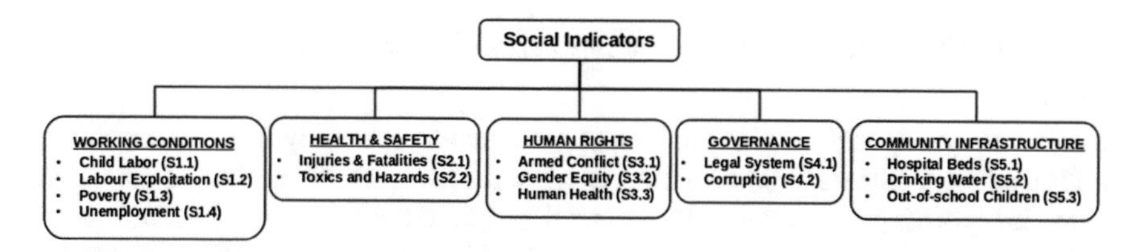

Fig. 3 Social indicators

The foundation for building the knowledge base was an analysis of inputs/outputs inventory. The environmental indicators considered in this study are carbon footprint, Acidification Potential, Photochemical Oxidation (POCP), Eutrophication Potential and water footprint, that summarizes the values of 7 sub-indicators (water, cooling, lake, river, turbine, unspecified and well). Values of these indicators have been established combining several information sources like manufacturer's performance handbook Caterpillar (2013), SimaPRO Pré Consultants (2016) with ecoinvent database and other EPDs EPD International (2015), and applying international standards AENOR (2012), International Organization for Standardization (2006a, b, c) guidelines.

Regarding to social impact, or Social Life Cycle Assessment (SLCA), methodologies are less developed than environmental ones. Thus, and according to the limited specialized bibliography data UNEP (2009), indicators shown in Fig. 3 have been selected. These indicators have been calculated using data from 253 countries obtained from TARIC database DataComex (2015). Another extra social indicator has been considered: job creation (S6). It is related to work hours needed to build the infrastructure and, unlike others, it is a positive indicator. Economic indicator measures the cost of developing the infrastructure, including line items for workforce, materials and machinery, as well as overheads.

MIVES Viñolas et al. (2009) is a decision support methodology suitable for comparison of variables with different measurement units. Other normalization methods could be used, e.g. Z-score, unity-based, etc. but one important feature of MIVES methodology is that evaluation model precedes the creation of alternatives. Thus, decisions are taken at the beginning, when the aspects taken into account and how to be assessed are defined. The advantage of this approach is that decision making is done without any influence of the assessments of the alternatives, preventing any kind of subjectivity. The steps of the MIVES methodology are:

- Decision delimit: setting system limits and boundary conditions.
- Decision-making tree: branching of decision making aspects.
- Value functions: such functions provide a rating in the interval [0-1].
- Weights assignment: i.e. relative importance of factors wrt its siblings.

3.2 Multi-objective Evolutionary Algorithms (MOEA)

Evolutionary Algorithms (EA) are adaptive methods, generally used in search and optimization problems of parameters, based on the principle of survival of the fittest. Genetic algorithms are search algorithms based on the mechanics of natural selection and natural genetics Goldberg (1989). In this case NSGA-II (Non-dominated Sorting Genetic Algorithm 2) method is used Deb et al. (2002). Best fitted individuals are randomly altered in the hope of improving their fitness to the next generation. Evaluation function provides a numerical measure of the goodness of a solution, setting whether individuals represent good solutions to the problem, or not.

Multi-objective programming seeks to provide efficient methods for decision-making on issues that usually include contradictory goals Coello (1998). Real problems require finding solutions that simultaneously meet multiple performance criteria or objectives, which can be contradictory Zitzler et al. (2000).

In this case, the problem is a Multi-objective Optimization Problem (MOP) because objective function has two parts: sustainability and constraints violation. Multi-objective evolutionary algorithms (MOEA) provide a set of non-dominated feasible solutions Deb (1999), i.e. satisfying all constraints, providing different planning alternatives for the rail infrastructure construction project being analyzed.

3.3 Indicator Evaluation

The total number of indicators is 27 including 11 environmental indicators, 15 social indicators and 1 economic indicator, as stated in Sect. 3.1.

After calculating indicators values, it is necessary to establish the evaluation methodology. Evolution of such methodologies has been uneven for the three pillars of sustainability: environmental, economic and social, and this has influenced the development of the evaluation method itself. By means of MIVES methodology it is possible to summarize the 27 indicators in 3 values: Environmental, Social and Economic impact.

A tool has been developed from this information compilation, applying data mining and computational intelligence techniques, e.g. multi-objective evolutionary algorithms. Starting from the three values provided by MIVES, these algorithms allow making alternative project units scheduling and showing different solutions, due to there could be more than one optimal solution.

4 Results

Based on the 27 indicators previously identified, a comprehensive compilation of basic information was performed, reviewing and analyzing more than 450 infrastructure and superstructure project units, see Table 2. The sum of both columns is higher

Table 2 Number of analyzed project units

Infrastructure	No.	Superstructure	No.
Earthwork	78	Preliminary topographic works	2
Drains	69	Materials supply, discharge and storage	22
Demolition and Deconstruction	15	Ballast spread out	2
Structures	133	Track and track devices mounting	20
Tunnels	61	Track adjustment and levelling	7
Service roads recovery	14	Tension release	2
Environm. integration, fences and closures	40	Welding	5
Signaling	31	Temporary/Auxiliary facilities	3
Railway installations	28	Various issues	2
Total	**469**	**Total**	**65**

Table 3 Example of project unit and its tasks

Task	Quantity	Units	Description	Type
	1	m^3	*Excavation at clearing site with use of explosive demolition*	
T1	0.004	h	Foreman	Workforce
T2	0.019	h	Skilled Workman	Workforce
T3	0.037	h	Unskilled Workman	Workforce
T4	0.224	kg	Dynamite with proportional share of detonator and fuse	Material
T5	0.019	h	Complete equipment of machinery for cut excavation	Machinery
T6	0.002	h	Bucket loader 375 HP, type CAT-988 or similar	Machinery
T7	0.012	h	Truck 400 HP, 32 T.	Machinery

because some project units are present in more than one category, e.g. ‘concrete sleeper mounting’ could be part of ‘track mounting’ or ‘track devices mounting’. Every project unit involves several tasks, as illustrated by Table 3. More than 520 tasks have been analyzed. As in the previous case, a task can be involved in more than one project unit.

Tables 3, 4 and 5 show an example of the information compiled for every project unit, including its tasks. Table 3 compiles the general information, while Tables 4 and 5 summarize environmental and social indicators, respectively.

Table 4 Environmental indicators compiled for every project unit and its tasks

Task	Carbon F.	Acidif.	POCP	Eutrop.	Water Footprint						
					Water	Cooling	Lake	River	Turbine	Unspec.	Well
T1	–	–	–	–	–	–	–	–	–	–	–
T2	–	–	–	–	–	–	–	–	–	–	–
T3	–	–	–	–	–	–	–	–	–	–	–
T4	1.11	0.01	0.00	0.00	–0.18	0.02	0.00	0.00	2.69	0.00	0.00
T5	0.15	0.00	0.00	0.00	–0.00	0.00	0.00	0.00	0.05	0.00	0.00
T6	0.31	0.00	0.00	0.00	–0.01	0.00	0.00	0.00	0.10	0.00	0.00
T7	1.77	0.01	0.00	0.00	–0.03	0.01	0.00	0.00	0.55	0.00	0.00
Total	**3.34**	**0.02**	**0.00**	**0.00**	**–0.22**	**0.03**	**0.00**	**0.00**	**3.39**	**0.00**	**0.00**

Table 5 Social indicators compiled for every project unit and its tasks

Task	Working conditions				H&S		Human rights			Govern.		Comm. Infr.			Jobs
	S1.1	S1.2	S1.3	S1.4	S2.1	S2.2	S3.1	S3.2	S3.3	S4.1	S4.2	S5.1	S5.2	S5.3	S6
T1	–	–	–	–	–	–	–	–	–	–	–	–	–	–	0.00
T2	–	–	–	–	–	–	–	–	–	–	–	–	–	–	0.02
T3	–	–	–	–	–	–	–	–	–	–	–	–	–	–	0.04
T4	0.01	0.04	0.06	0.03	0.02	0.00	0.04	0.06	0.06	0.07	0.14	0.06	0.22	0.03	–
T5	0.00	0.01	0.01	0.01	0.00	0.00	0.01	0.01	0.01	0.01	0.02	0.01	0.04	0.00	–
T6	0.00	0.01	0.02	0.01	0.01	0.00	0.01	0.03	0.02	0.02	0.04	0.02	0.07	0.01	–
T7	0.03	0.08	0.12	0.06	0.04	0.02	0.07	0.14	0.11	0.10	0.25	0.12	0.41	0.03	–
Total	**0.04**	**0.14**	**0.21**	**0.11**	**0.07**	**0.02**	**0.13**	**0.24**	**0.20**	**0.20**	**0.45**	**0.21**	**0.74**	**0.07**	**0.06**

Table 6 MIVES weighting factors, by category, for social indicators

Data source	Category	Weighting Factor (*WF*)	$\sum WF$
Indirect sources of information	(S1) Working conditions	0.1	0.5
	(S2) Health and Safety	0.1	
	(S3) Human rights	0.1	
	(S4) Governance	0.1	
	(S5) Community infrastructure	0.1	
Direct sources of information	(S6) Job creation	0.5	0.5

Units for Table 4 are kg CO_2 *eq* for carbon footprint, kg SO_2 *eq* for acidification, kg C_2H_4 *eq* for POCP, kg PO_4^{-3} *eq* for eutrophication and m^3 for all water footprints. Data in Table 5 are dimensionless due to MIVES methology.

Economic indicator depends on every construction company and the infrastructure owner requirements. So, the cost should be specified case by case. Workforce tasks have no impact on environmental and social indicators, except for 'S6 job creation', while material and machinery tasks are just the opposite case.

As explained in Sect. 3, MIVES is suitable for combining different variables. A weighting factor should be applied between the existing categories. For this, a priority relationship has been established that is heuristic and directly proportional to the data source. Since (S6) 'job creation' category is obtained directly from project data, it is assigned half of the weight, giving the other categories equal importance within the scope of 'indirect sources of information'. Therefore, the weighting between categories has been made according to the values shown in Table 6. On MOP tool, users can adjust these weights according to their own criteria.

Fig. 4 Footprints calculator home page

Weights are also assigned to indicators within each category, considering 3 factors: *severity*, i.e. hardness and social difficulty for a country; *duration*, time while the effects are evident; and *data variability*, i.e. short-term data volatility.

For instance, Health and Safety indicators (S2.1) 'injuries and fatalities' and (S2.2) 'toxics and hazards' are assigned 0.25 and 0.75 weights, respectively, considering that both indicators have *high* severity, S2.2 last longer than S2.1 and data variability is *medium* for both cases. A similar procedure has been applied to environmental indicators, due to at least three different sources of environmental footprints have been considered, as mentioned in Sect. 3.1. The economic indicator case is different, because cost is the only indicator and there is no need to apply any weight.

A tool has been developed as a web application, see Figs. 4 and 5. The user can select a project unit, assign the quantity and specify the configuration parameters according to project location. Once the project is completely defined, the tool will provide alternative solutions with equivalent project units or scheduling that optimize the environmental, social and economic impacts.

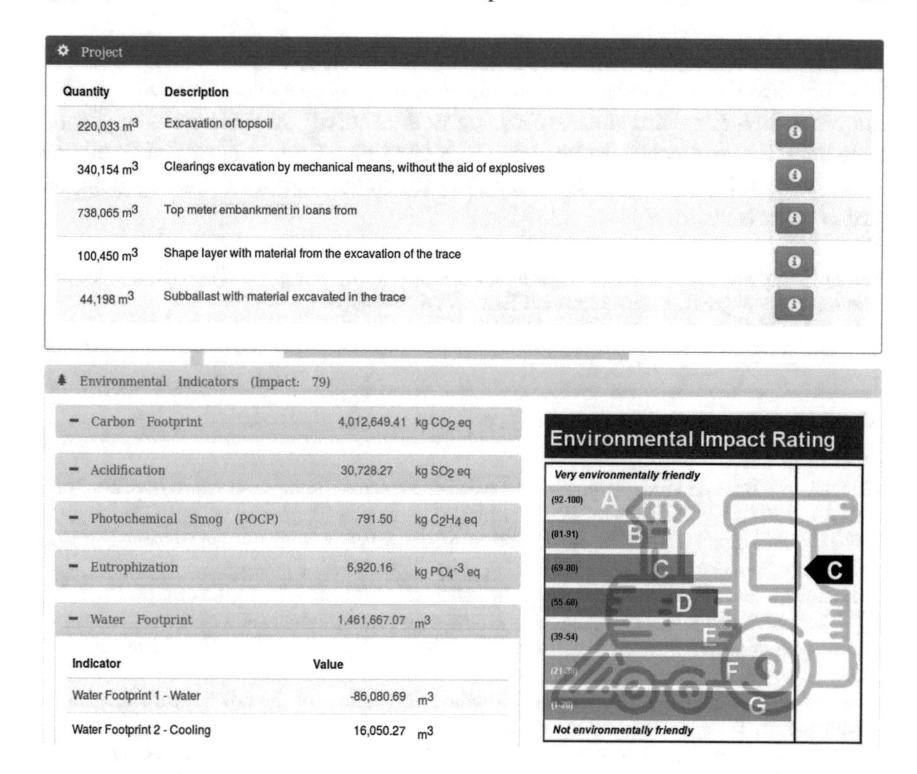

Fig. 5 Open access tool screenshot

5 Conclusions

This paper shows an in-depth analysis of rail infrastructure project units. A threefold (environmental, economic and social) analysis has been performed on more than 450 project units, considering more than 520 tasks. For every item, 27 indicators have been calculated and summarized in three values applying MIVES methodology. A multi-objective optimization has been performed on these three values, in order to find a trade-off solution.

Results have allowed the development of a series of environmental impact indicators, promoting the incorporation of environmental criteria for public bodies or bidders, prevailing the use of this or similar tools. An open version of the tool is available online, with all the information regarding LCA and Social LCA (SLCA) but the optimization capability.

Implementation of project units involve a multitude of decisions, from supplier selection to machinery or execution methods. The availability of a methodology to promote not only the calculation of the carbon and water footprints, but also the analysis of alternative project unit scheduling, would be a substantial boost in reducing emissions and impacts of railway infrastructures construction.

Further work could be done applying this methodology to phases not included in this study, see Fig. 1. Another field for improvements is on standardization of carbon and water footprints measures. According to the source, values can change for the same item. Some efforts have been made by creating a Carbon Footprint Databases Task Force OECC (2015), to reach a consensus carbon footprint value by item, but further work is necessary.

Acknowledgments This work has been partially supported by European Union LIFE+ financial instrument, under grant agreement no. LIFE12 ENV/ES/000686.

References

ADIF. (2008). Base de Precios tipo General para los Proyectos de infraestructura (BPGP), Tomo III (General type base prices for infrastructure projects, Vol. III). ADIF

AENOR: UNE EN 15804:2012 sustainability of construction works—environmental product declarations—core rules for the product category of construction products

Barrientos, F., Moral, A., Rodríguezguez, J., Martínez, C., Campo, F., Carnerero, R., Parra, M., Benítez, J., & Sainz, G. (2016, in press). Knowledge-based minimization of railway infrastructures environmental impact. *Transportation Research Procedia*. doi:10.1016/j.trpro.2016.05.032.

Botniabanan, A. B. (2010). Environmental product declaration for the rail infrastructure on the bothnia line. Technical report, Reg. no. S-P-00196 (2010)

Campo, F., Carnerero, R., Barrientos, F., Moral, A., Martínez, C., Rodríguez, J., Fernández, C., Cela, B., Benítez, J., & Parra, M. (2014). Life huellas project: Sustainability in the rail infrastructure construction sector—carbon and water footprints. In *4th Congress of Innovation on Sustainable Construction (CINCOS'14)*, Porto (Portugal)

Carbon Trust. (2011). Conversion factors: Energy and carbon conversions. 2011 update. Retrieved September 2015 from https://www.carbontrust.com/.

Caterpillar Inc. (2013). *Caterpillar performance handbook* (Vol. 43). Peoria, Illinois, U.S.A.

Coello, C. (1998). A comprehensive survey of evolutionary-based multiobjective optimization techniques. *Knowledge and Information Systems*, *1*, 269–308.

Deb, K. (1999). Multi-objective genetic algorithms: Problem difficulties and construction of test problems. *Evolutionary Computation*, *7*(3), 205–230.

Deb, K., Pratap, A., Agarwal, S., & Meyarivan, T. (2002). A fast and elitist multiobjective genetic algorithm: NSGAII. *IEEE Transactions on Evolutionary Computation*, *6*(2), 182–197.

EPD International AB. (2014). First pre-verified EPD® tool published. http://environdec.com/en/News-archive/First-pre-verified-EPD-Tool-published1/

EPD International AB. (2015). EPDs for railway transports. http://www.environdec.com/en/Articles/EPD/EPDs-for-railway-transports/

EUWI: European water initiative. (2002). Retrieved September 2015 from http://www.unep.org/GC/GCSS-VIII/EUWaterInitiative2.IWRM.doc.

Goldberg, D. (1989). *Genetic algorithms in search, optimization and machine learning*. Reading, Massachusetts: Addison-Wesley Publishing Company.

International Organization for Standardization. (2006a). ISO 14025:2006 environmental labels and declarations—type III environmental declarations—principles and procedures.

International Organization for Standardization. (2006b). ISO 14040:2006: Environmental management—life cycle assessment—principles and framework.

International Organization for Standardization. (2006c). ISO 14044:2006 environmental management—life cycle assessment—requirements and guidelines.

OECC. (2015). Oficina española del cambio climático (Spanish office for climate change) and LIFE HUELLAS project. http://www.life-huellas.eu/newsandevents.

Pré Consultants B.V. (2016). SimaPRO. https://simapro.com/

Secretaría de Estado de Comercio del Ministerio de Economía y Competitividad. (2015). DataComex: Estadísticas del comercio exterior español (TARIC: Spain foreign trade statistics). Retrieved February 2015 from http://datacomex.comercio.es/.

Simonsen, M. (2010). Indirect energy use. Retrieved September 2015 from http://sip.vestforsk.no/pdf/Felles/IndirectEnergyUse.pdf

Skinner, I., Essen, H., Smokers, R., & Hill, N. (2010). Towards the decarbonisation of EUs transport sector by 2050. Final report produced under the contract ENV.C.3/SER/2008/0053 between European commission directorate-general environment and AEA technology plc. Technical report. http://www.eutransportghg2050.eu.

Thinkstep: Gabi lca software. http://www.gabi-software.com/

UIC. (2011). International union of railways. http://www.uic.org

UNEP. (2009). Guidelines for social life cycle assessment of products; social hotspot database; ISO 21929-2: Draft on sustainability in buildings and civil engineering works—sustainability indicators—part 2: Framework for the development of indicators for civil engineering works.

Viñolas, B., Cortés, F., Marques, A., Josa, A., & Aguado, A. (2009). MIVES: Modelo integrado de valor para evaluaciones de sostenibilidad. In *Congrés Internacional de Mesura i Modelització de la Sostenibilitat. II Congrés Internacional de Mesura i Modelització de la Sostenibilitat*. Barcelona: Centro Internacional de Métodos Numéricos en Ingeniería (CIMNE) (pp. 1–24). Retrieved September 2015 from http://hdl.handle.net/2117/9704. (in Spanish)

Zitzler, E., Deb, K., & Thiele, L. (2000). Comparison of multiobjective evolutionary algorithms: Empirical results. *Evolutionary Computation*, *8*, 173–195.

Part VII
Health Systems

A Computational Intelligence Approach to Diabetes Mellitus and Air Quality Levels in Thessaloniki, Greece

Kostas Karatzas, Vassiliki Dourliou, Nikolaos Kakaletsis, Nikolaos Katsifarakis, Christos Savopoulos and Apostolos I. Hatzitolios

Abstract We employ Computational Intelligence (CI) methods to investigate possible associations between air pollution and Diabetes Mellitus (DM) in Thessaloniki, Greece. Models are developed for describing key DM parameters and for identifying environmental influences to patient status. On this basis new, more accurate models for the estimation of renal function levels are presented while a possible linkage is indicated concerning disease parameters and the quality of the atmospheric environment.

Keywords Computational intelligence · Air pollution · Diabetes mellitus

1 Introduction

The relationship between air pollution and increased risk of diabetes mellitus (DM) has been discussed in related studies (Chen et al. 2016; Papazafiropoulou et al. 2011; Solomini et al. 2015). DM is a complex, chronic illness characterized by hyperglycemia resulting from defects in insulin secretion, insulin action, or both. The chronic hyperglycemia of diabetes is associated with long-term damage, dysfunction, and failure of various organs, especially the eyes, kidneys, nerves, heart, and blood vessels (American Diabetes Association 2014). Air pollutants might deteriorate glucose metabolism and insulin sensitivity by oxidative stress and inflammation (Eze et al. 2016). The investigation of any dependency between Air

K. Karatzas (✉) · N. Katsifarakis
Department of Mechanical Engineering, Informatics Systems and Applications – Environmental Informatics Research Group, Aristotle University, 54124 Thessaloniki, Greece
e-mail: kkara@eng.auth.gr

V. Dourliou · N. Kakaletsis · C. Savopoulos · A.I. Hatzitolios
First Propaedeutic Department of Internal Medicine, Aristotle University, AHEPA Hospital, 54636 Thessaloniki, Greece

V. Wohlgemuth et al. (eds.), *Advances and New Trends in Environmental Informatics*, Progress in IS, DOI 10.1007/978-3-319-44711-7_20

Quality (AQ) and DM is still open (Tamayo et al. 2016) and thus contributions on this direction are of importance.

Coming to DM complications, a critical parameter describing its impacts to patients is the Glomerular Filtration Rate (GFR) that is the overall index of kidney function (National Kidney Foundation 2016). Several equations based mainly on serum creatinine values (another medical parameter estimated via blood tests) have been suggested, while also taking into account additional demographic and clinical variables, in order to more reliably estimate GFR. In the frame of the current study we aimed to:

(a) Develop a new model that could more effectively estimate the GFR levels, incorporating several laboratory and clinical characteristics as well as AQ levels if appropriate and
(b) Investigate the association between air pollution and metabolic/glycemic profile in adult patients with type 2 DM in the area of Thessaloniki, Greece.

2 Materials and Methods

2.1 Data Presentation

Data coming from 193 consecutive type 2 DM patients, followed-up at the Diabetes Outpatient Clinic of the First Propaedeutic Department of Internal Medicine, AHEPA Hospital, Thessaloniki, Greece, were used in this prospective study. Several (57) laboratory and clinical characteristics have been recorded as well as the home address for each patient. In addition, air quality data coming from the two monitoring networks covering the Greater Thessaloniki Area (GTA) were used, namely the network of the Ministry of Environment (7 stations, Prefecture of Central Macedonia 2016) as well as the network operated by the Department of Environment of the Municipality of Thessaloniki (7 stations, Thessaloniki Open Data 2016). All data values were numerical, while the transformation of the GFR values to categorical was based on the definition and classification of stages of chronic kidney disease of the National Kidney Foundation 2016 (Table 1).

Table 1 Stages of chronic kidney disease based on GFR levels. Severity ranges from low, stage 1, to maximum, stage 5 (National Kidney Foundation 2016)

Stage (from lower-1 up to higher-5)	GFR level
Stage 1	≥90
Stage 2	60–89
Stage 3	30–59
Stage 4	15–29
Stage 5	<15

2.2 Methods Presentation

The home addresses of patients were used for estimating the relevant latitude and longitude (Fig. 1) in order to calculate the Euclidean distance between patient residence (***x***) and the nearest AQ monitoring station (***y***), according to Eq. X.1:

$$d(s,y) = \sqrt{\sum_{i=1}^{2} (x_i - y_i)^2} \qquad \text{(X.1)}$$

This was done in order to take into account possible air pollution influences to patient status in terms of DM-related parameters, as the representativeness of AQ monitoring stations concerning atmospheric quality is considered to decrease with increased distance from their location (Spangl et al. 2007).

The time period of interest was 3 months from the date stamp of each patient record (i.e. from the time that all clinical and laboratory data were accessed per patient), as this is the time frame regarded as representative for estimating DM parameters like the glycated hemoglobin (HbA1c), a form of hemoglobin that is measured primarily to identify the 3 month average plasma glucose concentration. On this basis, each one of the 193 patient records was enriched with data describing the mean, max, min, standard deviation, 25th and 75th percentile values of the four main pollutants affecting AQ levels in Thessaloniki, Greece (Table 2), leading to a total of 24 AQ parameters. Thus a common data matrix ***X*** was formulated,

Fig. 1 Epidemiological map of the 193 consecutive type 2 DM patients

Table 2 Basic air pollutants in the Greater Thessaloniki Area (Voukantsis et al. 2011)

Pollutant	Basic characteristics
PM_{10}	Particulate matter (PM) are microscopic solid or liquid matter suspended in the Earth's atmosphere. PM_{10} is particulate matter with a mean aerodynamic diameter of 10 μm and is emitted by traffic, industrial processes and nature
NO_2	Nitrogen dioxide is an intermediate in the industrial synthesis of nitric acid, millions of tons of which are produced each year. It is commonly found in urban areas as a byproduct of combustion (internal combustion engines) and industrial emissions
CO	Carbon monoxide is spatially variable and short lived, having a role in the formation of ground-level ozone, while emitted by traffic and combustion processes
O_3	Ozone is a secondary pollutant formatted as a result of the changes in the chemical equilibrium in the atmosphere, caused by other pollutants like hydrocarbons, nitrogen oxides and carbon monoxide. It is commonly found away from dense traffic, thus sometimes characterizing the atmosphere of suburbs of major cities

including all patient (a total of 57) and AQ (a total of 24) values, consisting of 193 rows (one for each patient) and 81 columns (one for each parameter). The calculation and use of the percentile values was done on the basis of the hypothesis that a potential influence of AQ to DM parameters, if existing, might be based on a mechanism affected by a "threshold" of the air pollutants concentration and not only by their basic descriptive statistics.

Concerning the investigation of possible relationships between AQ and DM, the large number of parameters has led us to choose a method being capable of (i) reducing the dimensionality of the data matrix and (ii) identifying the most influential parameters in terms of their relevant importance, while preserving as much as possible of the variation present in the data set (Solimini et al. 2015). For this reason we selected Principal Components Analysis (PCA), aiming at identifying and prioritizing those factors that describe glycated hemoglobin (HbA1c), a key DM parameter, in the "best" possible way. "Best" in this context refers to the ability of the method to express the majority of the variance of the dataset in terms of the parameter of interest. PCA originates from multivariate statistical analysis and allows for the identification of the major drives (measured as a percentage of the overall variance representation) within a certain multidimensional data set and thus may be applied for data compression as well as for identifying patterns in data.

By applying the PCA, we estimate the eigenvalues λ_i, and the corresponding eigenvectors a_i, by solving the following equations respectively:

$$|A - \lambda I| = 0 \tag{X.2}$$

$$(\mathrm{A} - \lambda_i I) \cdot a_i = 0, \quad i = 1, \ldots, N \tag{X.3}$$

where A is the correlation coefficient matrix of the data matrix $\boldsymbol{X}$ including the 193 records of the N (here 81) parameters and. The result is a new set of uncorrelated variables, the principal components (PCs), sorted by the percentage of the original variance they account for. By selecting the most significant PCs we reduce the original dataset's dimensions and we thus select the most important parameters. We applied the PCA method with the aid of the Matlab computational environment, in order to identify the relevant contribution of each attribute to the calculated PCs.

2.3 Model Formulation

Regarding the modelling and forecasting of DM-related parameters, we focused on GFR as this is a critical parameter describing the severity of kidney malfunction, and we employed Weka version 3.7 (Hall et al. 2009) enriched with a number of additional regression and classification algorithms, in an effort to enhance the population of algorithms to be tested. More specifically, the algorithms employed were:

(a) Classification problem (GFR treated as categorical parameter). All available via Weka (more than 30) algorithms from classifier categories like Bayes, functions, meta-algorithms, rules, decision trees were tested. On the basis of their performance, results from the following are presented in the next chapter:

- *Averaged One Dependence Estimators (AODE),* originating from the Naive Bayes (NB) rule and a set of conditional independence assumptions (the weak point of the NB approach). AODE is a probabilistic classification learning method that strives to improve its accuracy by overcoming the deficiencies of its attribute independence assumption, via averaging over a space of alternative naive-Bayes-like models that have weaker (and hence less detrimental) independence assumptions than NB. The resulting algorithm is computationally efficient while delivering accurate classification (Webb et al. 2005).
- *MultiboostAB*, an algorithm belonging to the meta-algorithms category, combining different individual algorithms by optimizing their overall performance (Freund and Schapire 1997). MultiBoosting is an extension to the highly successful AdaBoost technique for forming decision committees. MultiBoosting can be viewed as combining AdaBoost with wagging (Webb 2000).

(b) Regression problem (GFR treated as a numerical parameter). Various (approx. 20) algorithms were tested, belonging to the categories of functions, Bayes, meta-algorithms, rules and decision trees. The best performing algorithm was found to be M5P, which is a decision tree algorithm. M5P is a reconstruction of Quinlan's M5 algorithm for inducing trees of regression models and combines a conventional decision tree with the possibility of linear regression functions at the nodes (Quinlan 1992). In M5P a model tree is used for numeric prediction and at each leaf a linear regression model is stored that predicts the class value of instances that reach the leaf. Two variations of the

Table 3 Value ranges of Cohen's Kappa according to Altman (1991)

Kappa statistic	Agreement strength
<0.20	Poor
0.210–0.40	Fair
0.41–0.60	Moderate
0.61–0.80	Good
0.81–1.00	Very Good

M5P algorithm were used, based on two different attribute selections, leading to two models, as reported in the results section.

The training and validation of the aforementioned models was made with the aid of a tenfold cross validation approach (Kovahi 1995). The evaluation metrics employed for choosing the best performing models were (a) the Pierson's correlation coefficient and the Root Mean Square Error (RMSE) for the regression-oriented modelling and (b) the Cohen's kappa (Cohen 1960, as described in Table 3), accompanied by the percentage of correctly classified instances, the Mean Absolute Error (MAE), the RMSE and the confusion matrix for the cases where GFR was treated as a categorical parameter (classification problem, Witten et al. 2011).

3 Results and Discussion

3.1 Relationship Between AQ and DM

The conducted analysis resulted in a number of PCs that were ranked according to the percentage of the overall data variance they describe. For each PC, the relevant participation of each attribute (i.e. parameter) of the initial dataset was described as a real value coefficient within the range $[-1, 1]$, all attributes having a coefficient value of the same sign thus considered to have a "direct" relationship while those having coefficients of different signs are considered to have an "inverse" relationship.

The PCA results indicate that the following parameters have the largest participation in the overall formation of the strongest PC (PC#1, approx. 21 % of the overall variance): *GFR*, *HDL cholesterol*, *total cholesterol* and the *25*th *percentile of Ozone*; these parameters demonstrate a similar behavior: the increase of the value of one of them is expected to occur in parallel with a similar increase in the value of the others. Furthermore, GFR demonstrates a different behavior in relation to *Cystatin*, *Urea*, *Creatinine Serum* and *Uric Acid*: when GFR is reduced (leading to worse kidney function) these parameters are increased. As high GFR values are associated with better kidney functionality in comparison to low values, its common pattern of behavior with the 25th percentile of Ozone might be related to the fact that *those patients in our study having a better kidney function are mostly living*

Table 4 Performance metrics of the two best models concerning the correct classification of the patients targeting the categorical GFR values

Indices	Algorithms	
	bayes.AODE	MultiBoostAB
Cohen's kappa	0.60	0.59
Correctly classified instances	82	82
MAE	0.21	0.19
RMSE	0.38	0.39
Confusion matrix	a b <– classified as 114 18	a = high 16 45	b = low	a b <– classified as 117 15	a = high 19 42	b = low

in areas with low traffic, away from the city center, i.e. areas with lower NOx emissions. As low GFR is a sign of problematic DM control, reflected by high HbA1c values, the aforementioned result is in line with the (weak) inverse association of accumulated ozone with HbA1c reported by Tamayo et al. 2016.

In addition, the following parameters demonstrate a strong participation in the 1st PC and a strong association: *HbA1c*, *Total Cholesterol*, *Triglycerides*, *Glucose* and the *25th percentile of CO*. This result suggest that the value of HbA1c may be influenced by the 25th percentile of Carbon Monoxide, a pollutant mainly attributed to traffic, as already discussed in some of the studies reviewed by Teichert and Herder (2016).

3.2 *Modelling of GFR as a Categorical Parameter*

GFR modelling initially focused on the 5 levels reported in Table 1, yet with poor results. In order to increase the validity of the modelling approach, the number of GFR value ranges was decreased from five to two,[1] on the basis of a medically accepted distinction between low and high levels of kidney malfunction, thus dividing the initial set of 193 patients to 61 having high levels of kidney malfunction (and thus low GFR values), and 132 patients having low levels of kidney malfunction (and thus high GFR values). Results for the best two algorithms are reported in Table 4 and discussed below:

(i) The bayes.AODE algorithm correctly classified 159 out of the 193 patients, while the MultiboostAB algorithm correctly classified 159 out of the 193 patients in terms of low or high GFR levels.

[1] 1st *case (low levels of kidney malfunction)*: corresponds to stage 1 and stage 2 (GFR >= 60) 2nd *case (high levels of kidney malfunction)*: corresponds to stages 3, 4 and 5 (GFR < 60).

Table 5 Comparison of the four available GFR models with the newly developed models for the numerical estimation of GFR

Statistics	MDRD GFR	CKD-EPI creatinine (2009)	CKD-EPI creatinine-cystatin C (2012)	CKD-EPI cystatin C (2012)	M5P model #1	M5P model #2
R	0.76	0.79	0.78	0.69	0.829	0.828
RMSE	22.70	10.70	0.30	8.30	16.398	6.103

(ii) Both algorithms demonstrate a moderate to good performance in terms of the Cohen's kappa, and an equivalent performance in terms of correctly classified instances, MAE and RMSE.

(iii) AODE leads to slightly better results when focusing on the estimation of the (more critical) low GFR values (correctly classified 45 out of 61 vs. 42 out of 61)

(iv) MultiboostAB is slightly better in correctly classifying the (less critical) high GFR values (117 out of 132 in comparison to 114 out of 162 for AODE).

3.3 *Modelling of GFR as a Numerical Value*

Several models exist concerning the arithmetic estimation of the GFR on the basis of other laboratory and medical data (Levey et al. 1999, 2006, 2009; Stevens et al. 2008; Inker et al. 2011, 2012). Our aim was to make use of CI algorithms in order to develop models that outperform existing ones. For this reason we tested a large number of algorithms, and we compared their performance with the performance of four standard GFR models made available via literature, in terms of Pierson's correlation coefficient and RMSE. Table 5 presents the performance of the two best models that we developed, together with the performance of the four standard GFR models.

It is evident that the new models, based on the M5P algorithm, have a better correlation coefficient by ~5 % (0.829 in comparison to 0.79), while their RMSE is the 2nd lower in comparison to the existing ones (6.1). Thus, the new models slightly outperform standard ones and provide with a promising outcome that needs to be further developed, tested and evaluated.

4 Conclusions

The analysis of patient data concerning the association between DM and AQ for the area of Thessaloniki, Greece, suggests complicated relationships between the two. Thus AQ parameters like Ozone (that correspond to types of residential areas with low-traffic related air emissions, away from the city center) seem to have a "positive" effect on GFR (a DM-dependent parameter), while other AQ parameters like CO (associated with high traffic-induced air emissions) seem to have a "negative" effect on HbA1c. These results suggest that further investigation is required concerning DM and AQ associations. In addition, the computational experiments towards the development of critical DM parameter models as in the case of GFR, suggest that (a) it is possible to come up with models that can be used to support the categorization of the GFR-related kidney conditions in more than 80 % of the patients examined and (b) it is also possible to make use of CI algorithms for developing models that are more accurate in the estimation of the numerical value of the GFR parameter in comparison to existing ones, and thus be more useful from the point of view of patient management, regarding the same kidney operation parameter (GFR). Current research outcomes indicate that it is worthwhile to examine the above findings in a larger cohort, taking into account limitations like the selection bias of patients to be included in such studies.

References

Altman, D. G. (1991). *Practical statistics for medical research*. London: Chapman and Hall.

American Diabetes Association. (2014). Diagnosis and classification of diabetes mellitus. *Diabetes Care, 37*(Suppl 1), S81–S90. doi:10.2337/dc14-S081.

Chen, Z., et al. (2016). Ambient xin Mexican Americans. *Diabetes Care,* pii: dc151795. Retrieved June 08, 2016 from http://dx.doi.org/10.2337/dc15-1795.

Cohen, J. (1960). A coefficient of agreement for nominal scales. *Educational and Psychological Measurement, 20*(1), 37–46. doi:10.1177/001316446002000104.

Eze, et al. (2016). A common functional variant on the pro-inflammatory Interleukin-6 gene may modify the association between long-term PM10 exposure and diabetes. *Environmental Health, 15*, 39. doi:10.1186/s12940-016-0120-5.

Freund, Y., & Schapire, R.E. (1997). A decision-theoretic generalization of on-line learning and an application to boosting. *Journal of Computer and System Sciences, 55*(1), 119–139. doi: 10.1006/jcss.1997.1504.

Hall, M., et al. (2009). The WEKA data mining software: An update. *SIGKDD Explorations, 11* (1), 10–18. Retrieved June 08, 2016 from http://www.kdd.org/exploration_files/p2V11n1.pdf.

Inker, L. A., et al. (2011). Expressing the CKD-EPI (chronic kidney disease epidemiology collaboration) cystatin C equations for estimating GFR with standardized serum cystatin C values. *American Journal of Kidney Diseases, 58*(4), 682–684. doi:10.1053/j.ajkd.2011.05.019.

Inker, et al. (2012). Estimating glomerular filtration rate from serum creatinine and cystatin C. *New England Journal of Medicine, 367*(1), 20–29. doi:10.1056/NEJMoa1114248.

Kohavi, R. (1995). A study of cross-validation and bootstrap for accuracy estimation and model selection. In *Proceedings of the Fourteenth International Joint Conference on Artificial Intelligence* (Vol. 2, No. 12, pp. 1137–1143). San Mateo, CA: Morgan Kaufmann.

Levey, A. S., et al. (1999). A more accurate method to estimate glomerular filtration rate from serum creatinine: A new prediction equation. Modification of Diet in Renal Disease Study Group. *Annals of Internal Medicine, 130*, 461–470.

Levey, A. S., et al. (2006). Using standardized serum creatinine values in the modification of diet in renal disease study equation for estimating glomerular filtration rate. *Annals of Internal Medicine, 145*, 247–54.

Levey, et al. (2009). A new equation to estimate glomerular filtration rate. *Annals of Internal Medicine, 150*, 604–612.

National Kidney Foundation: Glomerular filtration rate. Retrieved June 08, 2016 from https://www.kidney.org/professionals/kdoqi/gfr.

Papazafiropoulou, A., Kardara, M., & Pappas, S. (2011). Environmental pollution and diabetes mellitus. *Recent Patents on Biomarkers, 1*, 44–48. doi:10.2174/2210309011101010044.

Prefecture of Central Macedonia: Air quality report 2013. Retrieved June 08, 2016 from http://www.pkm.gov.gr/default.aspx?lang=el-GR&page=507.

Quinlan, R. (1992). Learning with Continuous Classes. In A. Adams & L. Sterling (Eds.), *Proc. AI'92, 5th Australian Joint Conference on Artificial Intelligence* pp. 343–348. Singapore: World Scientific.

Solimini, A., D'Addario, M., & Villari, P. (2015). Ecological correlation between diabetes hospitalizations and fine particulate matter in Italian provinces. *BMC Public Health, 15*, 708. doi:10.1186/s12889-015-2018-5.

Spangl, W., Schneider, J., Moosmann, L., & Nagl, C. (2007). Representativeness and classification of air quality monitoring stations, draft final report. Retrieved June 08, 2016 from http://ec.europa.eu/environment/air/quality/legislation/pdf/report_uba.pdf.

Stevens, L.A., et al.: Estimating GFR using serum cystatin C alone and in combination with serum creatinine: A pooled analysis of 3,418 individuals with CKD. *American Journal of Kidney Diseases, 51*, 395–406 (2008). doi:10.1053/j.ajkd.2007.11.018.

Teichert, T., & Herder, Ch. (2016). Air Pollution, Subclinical Inflammation and the Risk of Type 2 Diabetes. In: Ch. Esser (Ed.), *Environmental Influences on the Immune System* pp. 243–271. Springer, Vienna. doi: 10.1007/978-3-7091-1890-0_11.

Thessaloniki Open Data: Thessaloniki municipality air quality measurements 1989–1999, 2000–2009, 2010–2013. Retrieved June 08, 2016 from http://opendata.thessaloniki.gr/.

Tamayo, T., et al. (2016). No adverse effect of outdoor air pollution on HbA1c in children and young adults with type 1 diabetes. *International Journal of Hygiene and Environmental Health, 219*(4–5), 349–355. doi:10.1016/j.ijheh.2016.02.002.

Voukantsis, D., Karatzas, K., Kukkonen, J., Räsänen, T., Karppinen, A., & Kolehmainen, M. (2011). Intercomparison of air quality data using principal component analysis, and forecasting of PM10 and PM2.5 concentrations using artificial neural networks, in Thessaloniki and Helsinki. *Science of the Total Environment, 409*, 1266–1276. doi:10.1016/j.scitotenv.2010.12.039.

Webb, G. I. (2000). MultiBoosting: A technique for combining boosting and wagging. *Machine Learning, 40*, 159–196. doi:10.1023/A:1007659514849.

Webb, G. I., Boughton, J., & Wang, Z. (2005). Not so naive Bayes: Aggregating one-dependence estimators. *Machine Learning, 58*(1), 5–24. doi:10.1007/s10994-005-4258-6.

Witten, I., Frank, E., & Hall, M. (2011). *Data mining: Practical Machine Learning Tools and Techniques*. Morgan Kaufmann Publishers. ISBN: 978-0-12-374856-0.

Aggregation and Measurement of Social Sustainability and Social Capital with a Focus on Human Health

Andi H. Widok and Volker Wohlgemuth

Abstract This paper addresses theoretical differences between the perceived content of social sustainability and different socially attributed capital stocks, as for example, rather transaction oriented so called social capital, human capital or societal capital. It does so, by mapping different aspects from existing capital approaches to social sustainability, in order to resolve theoretical misconceptions. In addition, the review is intended to highlight different approaches for the measurement and aggregation of social aspects. Consequently the balance within sustainability approaches is addressed and strengthened by elaborating ways to formalize the social perspective. Furthermore, technical solutions considering the resulting impact assessment of social aspects are discussed and in order to demonstrate the significance of the presented ideas, an own application of a threshold based aggregation method, in a manufacturing simulation software, is presented.

Keywords Social sustainability · Social capital · Occupational health and safety (OHS)

1 Introduction and Motivation

For a few years now research has been done at the HTW Berlin to integrate a social dimension to the simulation of manufacturing entities (Widok and Wohlgemuth 2013, 2014, 2015, 2016). In that regard, measurement and aggregation approaches

A.H. Widok (✉) · V. Wohlgemuth
Industrial Environmental Informatics Unit, HTW Berlin, Wilhelminenhof Str. 75 A, 12459 Berlin, Germany
e-mail: a.widok@htw-berlin.de

V. Wohlgemuth
e-mail: volker.wohlgemuth@htw-berlin.de

V. Wohlgemuth et al. (eds.), *Advances and New Trends in Environmental Informatics*, Progress in IS, DOI 10.1007/978-3-319-44711-7_21

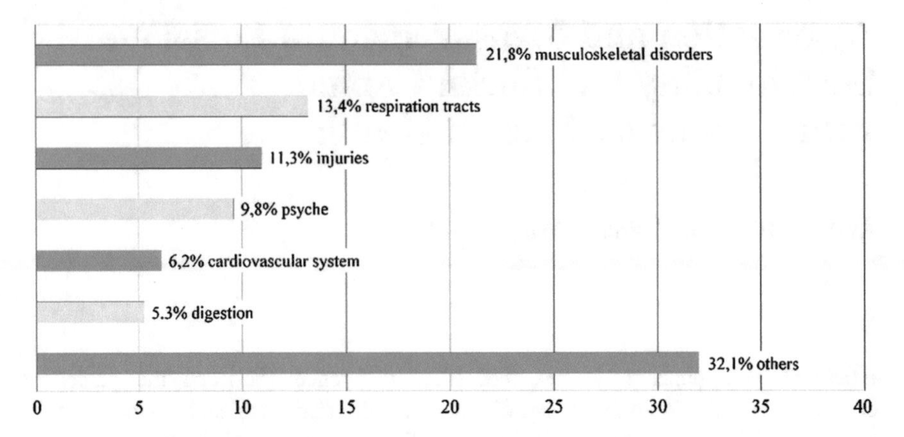

Fig. 1 Proportion of inability to work days based on relevant effects (cf. Badura et al. 2014)

of social aspects have been of special interest, in order to formalize social impacts and implement technical systems able to model and simulate these. While the implementation of a system, with the ability to model and simulate social influences, was challenging, the methodical research of valid aggregation methods was even more so and mainly intended to facilitate a quick way to compare processes based on key figures, representing the three sustainability dimensions. This paper discusses the chosen approach.

To elaborate on our motivation, one can consider current data of inability to work days in the industry; a significant part of these is attributed to musculoskeletal disorders, which are often a result of the exceedance of biomechanical exposure thresholds (Badura et al. 2010, 2014; Neumann et al. 2006). In addition, a strong rise of psychological factors is to be acknowledged (Busch et al. 2014). See Fig. 1 for a brief overview in Germany:

In that regard, it has to be recognized that the proof of unspecific, work related impacts is, in individual cases, especially challenging, as there are a lot of correlations to be considered (Biendarra and Weeren 2009). For example skin related, or allergic reactions have a variety of possible influence factors, so do respiratory diseases (cf. Biendarra and Weeren 2009). Psychological influences, as for example stress, have to be considered as well, especially as contrary to physical influences, their impact is increasing in the last decades (Busch et al. 2014; Bamberg 1991, 2003).

Being perceptive of these influences on health, the main goal of our social domain for manufacturing systems, was to be able to quantify similar health related aspects, while in the same time addressing the question about the place of them in a social sustainability perspective. The second goal follows the intention to allow a qualification of measures by attributed sustainability indicators. The following section will elaborate on this, as well as on related literature and current approaches.

2 Literature Review and Theoretical Discussion of Social Capital, Social Sustainability and Aggregation Approaches

The number of current approaches that try to measure or at least evaluate social aspects is vast, while health is only one of the considered aspects. Author Dubielzig argues that the great number of approaches itself is already an indication that the validity of the current assessments has to be questioned (Dubielzig 2009).

In order to clarify what currently is considered to be measurable and attributed to the social sustainability perspective, social capital initiatives were considered. Capital approaches are a rather established way of measuring and aggregating related subject matter, i.e., attributed "capital", while being in accordance with many sustainability definitions, that there is a underlying "capital" that should be protected for further generations (Conrad 2007; Grootaert and van Bastelaer 2002). In that regard, it has to be acknowledged that different authors see the application of such assessment approaches (at least considering environmental and social implication) critically and argue that there is a need for less economically oriented/originated methods (von Hauff 2015; Pezzey and Toman 2002, cf. Möller 2010).

As mentioned in the introduction the underlying capital of the social pillar of sustainability does not correspond with the notion of what is considered social capital. This is based on long theoretical discussions considering the function and content of social capital itself, which is related to the interactions between individuals, rather than to for example their health (which is attributed to human capital). Author Conrad dates these discussions back to the 1916s (Conrad 2007), and while it is noteworthy that especially during the two world wars, much research has been conducted considering the productivity of individuals (with accorded realizations that working less can actually create higher output) (cf. Conrad 2007), the most referenced and likely most relevant contributions to the definition and understanding of social capital have been provided by Bourdieu (1986), Coleman (1988), Putnam (1995) and Fukuyama (2001). All four have formulated specific ideas with relevance to the definition and attributed subject matter. Without going into too much details (a summary can be found in Grootaert and van Bastelaer 2002), for the following social capital will be understood in accordance to Putnam's definition that specifies

> features of social organization such as networks, norms, and social trust that facilitate coordination and cooperation for human benefit (Putnam 1995)

and Fukuyama's "addition" that

> these norms must be instantiated in an actual human relationship (…) (Fukuyama 2001).

This definition implies that what is considered social capital does not encompass human health, which is rather attributed to, so called, human capital (cf. Badura et al. 2010). In this respect, it also has to be addressed, that not all definitions of human capital highlight human well-being; some rather focus on the capacity of execution of basic functions, as well as knowledge, in terms of beneficial abilities for accorded organizations (cf. critically reviews due to economic orientation). Nonetheless one can come to the realization that what is considered the social pillar of sustainability, actually tries to encompass social impacts and by doing so addresses not only the above mentioned social capital, but also human capital and societal/institutional capital (which can be understood in reference to Spangenberg's definition of an institutionalized dimension of sustainability (cf. Spangenberg 1997). Furthermore, it addresses the change of nature capital and the resulting impact it has on humans (measured in human/social capital, or via social impact assessment SIA) and their societies (societal/structural capital), as well as changes of economical capital with similar respective impacts.

In summary to clarify why the existing definitions of social capital were not simply adapted for our sustainability aggregation approach, one has to consider four conceptual difficulties. First, social capital is not in correspondence with the definition of the underlying capital of social sustainability (rather contrary to economic sustainability and economic capital, as well as environmental sustainability and nature capital). Second, the definitions are still rather incoherent in the literature and consequently so are the accorded indicators as well as the resulting assessment approaches (cf. Dubielzig 2009). Third, the transaction oriented social capital is very hard to resolve formally, due to a variety of unknown correlations (cf. Dubielzig 2009; Baumanns 2009), even though statistical examples exist (Baumanns 2009). In comparison, for human capital alone, there are a few different measurement and aggregation approaches in the given scope of research (Neumann et al. 2006; Takala et al. 2013). Lastly, some definitions of social capital strongly highlight the (economical) transaction enabling nature of social capital, i.e., the solely perception of social capital as access building element, enabling economical gain, rather than the focus on well-being of humans. This economic understanding of social capital is seen critically by numerous authors, see also Badura et al. (2010), Dubielzig (2009), Bourdieu (1986), and Fukuyama (2001).

Upon acceptance of the very different aspects of social sustainability, consequently questions arose how these aspects could be compared and aggregated. While the capital approaches enabled a first categorization, the aggregation of attributed impacts was still unresolved and resulted in a stronger research focus on aggregation methods, regarding social sustainability and sustainability in general, which will be elaborated in the following.

In order to understand measurement and aggregation methods, one should first consider the main strategies of achieving/strengthening sustainability. Many authors make in this regard the differentiation between three strategies, namely efficiency, sufficiency and consistency strategies (Thiede 2012; Langer 2011), (cf. Dubielzig 2009). Figure 2 visualizes these, in respect to their impact in the economic and environmental dimensions:

In accordance with the most common definitions of sustainability, the different strategies focus on the preservation of underlying capital, while in the same time acknowledging the effects, different measures have in the different dimensions (as

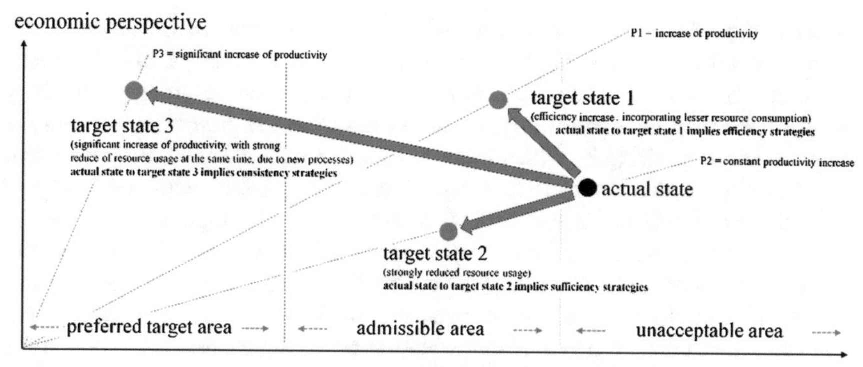

Fig. 2 Strategies of sustainability (cf. Thiede 2012)

depicted above for the environmental and economic dimension) (Thiede 2012; Langer 2011). In order to formulate such a relation, as the one above, there is a need for an indicator which represents the aggregated value of the given dimension and for an object under observation (which may be a single process step, a measure or a life cycle). For the economic dimension, this is naturally a financial value, which represents the affected financial capital under observation. In the environmental perspective a lot of research has been conducted over the past decades, with a few resulting established aggregation methods. For example LCA approaches have established themselves for the evaluation of different process steps in the lifecycle of, e.g., products. The resulting environmental impacts are usually measured in CO_2 equivalents, and aggregated to form a single value in the end, such as for example the global warming potential (GWP), or the environmental impact points ("Umweltbelastungspunkte" UBP). While the CO_2 equivalents should be questioned regarding their missing qualification of biodiversity, or some forms of land use impact and accorded impact on nature capital, the general concept of creating reference values, is supposed to help with an easier qualification of processes and hence reinforce sustainable measures, according to the strategies formulated above.

For the social dimension this form of aggregated value is still lacking conceptual clarity (Omann and Spangenberg 2002; Colantonio 2007) and consequently acceptance among involved stakeholders. This is mainly based on the abstract, non-material, and so called unmeasurable nature of many social effects (cf. Dubielzig 2009; Badura et al. 2010). While we do not argue, that social aspects are harder to measure, we intuitively feel that their immaturity considering data quality and absence of correlation knowledge, is comparable as to what used to be the case for environmental data, two decades earlier. Analysis of aspects that contribute to this factor can be found in Dubielzig (2009), Omann and Spangenberg (2002), and Benoît and Vickery-Niederman (2010).

To give but one example about the main difficulty, one can consider three different social impacts, the first consisting of a physical task influence on a single human being, which may, to an extent, have impact on his health. The second could be the interaction between the same human being and a group of his coworkers, which may or may not have influences on his motivation and resulting productivity and the third may be a social impact that results from the production process the human being is working in, which has influences on nature capital in a region and this possible degradation of nature capital has a correlation with the health of regional human beings. These three very different social impacts should all be addressed when considering, for example, the manufacturing entity in this scenario. Consequently the question arises, on how to formulate an aggregation of these impacts, in order to come to a similar value as the GWP for the social dimension.

As mentioned there is a variety of methods surrounding the assessment and more specifically, the management of social criteria. Generally these are categorized as International Policy Frameworks, Codes of Conduct and Principles, Sustainability Reporting Frameworks, SR Implementation Guidelines, Auditing and Monitoring Frameworks and Financial Indices (Benoît and Vickery-Niederman 2010; Dubielzig 2009).

The most prominent examples for standards are likely the renewed ISO 26000 (Corporate Social Responsibility (CSR) specification and organizational guidelines) (DIN Deutsches Institut für Normung 2011), and the GRI G4 sustainability guidelines (Global Reporting Initiative 2013). It is noteworthy in this context that the ISO 26000 understands and clearly describes itself as enabler of sustainability (DIN Deutsches Institut für Normung 2011) and thus understands the introduction of measures, which lead to a strengthening of social aspects, as measures that are directly linked to a more sustainable functioning of organizations. This is in accordance with the understanding of the depicted strategies in Sect. 2.2. Furthermore, the ISO 26000 specifies, in accordance with other social assessment methods, such as social impact assessment (SIA) and human rights impact assessment (HRIA), the definition of stakeholders, as general approach to associate impacts to different groups and to assess relevant correlations (Benoît and Vickery-Niederman 2010; Jørgensen et al. 2008). This concept is comparable with the scope definition and resulting stakeholder definition of S-LCA (which also points to the conceptual importance of Fukuyama's "circles of trust" (Fukuyama 2001)). Benoît and Vickery-Niedermann highlight in this regard however, that good practices of SIA imply quantitative, as well as qualitative indicators, as SIA normally goes beyond only quantifiable indicators (cf. Benoît and Vickery-Niederman 2010. This also highlights the problems of quantification for some social impacts and thus leads again to the question of aggregation methods.

One approach that addresses the combination of indicators that are hard to compare is the benefit-analysis, even though it is often used from an economic point of view, i.e., as cost-benefit-analysis. The main idea behind this approach is that the weighting of categories and indicators is done manually and represents subjective qualifications, in reference to the object of investigation (VDI (Verein Deutscher Ingenieure) 2013). Similar aggregation methods are used in numerous assessment framework approaches (Odeh 2013). While the general qualification, according to the object of investigation, is usually implied, the general approach does not represent a comparable concept, considering the aggregation of social aspects, as individual weightings for different cases will not correspond to each other. In that regard, other methods were researched, especially in the occupational health and safety sector. The resulting findings and the chosen aggregation approach will be presented in the following.

3 Threshold Based Aggregation Method

Occupational health and safety (OHS) has a long tradition in addressing aspects of human health, which were/are hardly quantifiable. Especially psychological impacts are, to this day, usually only qualified in terms of positive or negative impact on the human psyche (with exceptions and possible gradations). Due to this long tradition and familiarity of OHS with hardly quantifiable indicators, a multitude of OHS assessment methods were researched for similarities in measurement and consequent handling of data, see Neumann et al. (2006), (Fritz 2001), and Westgaard and Winkel (2011) for examples.

As a result, the commonly incorporated definition of normative values and individual influence variables, appeared similar to the needed formulation of normative values, for the qualification of measures, regarding the sustainability principles of long-term orientation and holistic orientation (Widok and Wohlgemuth 2013). To recapitulate this, we argued in a previous paper, that the simple aggregation and/or simulation and comparison of two processes or two version of the same process with different parametrizations, is not enough for a qualification with regard to the sustainability of these processes, as the qualification of what is sustainable logically changes over time. In the social dimension it is furthermore not only a shift in the time dimension, but also culture (Badura et al. 2014), (cf. Rozlina et al. 2012) and multi-polarity (cf. Dubielzig 2009) (in respect to different organizational influences). Hence the comparison should always include the definition of normative (target) values, to which the aggregated data of processes has to be compared (cf. Widok and Wohlgemuth 2013). This is also relevant for processes that do not share the totality of (identic) indicators, as the definition of normative values for each process allows a clearer comparison, contrary to the exclusion of not corresponding indicators or the reference to the total values.

To clarify what is meant by the description of an OHS oriented approach, one can consider Table 1 and Fig. 3, which are taken from the Federal Institute for Occupational Safety and Health (BAuA) and relay a general (simplified) categorization of threshold values considering lifting processes performed by human beings. The points related to the threshold values are calculated based on four controlling characteristics, see Fig. 3 and (Bundesanstalt für Arbeitschutz und Arbeitsmedizin und Länderausschuss für Arbeitschutz und Sicherheitstechnik (BAuA) 2001).

Table 1 Rough qualification of lifting impact based on the key indicator method (cf. Bundesanstalt für Arbeitschutz und Arbeitsmedizin und Länderausschuss für Arbeitschutz und Sicherheitstechnik (BAuA) 2001)

Risk factor	Threshold value	Description
1	<10	Low physical stress, danger to health is unlikely
2	10–<25	Elevated physical stress, danger to health is possible for individuals with reduced resilience
3	25–<50	Strongly elevated physical stress, danger to health is likely for normal individuals
4	>50	High physical stress, overload of individuals is very likely

Threshhold value perception and resulting points, following the four controlling characteristics duration/frequency, load, body posture and performance conditions (including rough categorization):

Fig. 3 Individual relativity between influence and threshold value (cf. Bundesanstalt für Arbeitschutz und Arbeitsmedizin und Länderausschuss für Arbeitschutz und Sicherheitstechnik (BAuA) 2001)

Considering this categorization and more importantly the qualification of processes or measures, in correspondence to threshold values, we deduced, that different social impacts on human health would be qualified with a similar method, just different normative values and calculation approaches for the threshold values. Consequently the deviation from threshold values would be comparable also for very different influence, i.e., impact categories, which basically lead to the implementation of a pre-orientation of normative value building. This means, that normative values, depicting desired states, had to be defined in accordance to the definition of social influences. The different influences were then comparable based on their deviation from the desired threshold values, furthermore depending on the number of individuals involved, they could be weighted accordingly.

To test the aggregation method, we have implemented a simplified version in our manufacturing simulation software. This was done, in order to quantify health related influences during the task, individuals (human resources) have to perform, for the working of the machines or processes. In addition, also transaction oriented influences can be modelled in the software, as the influence building allows for the definition of influences that occur, either if human resources are assigned to the same workplaces, or if they are otherwise modeled as related to each other.

The aggregation process itself is performed once the simulation finishes and results are available. Different adaptations of the simulation software had to be made to facilitate this additional perspective, see Widok and Wohlgemuth (2014, 2016). Interaction based influences have to be defined two ways, i.e., a single influence is always only working one way, as the impact of the influence is naturally different for the two (or more) individuals in question. This also means, that for the modelling of interaction based influences, which should work both ways, two influences have to be modelled.

Finally there is the integration of social life cycle assessment (S-LCA) data, which would enable an even larger picture on the incorporation of social impacts. While the general technology to integrate such data is already implemented, the data itself is still lacking and thus no such experiments were yet realized. To clarify this point, the software can already integrate environmental LCA data (E-LCA data) of processes that lead up to the manufacturing systems, through accorded materials, substances and or energy, which could also be done with S-LCA data. The currently used database however does not offer the S-LCA data in the current version. This will however change in the future (as far as we know) and thus enable the integration of social impact data from the accorded materials and energy mixes. This will likely pose new questions regarding the aggregation of the impact values, but also enable an even broader perspective on the sustainability of the analyzed manufacturing system.

4 Conclusion, Discussion and Outlook

We understand that there are more sophisticated simulation tools available for the calculation and attribution of ergonomic criteria. While our simulation software is not equally sophisticated in regards of the measurement and attribution of physical aspects, a free definition of social influences is possible and thus statistical data, as is provided for example by Baumanns and Münch (2009), (cf. Badura et al. 2014) for interaction oriented social capital can possibly be used to create more complex models, that go beyond the analysis of only ergonomic aspects.

Considering the threshold value aggregation, one has to acknowledge that the definition of threshold values is also subjective and hence comparable to the benefit-analysis. However there is a strong difference, as at least for physical exposures and substance interactions, there are specific threshold values from multiple organizations, which give the accorded values a stronger scientific foundation than the prioritization based on subjective preferences or subjectively perceived importance.

In the future the modeling and simulation approach still has to be verified by conducting more use cases with the software as it is. While the presented functioning is already a big step towards the simulation of social aspects in manufacturing entities, the validity of accorded results and of the quality of the approach itself is still in question. In that regard, our current focus will be to conduct more case studies with the software. The future integration of S-LCA data will, as described, add an additional dimension to the assessment of social impacts and may lead to an even broader understanding of the impact of manufacturing processes on sustainability.

References

Badura, B., Walter, U., & Hehlmann, T. (2010). *Betriebliche Gesundheitspolitik—Der Weg zur gesunden Organisation*. Bielefeld/Bremen: Deutschland, Springer-Verlag.

Badura, B., Ducki, A., Schröder, H., Klose, J., & Meyer, M. (Eds.) (2014). *Fehlzeiten-Report 2014 —Erfolgreiche Unternehmen von morgen—gesunde Zukunft heute gestalten*. Springer-Verlag.

Bamberg, E. (1991). Arbeit, Freizeit und Familie. In S. Greif, E. Bamberg, & N. Semmer (Eds.), *Psychischer Streß* (pp. 201–221). Hogrefe Verlag, Germany: Göttingen.

Bamberg, E. (2003). Organisationsberatung aus der Perspektive der Arbeitspsychologie. In K.-C. Hamborg & H. Holling (Eds.) *Innovative Personal- u. Organisationsentwicklung* (S. 355–380). Germany: Göttingen. Hogrefe Verlag.

Baumanns, R. (2009). Unternehmenserfolg durch betriebliches Gesundheitsmanagement, Nutzen für Unternehmer und Mitarbeiter. In: B. Badura, U. Walter, & T. Hehlmann (Eds.) (2010): Betriebliche Gesundheitspolitik - Der Weg zur gesunden Organisation.

Benoît, C., & Vickery-Niederman, G. (2010). Social sustainability assessment literature review. In *Sustainability Consortium White Paper #102*. Durham, NH, USA.

Biendarra, I., & Weeren, M. (Eds.). (2009). *Gesundheit - Gesundheiten?: Eine Orientierungshilfe*. Germany: Königshausen u. Neumann Verlag.

Bourdieu, P. (1986). The forms of capital. In J. Richardson, *Handbook of Theory and Research for the Sociology of Education* (pp. 46–58).

Bundesanstalt für Arbeitschutz und Arbeitsmedizin und Länderausschuss für Arbeitschutz und Sicherheitstechnik (BAuA) (2001). Leitmerkmalmethode zur Beurteilung von Heben, Halten, Tragen.

Busch, C., Cao, P., Clasen, J., & Deci, N. (2014). *Betriebliches Gesundheitsmanagement bei Kultureller Vielfalt*. Germany: Springer Verlag.

Colantonio, A. (2007). Measuring social sustainability: best practice from urban renewal in the EU, In *2007/01: EIBURS Working Paper Series*.

Coleman, J. S. (1988). Social capital in the creation of human capital. *The American Journal of Sociology, 94*, S95–S120.

Conrad, D. (2007). Defining social capital. *Electronic Journal of Sociology*, pp. 1–5.

DIN Deutsches Institut für Normung e. V. (2011). Guidance on social responsibility (ISO 26000:2010), Beuth Verlag GmbH, Berlin, Germany.

Dubielzig, F. (2009). *Sozio Controlling in Unternehmen*. Lüneburg, Germany: Das Management erfolgsrelevanter sozialgesellschaftlicher Themen in der Praxis.

Fritz, M. (2001). Simulation der Druckkräfte in den Beingelenken und der Wirbelsäue bei praxisnahen Schwingungsbelastungen mit Hilfe eines biomechanischen Modells, In *Zeitschrift für Arbeitswissenschaften* (Vol. 55, No. 3, pp. 154–161).

Fukuyama, F. (2001). Social capital, civil society and development. *Third world quarterly, 22*(1), 7–20.

Global Reporting Initiative (GRI). (2013). *Sustainability reporting guidelines - reporting principles and standard disclosures*. Amsterdam, The Netherlands: Global Reporting Initiative.

Grootaert, C., & van Bastelaer, T. (Eds.). (2002). *Understanding and measuring social-capital—a multidisciplinary tool for practitioners*. Washingtong, DC, USA: The World Bank.

Jørgensen, A., Le Bocq, A., Narzarkina, L., & Hauschild, M. (2008). Methodologies for social life cycle assessment. *I Journal of Life Cycle Assessment, 13*(2), 96–103.

Langer, U. (2011). Unternehmen und Nachhaltigkeit Analyse und Weiterentwicklung aus der Perspektive der wissensbasierten Theorie der Unternehmung, Ph.D. Thesis, Germany.

Möller, A. (2010). Software-Unterstützung für Routine im betrieblichen Umweltschutz, In *Proceedings of the EnviroInfo 2010*. Cologne, Bonn, Germany: Shaker Verlag.

Neumann, W. P., Winkel, J., Medbo, L., Magneberg, R., & Mathiassen, S. E. (2006). Production system design elements influencing productivity and, In *International Journal of Operations & Production Management* (Vol. 26, No. 8, pp. 904–923). Emerald Group Publishing.

Odeh, K. (2013). Framework for assessing Environmental, Social and Economic Sustainability of ICT Ogrnizations, Ph.D. Thesis.

Omann, I., & Spangenberg, J. H. (2002). Assessing social sustainability—the social dimension of sustainability in a socio-economic scenario. In *Proceedings of the 7th Biennial Conference of the International Society for Ecological Economics*. Sousse, Tunisia.

Pezzey, J. C. V., & Toman, M. A. (2002). The economics of sustainability: a review of journal articles. In *Resources of the Future Discussion Papers 02–03*. Resources for the Future.

Putnam, R. D. (1995). Bowling Alone: America's Declining Social Capital. *Journal of Democracy*, 65–78.

Rozlina, S., Awaluddin, M. S., Hamnid, S. H. S., & Norhayati, Z. (2012). Perceptions of ergonomics importance at workplace and safety culture amongst Safety & Health (SH) Practitioners in Malaysia. In *Proceedings of the WCE 2012* (Vol I).

Spangenberg, J. H. (1997). Prisma der Nachhaltigkeit); Nr. UM-631/97; Wuppertal Institut, In: Kleine, A. (2009). Operationalisierung einer Nachhaltigkeitsstrategie, Gabler, GWV Fachverlage GmbH, Wiesbaden, Germany.

Takala, E.-P., Pehkonen, I., Forsman, M., Hansson, G.-Å., Mathiassen, S. E., Neumann, W. P., et al. (2013). Systematic evaluation of observational methods assessing biomechanical exposures at work. *Scandinavian Journal of Work, Environment & Health, 36*(1), 3–24.

Thiede, S. (2012). *Energy efficiency in manufacturing systems*. Berlin Heidelberg, Germany: Springer Verlag. ISBN 978-3-642-25914-2.

VDI (Verein Deutscher Ingenieure) (2013). VDI 3633 - VDI-Richtlinien - Simulation of systems in materials handling, logistics and production – Terms and definitions, Beuth Verlag.

von Hauff, M. (2015). Nachhaltiges Wachstum—ein anderer Weg. In *WISU - Das Wirtschaftsstudium* (Vol. 7, pp. 310–315).

Westgaard, R. H., & Winkel, J. (2011). Occupational musculoskeletal and mental health: significance of rationalization and opportunities to create sustainable production systems—A systematic review. *Applied Ergonomics, 42*(2011), 261–296.

Widok, A. H., & Wohlgemuth, V. (2013). Simulating sutainability. In *Proceedings of the 27. Conference on Environmental Informatics* (S. 514–522). Aachen, Germany: Shaker Verlag.

Widok, A. H., & Wohlgemuth, V. (2014). Technical concept of a software component for social sustainability in a software for sustainability simulation of manufacturing companies. *Proceedings of the SEDSE 2014* (pp. 75–81). Genova, Italy: Dime Università di Genova.

Widok, A. H., & Wohlgemuth, V. (2015). Definition of social sustainability criteria for the simulation of OHS in manufacturing entities. In *EnviroInfo & ICT4S* (S. 7–14). Denmark.

Widok, A. H., Wohlgemuth, V. (2016). Integration of a social domain in a manufacturing simulation software. In *International Journal of Service and Computing Oriented Manufacturing* (Vol. 2, No. 2, pp. 138–154).

Optimal Noise Filtering of Sensory Array Gaseous Air Pollution Measurements

Barak Fishbain, Shai Moshenberg and Uri Lerner

Abstract One of the fundamental components in assessing air quality is continuous monitoring. However, all measuring devices are bound to sensing noise. Commonly the noise is assumed to have zero mean and, thus, is removed by averaging data over temporal windows. Generally speaking, the larger the window, the better the noise removal. This operation, however, which corresponds to low pass filtering, might result in loss of real abrupt changes in the signal. Therefore, the need arises to set the window size so it optimally removes noise with minimum corruption of real data. This article presents a mathematical model for finding the optimal averaging window size. The suggested method is based on the assumption that while real measured physical phenomenon affects the measurements of all collocated sensors, sensing noise manifests itself independently in each of the sensors. Hence, the smallest window size which presents the highest correlation between the collocated sensors, is deemed as optimal. The results presented here show the great potential of the method in air quality measurements.

Keywords Air pollution measurements · Noise filtering · Micro sensing units

1 Introduction

Air quality has a tremendous effect on public health and the environment (Künzli et al. 2000). Many studies have associated various adverse effects to general air pollution and its specific components such as nitrogen dioxide (NO_2), ozone (O_3) carbon monoxide (CO) and particular matter (PM), to name a few (Kampa and Castanas

B. Fishbain (✉) · S. Moshenberg · U. Lerner
Faculty of Civil & Environmental Engineering,
Technion - Israel Institute of Technology, 32000 Haifa, Israel
e-mail: fishbain@technion.ac.il

S. Moshenberg
e-mail: shaisho@tx.technion.ac.il

U. Lerner
e-mail: uriler@technion.ac.il

V. Wohlgemuth et al. (eds.), *Advances and New Trends in Environmental Informatics*, Progress in IS, DOI 10.1007/978-3-319-44711-7_22

2008). These pollutants, for example, affect the respiratory system, the cardiovascular system and other systems in the human body (Laumbach and Kipen 2012). Some of the pollution is due to natural phenomena and some due to anthropogenic activity (Robinson and Robbins 1970; Cullis and Hirschler 1980). Regardless of its sources, air pollution undergoes a set of chemical processes in the atmosphere, depending on initial concentration and ambient conditions. The large number of sources and the intricateness of the chemical processes, lead to the creation of complex scenarios, displaying highly variable spatial and temporal pollution patterns rendering the analysis of air-pollution and its effects as a challenging task (Nazaroff and Alvarez-Cohen 2001; Levy et al. 2014).

One of the primary tools for assessing air-pollution patterns is continuous monitoring of pollutants' ambient levels. To accomplish that, numerous chemical-physical methods have been developed and standardized Air Quality Monitoring (AQM) station networks have been spread around the world. However, as any other sensor, these AQM stations are bound to measurement errors due to sensors' and circuitry noise. This noise limits AQM's capability to accurately capture ambient pollution levels and thus, hinders the study of air-pollution (Duyzer et al. 2015). With the growing usage of Micro Sensing Units (MSUs) for measuring ambient pollutants' levels (Künzli et al. 2000; Kampa and Castanas 2008; Mead et al. 2013; Williams et al. 2013; Moltchanov et al. 2015; Lerner et al. 2015), this problem increases as MSUs are more error prone than the standard measuring equipment (Tchepel and Borrego 2010; Mead et al. 2013; Williams et al. 2013; Moltchanov et al. 2015; Lerner et al. 2015). Thus, in order to better utilize the sensing equipment, noise must be effectively filtered out.

Filtering the noise out requires full characterization of either the noise or the signal. The statistical properties of the sensing noise may be known from the certification of the monitoring system. However, in many applications these data are unavailable. Further, it was shown that MSUs' accuracy, i.e. sensory noise level, varies over time, which makes any characterization futile, as it is valid for only a limited time period (Künzli et al. 2000; Gupta et al. 2011; US Environmental Protection Agency 2012; Mead et al. 2013; Moltchanov et al. 2015).

Sensing noise is often characterized as Additive White Gaussian Noise (AWGN) (Schwartz and Marcus 1990; Rao and Zurbenko 1994; Varotsos et al. 2005). Thus, for x_i, the true pollutant's ambient level and ε_i, the noise at time step i, the measurement y_i is given by: $y_i = x_i + \varepsilon_i$, where ε_i is a normally distributed random variable with zero mean and unknown variance (Wu and Huang 2009).

Realistically, changes in the composition of the atmosphere happen over relatively long period of time when compared to the sampling rate, i.e. order of tens of minutes with respect to the sampling rates of tens of seconds (Rao and Zurbenko 1994; Wang et al. 2003; Peng et al. 2006). Even when considering photochemistry in hot regions, a global change in air-pollution composition takes a much longer time than the sampling rate (Leighton 2012; Weinstein et al. 2016). Combined with the assumption of AWGN, noise may be filtered out by averaging the signal over a temporal sliding window, i.e., replacing each measurement, y_i, with the computed average of samples within a temporal window centered at i (Schwartz and Marcus 1990). This procedure is also known as Kolmogorov-Zurbenko (KZ) (Zurbenko 1986) or Sinc filtering (Yaroslavsky 2014), and for a window size of $2K + 1$ is given by:

$$y_i = \frac{1}{2K+1} \sum_{j=-K}^{K} y_{i+j} \tag{1}$$

The KZ operator essentially suppresses abrupt changes in the signal. The larger the temporal window is, the smoother the output signal (Yaroslavsky 2014). This is equivalent to removing higher frequencies of the signal, thus, low-pass filtering (Zurbenko 1986; Schwartz and Marcus 1990; Yaroslavsky 2014).

To this end, low-pass filtering in its simplest form means zeroing all signal's frequency coefficients above a given frequency, called the cut-off frequency. The larger the window size, the lower the cut-off frequency.

Previous analyses suggested to set the cutoff frequency so it maximizes the coefficient of determination, R^2, of a regression model associating mortality (Peng et al. 2006) or temperature (Rao and Zurbenko 1994) with air-pollution measurements. In both cases, the temporal window size found was considerably large (order of days), heavily smoothing the signals. This outcome is expected as signal's temporal local variations, whether originated from genuine signal's fluctuations or from noise, degrade R^2 value. Thus, removing these perturbations improves the regression model, but deteriorates the signal's high frequencies.

Therefore, using such a filter for noise filtering calls for a method to determine the ideal cutoff frequency or the size of the temporal window, so it eliminates as much noise as possible, while preserving real data. Here we present a mathematical model to optimally set the window's size.

2 Materials and Methods

2.1 *Optimal Filtering Window Size*

Typically, as the level of noise increases, the correlation between the real signal and the measured signal decreases (Fishbain et al. 2008). Assuming that the noise affects each sensor independently, if a pollution signal is measured by two separate collocated devices, the correlation between them is expected to decrease as the noise level grows. This is illustrated in Fig. 1, where Fig. 1a depicts real-life NO_2 time series, A_N, acquired between January 1st and December 31st, 2010 (16,949 samples) by a standard AQM station located at the heart of the Haifa industrial/commercial area (LAT/LON: 32.78919/35.04038)—see (Moltchanov et al. 2015) for more details on the study area. A_N's maximum measurement was 48 [ppb], its average was 4.67 [ppb] and its standard deviation was 5.98. From this signal a synthetic noisy signal, S_k, is generated by adding random AWGN, ε_σ, with zero mean and standard deviation, so the signal to noise ratio (SNR) is 5. The process is then repeated with

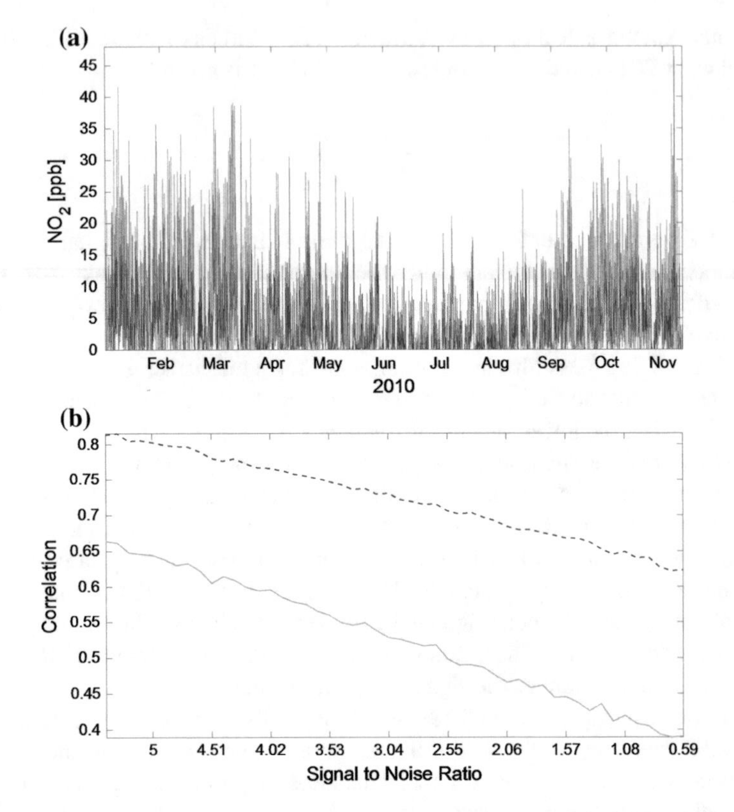

Fig. 1 Correlation coefficient as a function of noise standard deviation. **a** Real-life NO_2 time series acquired between January 1st and December 31st, 2010 (16,949 samples) by a standard AQM station located at the heart of the Haifa industrial/commercial area (LAT/LON: 32.78919/35.04038). **b** Correlation between two sets of synthetic noisy signals as a function of added noise characteristics (*solid-red*) and between the original signal and one of the synthetic signals (*dashed-blue*)

SNR=$\{4.9, 4.8, \ldots, 0.1\}$. Hence, a set of fifty signals with noise increasing standard deviations is created. Two such sets are used here.

Figure 1b shows the correlation between the two sets of synthetic noisy signals as a function of the added AWGN's standard deviation (solid-red) and between the original signal, A_N, and one of the synthetic signals (dashed-blue). It is evident that indeed the correlation drops as the noise level increases.

Signal's energy is a characteristic used in signal processing for quantifying the amount of data within a signal. For a continuous signal, $p(t)$, the energy is given by:

$$E = \int_{-\infty}^{\infty} |p(t)|^2 dt \qquad (2)$$

Following the Parseval's theorem (Boas 1966), the energy of a signal is equal to the energy of its frequency transform, $P(\omega)$:

$$E = \int_{-\infty}^{\infty} |P(t)|^2 d\omega \tag{3}$$

Hence, the function $|P(\omega)|^2$ represents the energy distribution in the frequency domain. Applying discrete sampling, (3) becomes:

$$E = \sum_{\omega=1}^{N} |P(\omega)|^2 \tag{4}$$

Given (4) and the notion presented in Fig. 1, the optimization goal is to find the highest cut-off frequency such that as little information, i.e., energy, in the higher frequencies is removed, while the correlation between two collocated sensors reaches its maximum.

This is the essence of the suggested filtering scheme. For removing AWGN, a temporal window is suggested. For finding the optimal window's size, one should balance between the window size which presents the highest correlation between two sensors measuring the same physical phenomenon, and by evaluating the signal's spectrum in search of a cut-off frequency, which removes as little as possible of a signal's energy, i.e., information.

The same physical phenomenon can be identically measured when the sensors are collocated (Mead et al. 2013; Moltchanov et al. 2015; Williams et al. 2013). This mode of operation is applicable mainly when MSUs are in use. Due to MSUs' inherent limitations, collocating is currently the common practice (Fishbain and Moreno-Centeno 2016; Lerner et al. 2015; Mead et al. 2013; Moltchanov et al. 2015; Williams et al. 2013). When the sensors are not collocated, measuring the same phenomenon can be achieved when it is uniform in all measuring points (Moltchanov et al. 2015).

2.2 *Frequency Representation*

In this study the transformation of the pollutants' time-series to the frequency domain is executed through the 2nd Discrete Cosine Transform (DCT). The DCT is well documented to have high energy-compaction, i.e., most of the signal's energy, in the frequency domain, lays with a small number of low-frequencies coefficients (Zurbenko 1986). Using DCT increases the amount of information in the lower frequencies, limiting true signal's information in the higher frequencies. For a pollutant time series, A_N, that is composited of N data points—a_k, the frequency coefficient, α_r is given by:

$$\alpha_r = \frac{2}{\sqrt{2N}} \sum_{k=0}^{N-1} \left(\frac{a_k \cos\left[\pi\left(k+\frac{1}{2}\right)\right]}{N} r \right) \tag{5}$$

2.3 Data

For demonstrating the suggested filtering scheme, two Air-Quality MSU pods (AQMesh 2015) were placed near an AQM station in Haifa, Israel (Lat:32.78741, Lon: 35.02119, height above ground level: 12 [m], height above sea level: 208 [m]). Each AQMesh unit was equipped with five environmental sensors: NO, NO_2, O_3, atmospheric pressure (AP), and relative humidity (RH). Additionally, the AQMesh measured the unit's internal temperature (Temp). Each pod has its own battery and communication device, wirelessly-transmitting the measurements to a central server every 15 min.

In order to compare the AQM and the MSU measurements, the time resolution of both should be the same. If that is not the case, the time series with the fine temporal resolution is aggregated so it fits the coarser resolution. The MSU measurements were acquired at a 15 min resolution, while the AQM time-series had a 30 min resolution. Hence, MSU measurements were averaged (without overlapping) to produce a time-series that corresponds to the AQM temporal resolution.

3 Results

For simulating true sensors' data post-processing, the measured signals of the two MSUs were low-pass filtered by averaging, with no overlapping windows and decreasing filter size, i.e., lowering the cut-off frequency at each iteration. For each window's size the correlation between the two averaged sequences was calculated. As seen in Fig. 2 for O_3, there is a peak at around 500 min. Also evident is that the variance of the correlation increases with the window size. This is attributed to the smaller number of window's positions, which decreases with the window's size.

The DCT transformation of the ozone time series is plotted in Fig. 3. Setting the cut-off frequency so 90 % of signal's energy is preserved, the cut-off frequency was found to be 53 [1/min]. This is equivalent to averaging the signal over 672 min. Evaluating this result with respect to Fig. 2, this value is sufficiently close to the highest correlation (found around 500 min) and thus noise can be filtered out without compromising on the correlation between the two signals. The 11 h average that was found by the suggested method agrees with the National Ambient Air Quality Standards (NAAQS) of the United States Environmental Protection Agency (US-EPA), which suggests an 8 h average (US Environmental Protection Agency 2012) for monitoring ozone.

Figure 4 illustrates the filtered signal (in red) versus the original noisy signal, in blue. It is noticeable that the filtered signal manages to describe the measurements truthfully, while giving a smoother behavior, without peaking at extreme high or low values.

The same process was performed on an NO_2 signal and is described in Figs. 5, 6 and 7. The cut-off frequency was obtained at 2,657. The 2,657 [1/ min] cut-off

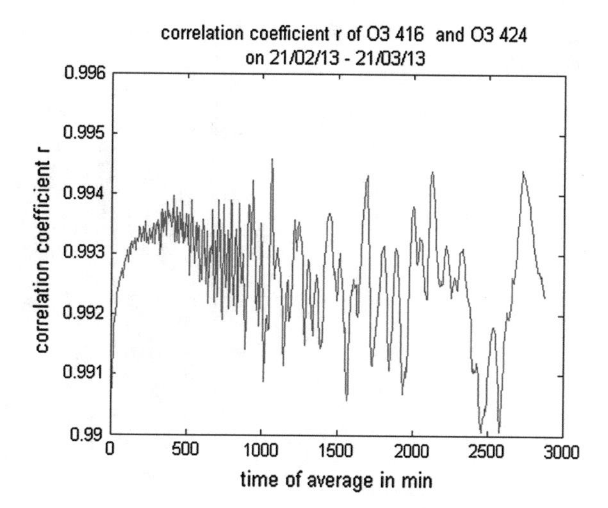

Fig. 2 Correlation between two MSUs as a function of the averaging temporal window size for O_3 measurements

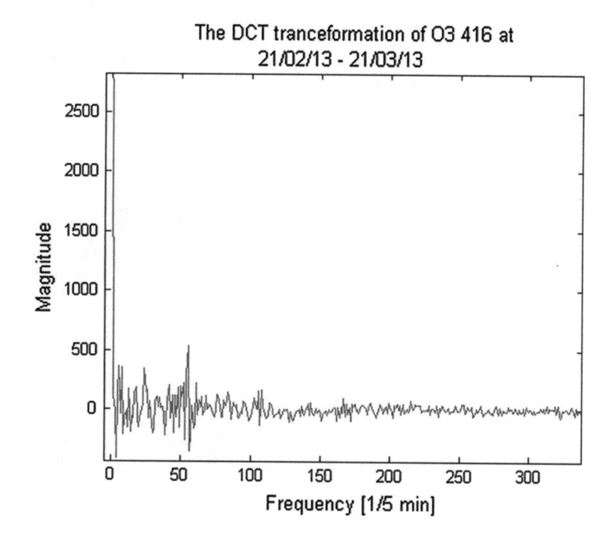

Fig. 3 DCT transformation of the O_3 time series

is equivalent to averaging over a temporal window of 15 min. The correlation, is highest when averaging the signal over a window of 50 min (Fig. 5). The US-EPA NAAQS for NO_2 is one hour (US Environmental Protection Agency 2012), which is close to the window suggested by our method.

In Fig. 7 the original noisy NO_2 signal can be seen in blue, and the filtered signal is in red, and again, it is evident that the filtered signal changes more gradually over time, and presents lower noise level.

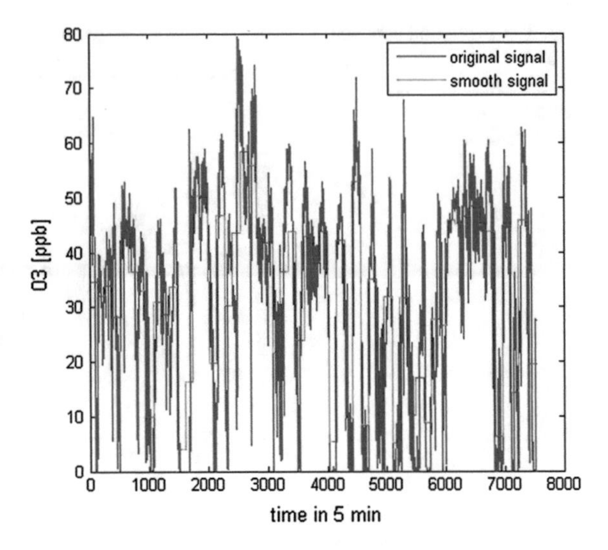

Fig. 4 DCT transformation of the O_3 time series

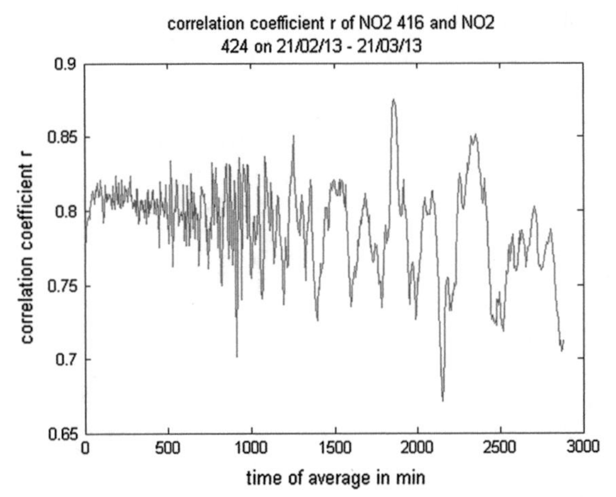

Fig. 5 Correlation between two MSUs as a function of the averaging temporal window size for NO_2 measurements

4 Conclusion

A methodology for finding the optimal averaging window size for noise removal in air-quality time series is suggested. The window's size is set by balancing between two criteria: maximum correlation between two signals obtained by collocated sensors, and applying a low-pass filter with the highest cut-off frequency. Using this method, the noise affecting the quality of the air pollution signal can be filtered out based on the actual measurement taken (and not by a common rule of thumb), thus giving a better assessment of the monitored signal, improving understanding of the environment.

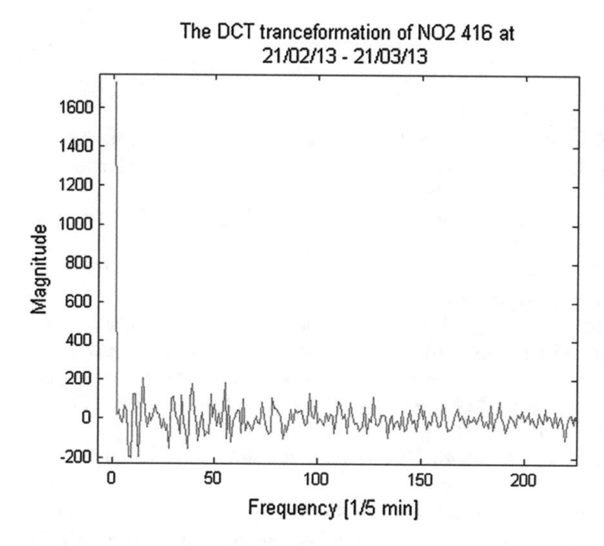

Fig. 6 DCT transformation of the NO_2 time series

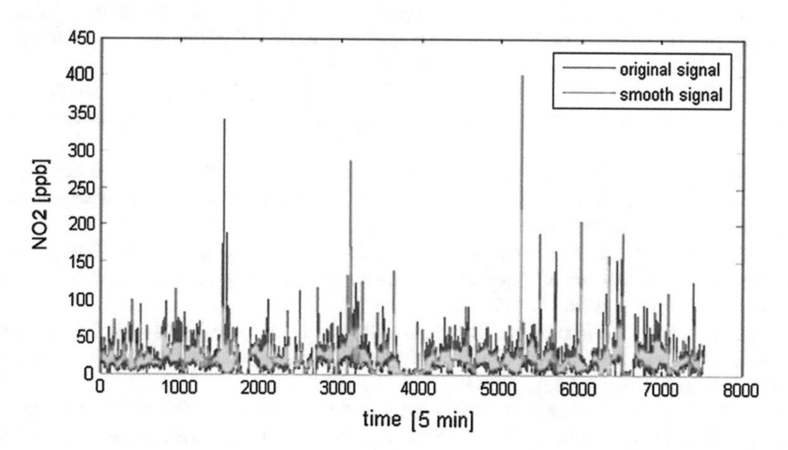

Fig. 7 Original (*blue*) and filtered (*red*) nitrogen-dioxide signal

More research regarding the optimal percent of energy preserved is needed. We assumed that disregarding 10 % of the energy from a long signal would not overly degrade the signal but a guiding methodology is needed. Further studies, which implement the suggested method on different pollutants acquired from different places would also be beneficial in supporting further the argument of the suitability of the method for the general case.

Acknowledgments This work was partially supported by the 7th European Framework Program (FP7) ENV.2012.6.5-1, grant agreement no. 308524 (CITI-SENSE), the Technion Center of Excellence in Exposure Science and Environmental Health (TCEEH) and the Environmental Health Fund (EHF).

References

AQMesh. (2015). Aqmesh website.
Boas, M. L. (1966). *Mathematical methods in the physical sciences* (vol. 2). New York: Wiley.
Cullis, C., Hirschler, M. (1980). Atmospheric sulphur: Natural and man-made sources. *Atmospheric Environment (1967)*, *14*(11), 1263–1278.
Duyzer, J., van den Hout, D., Zandveld, P., & van Ratingen, S. (2015). Representativeness of air quality monitoring networks. *Atmospheric Environment*, *104*, 88–101.
Fishbain, B. & Moreno-Centeno, E. (2016). Self calibrated wireless distributed environmental sensory networks. *Scientific reports*, 6.
Fishbain, B., Yaroslavsky, L. P., & Ideses, I. (2008). Spatial, temporal, and interchannel image data fusion for long-distance terrestrial observation systems. *Advances in Optical Technologies*.
Gupta, M., Shum, L. V., Bodanese, E., & Hailes, S. (2011). Design and evaluation of an adaptive sampling strategy for a wireless air pollution sensor network. In *2011 IEEE 36th Conference on Local Computer Networks (LCN)* (pp. 1003–1010). IEEE.
Kampa, M., & Castanas, E. (2008). Human health effects of air pollution. *Environmental Pollution*, *151*(2), 362–367.
Künzli, N., Kaiser, R., Medina, S., Studnicka, M., Chanel, O., Filliger, P., et al. (2000). Public-health impact of outdoor and traffic-related air pollution: a european assessment. *The Lancet*, *356*(9232), 795–801.
Laumbach, R. J., & Kipen, H. M. (2012). Respiratory health effects of air pollution: Update on biomass smoke and traffic pollution. *Journal of Allergy and Clinical Immunology*, *129*(1), 3–11.
Leighton, P. (2012). *Photochemistry of air pollution*. Elsevier.
Lerner, U., Yacobi, T., Levy, I., Moltchanov, S. A., Cole-Hunter, T., & Fishbain, B. (2015). The effect of ego-motion on environmental monitoring. *Science of the Total Environment*, *533*, 8–16.
Levy, I., Mihele, C., Lu, G., Narayan, J., & Brook, J. R. (2014). Evaluating multipollutant exposure and urban air quality: Pollutant interrelationships, neighborhood variability, and nitrogen dioxide as a proxy pollutant. *Environmental Health Perspectives*, *122*(1), 65.
Mead, M., Popoola, O., Stewart, G., Landshoff, P., Calleja, M., Hayes, M., et al. (2013). The use of electrochemical sensors for monitoring urban air quality in low-cost, high-density networks. *Atmospheric Environment*, *70*, 186–203.
Moltchanov, S., Levy, I., Etzion, Y., Lerner, U., Broday, D. M., & Fishbain, B. (2015). On the feasibility of measuring urban air pollution by wireless distributed sensor networks. *Science of The Total Environment*, *502*, 537–547.
Nazaroff, W., & Alvarez-Cohen, L. (2001). *Environmental Engineering Science*. John Wiley.
Peng, R. D., Dominici, F., & Louis, T. A. (2006). Model choice in time series studies of air pollution and mortality. *Journal of the Royal Statistical Society: Series A (Statistics in Society)*, *169*(2), 179–203.
Rao, S. T., & Zurbenko, I. G. (1994). Detecting and tracking changes in ozone air quality. *Air & waste*, *44*(9), 1089–1092.
Robinson, E., & Robbins, R. C. (1970). Gaseous nitrogen compound pollutants from urban and natural sources. *Journal of the Air Pollution Control Association*, *20*(5), 303–306.
Schwartz, J., & Marcus, A. (1990). Mortality and air pollution j london: A time series analysis. *American Journal of Epidemiology*, *131*(1), 185–194.
Tchepel, O., & Borrego, C. (2010). Frequency analysis of air quality time series for traffic related pollutants. *Journal of Environmental Monitoring*, *12*(2), 544–550.
US Environmental Protection Agency. (2012). Epa national ambient air quality standards. Technical report.
Varotsos, C., Ondov, J., & Efstathiou, M. (2005). Scaling properties of air pollution in Athens, Greece and Baltimore, Maryland. *Atmospheric Environment*, *39*(22), 4041–4047.

Wang, T., Poon, C., Kwok, Y., & Li, Y. (2003). Characterizing the temporal variability and emission patterns of pollution plumes in the pearl river delta of China. *Atmospheric Environment, 37*(25), 3539–3550.

Weinstein, B., Steyn, D., & Jackson, P. (2016). Modelling photochemical air pollutants from industrial emissions in a constrained coastal valley with complex terrain. In *Air pollution modeling and its application XXIV* (pp. 289–294). Springer.

Williams, D. E., Henshaw, G. S., Bart, M., Laing, G., Wagner, J., Naisbitt, S., et al. (2013). Validation of low-cost ozone measurement instruments suitable for use in an air-quality monitoring network. *Measurement Science and Technology, 24*(6), 065803.

Wu, Z., & Huang, N. E. (2009). Ensemble empirical mode decomposition: A noise-assisted data analysis method. *Advances in Adaptive Data Analysis, 1*(01), 1–41.

Yaroslavsky, L. (2014). Signal Resotoration by means of linear filtering. In *Digital signal processing in experimental research—Fast transform methods in digital signal processing* (pp. 67–80). Sharjah, U.A.E: Bentham Science Publishers Ltd.

Zurbenko, I. (1986). *The spectral analysis of time series*.North-Holland, Inc.: Elsevier.

Part VIII
Frameworks, Platforms, Portals

Generic Web Framework for Environmental Data Visualization

Eric Braun, Clemens Düpmeier, Daniel Kimmig, Wolfgang Schillinger and Kurt Weissenbach

Abstract The Web and the growing popularity of Internet of Things show a large need for data visualization. This paper presents both a concept and an implemented prototype for a framework that supports a user in creating data visualizations on the Web. The concept is based on a microservice architecture using lightweight REST based service interfaces for communication. The prototype is evaluated using environmental data gathered by the State Office for the Environment, Measurements and Nature Conservation of the Federal State of Baden-Württemberg (LUBW). The main advantages of the framework are high customizability and no requirements for programming skills for using a large part of the framework.

Keywords Data visualization · Web component · Microservice · REST

E. Braun (✉) · C. Düpmeier
Karlsruhe Institute of Technology (KIT), 76344 Eggenstein-Leopoldshafen, Germany
e-mail: eric.braun2@kit.edu

C. Düpmeier
e-mail: clemens.duepmeier@kit.edu

D. Kimmig · W. Schillinger
State Office for the Environment, Measurements and Nature Conservation
of the Federal State of Baden-Württemberg (LUBW), 76185 Karlsruhe, Germany
e-mail: daniel.kimmig@lubw.bwl.de

W. Schillinger
e-mail: wolfgang.schillinger@lubw.bwl.de

K. Weissenbach
Ministry of the Environment, Climate Protection and the Energy
Sector Baden-Württemberg, 70182 Stuttgart, Germany
e-mail: kurt.weissenbach@um.bwl.de

V. Wohlgemuth et al. (eds.), *Advances and New Trends in Environmental Informatics*, Progress in IS, DOI 10.1007/978-3-319-44711-7_23

1 Introduction

Nowadays, the world is acquiring more data than ever before. This data is collected from various sources like customers, mobile devices, and sensors. The amount of sensors has increased by multiple magnitudes because the Internet of Things became more and more relevant. The Web-based Information System (WebIS) group at the KIT (Karlsruhe Institute of Technology) works on modern technologies to acquire, store, analyze, and visualize such data. Cooperation with LUBW allows applying and evaluating these new concepts in concrete Environmental Information System (EIS) projects. The LUBW among other things gathers data about the deposition of pollutants into the soil, the emission of pollutants into the air, and water levels which need to be visualized on the Web. This paper focuses on pollutants that are measured in the air. These types of data are typically structured as time series. The time series data is often complemented by metadata, e.g. information about the measuring station and measurement equipment which gathers the time series. This is often important for the visualization of the data as it determines some implicit semantics of the data, e.g. the units for the measurement data or data quality.

A lot of visualizations on the Web (and this is also the case in EIS systems build by the LUBW) are still static images that have multiple disadvantages compared to a responsive HTML5 Web visualization. A static image has to be updated manually and is not responsive because the resolution cannot adapt to the viewport of the browser. This results for example in static chart scales or values that are not accordingly aggregated. Furthermore, the image cannot display data that is changing over time in a dynamic way. In addition, static images do not provide any user interaction at all. In contrast, an HTML5 visualization, which is managed with the visualization framework, features dynamic resolutions and multiple possibilities for user interaction on desktop computers and mobile devices.

Because of this a recently developed concept of the WebIS group describes a framework which allows the usage of modern highly customizable and dynamic HTML5 visualizations while supporting a variety of different data sources. The goal of the framework is allowing a user to create a visualization of such data sets without any programming knowledge. The visualization has to be implemented using Web technologies at some point but the requirement is that the implementation of new visualization code can be achieved apart and that this code can be reused by users who have no programming skills. In addition, the framework supports that visualizations can be highly customizable in layout, style, and behavior. This allows the user to configure the visualization to his needs.

The next chapters describe the main concepts of this generic visualization framework in more detail.

2 Microservice Based Architecture

The Web framework for data visualization is designed as a microservice based architecture as defined by Newman (2015). It can be divided into a backend part consisting of several microservices and a frontend part including different graphical user interfaces (GUI) for configuration as well as a visualization component that displays the actual visualization.

The backend is composed of three microservices using relational databases for data persistence. As a microservice is meant to have one single responsibility, one microservice manages data sources, a second microservice manages different visualization templates, and a third microservice manages instances of the concrete visualization. The architecture is outlined in Fig. 1 and in the following this architecture will be described in more detail.

The *Data Source Service* manages data sources that provide data to be visualized. The format of the data has to be specified for every data source in order to map the data to a visualization. The concept of the framework describes the usage of JSON and CSV formats that can be either read from a file or from a URL providing a response in one of these formats. Additionally, data can be read from a relational database with specified queries to access the data. A data source address can include parameters which change what data are served. These parameters can be used in the file path, URL or a relational database query accordingly to the format that is used. E.g., such a parameter can set the resolution of a time series data set (e.g. one value per minute, hour, day, etc.).

The *Visualization Template Service* manages visualization templates which semantically describe different types of visualization implementations, e.g. from what location the code of the visualization can be served, used parameters in this code and additional information about this type of visualization implementation.

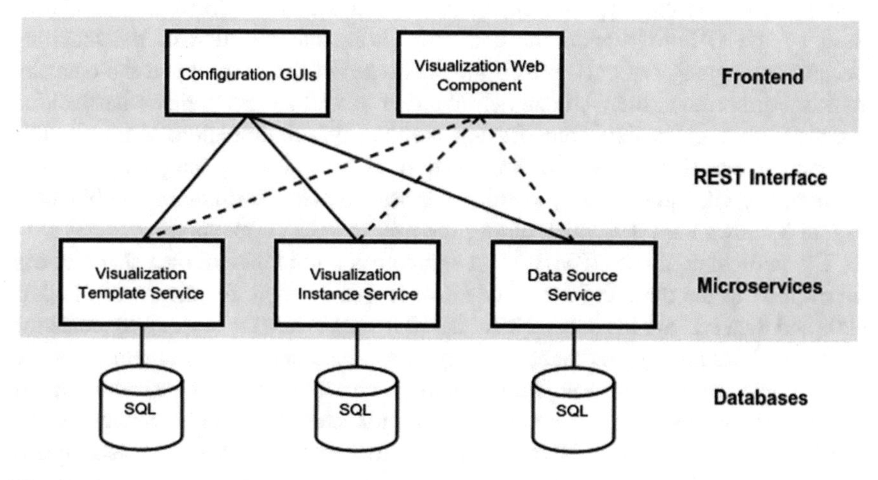

Fig. 1 The architecture of the framework

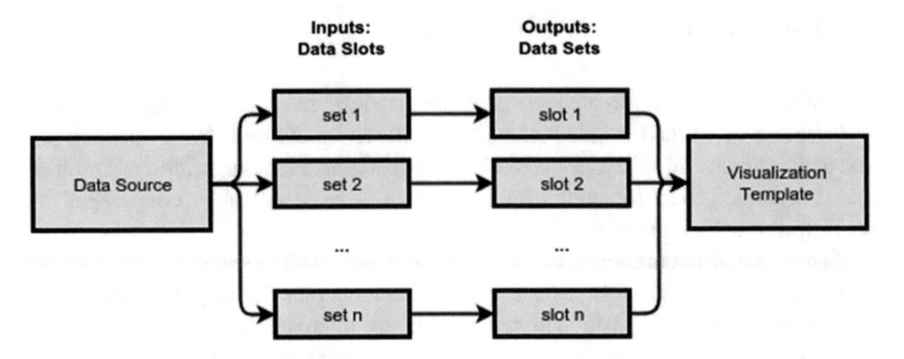

Fig. 2 Concept of the mapping between a data source and a visualization template

Such a template is a generic blueprint for a visualization which can be instantiated many times with different values for each parameter and with different data sources. This leads to the last microservice: the *Visualization Instance Service*.

The *Visualization Instance Service* manages every concrete visualization as instances that can be shown on a web page. The instance consists of a mapping between different data sources and a visualization template to form a complete visualization. Figure 2 shows the concept of such a mapping. Each visualization template defines different *data slots* that can be connected with a data source. A data source has similar connection ports named *data sets*. The user can map each data slot to a data set that is available for a specific data source.

Figure 2 shows a mapping between one data source and a visualization template but it is also possible to connect multiple data sources to a single visualization template. The number of data sets which can be connected to a single data slot can be configured. This configuration can be used to help the user with multiple mappings. Additionally, the users managing data sources and visualization templates can set the data type for each data set and data slot. This data type can be used by the GUI to support the end user during the selection of the mapping because a mapping can only exist if the data types of the data set and the data slot match. Furthermore, the GUI can help the user to find an appropriate visualization template for a given data source using the data type information. If no possible mapping exists for a specific combination of visualization template and data sources, the GUI immediately notifies the user to select a different combination. Figure 3 shows the GUI which allows the user to select a mapping.

The particulate matter (PM10) data source has three different data sets that can be mapped to the three different data slots. In this example the names of the data slots and data sets are the same which makes it even easier for the user to create the mapping. Additionally, the user can change every parameter that is supported by the selected visualization template. Furthermore, parameters for each data source can be changed individually. Those two groups of parameters can also be changed using JavaScript code during the display of the visualization. This leads to the second part of the framework—the frontend.

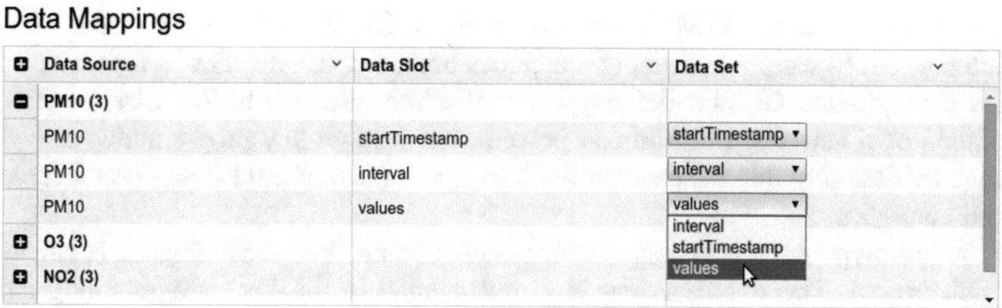

Fig. 3 User selection of a data mapping

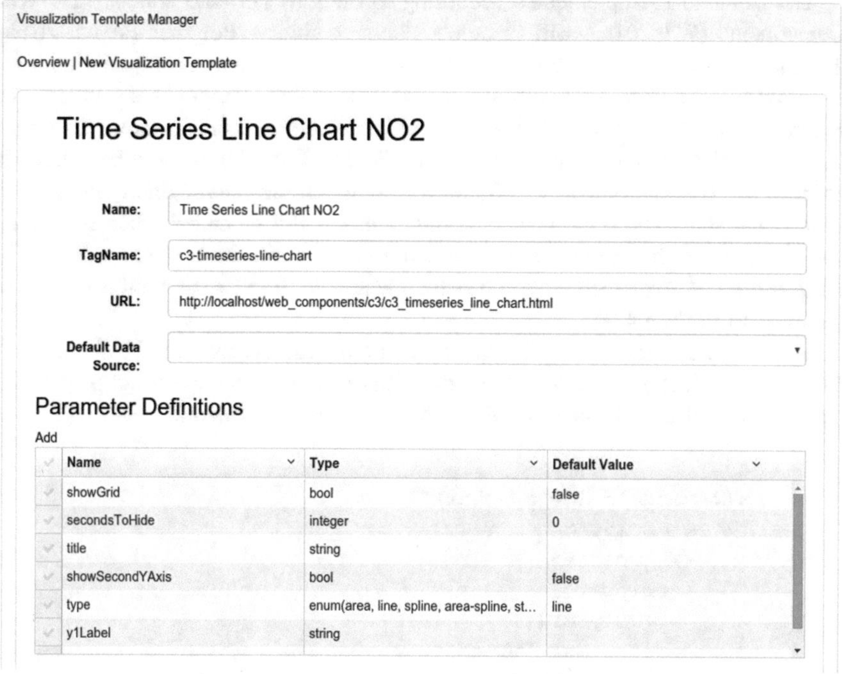

Fig. 4 A part of the visualization template GUI

The frontend consists of a set of configuration GUIs, one for every service. The GUIs allow different types of users to add and manage data sources, visualization templates and visualization instances. Figure 4 displays a part of the visualization template GUI.

A second part of the frontend is the *Visualization Web Component*. It is a HTML5 Web Component implemented in JavaScript code which shows the actual visualization that is defined by a visualization instance. The Visualization Web Component can be easily embedded into a website using only a few lines of code which is common for HTML5 Web Components. The WebIS group is working on a Liferay (2016) wrapper portlet that includes both, the visualization component and the configuration GUI for defining a visualization instance. In that case, the integration of a new visualization can be achieved without any programming knowledge by dragging this wrapper portlet into a web page of the Liferay portal server and configuring it.

A prototype for this concept, which uses modern Web technologies, is already implemented. The microservices are implemented in the Java programming language with the help of the Spring Boot Framework (Spring Boot 2016) which is one of the leading microservice frameworks. The services are each connected to a MySQL database to persist the configurations.

The frontend is implemented according to the new HTML5 standard for Web Components (W3 2016) with Google's Polymer library (Polymer Project 2016), that adds syntactic sugar to write 'cleaner code' and providing a compatibility layer for browsers which haven't fully implemented the Web Components standard yet. Figure 5 demonstrates a visualization template's source code. It shows an empty Polymer Web Component with three additional functions that have to be defined in order to use the framework. The framework will call the *visualization_init* method once at the start. The *visualization_update* method will be called when something has changed. Both parameters *data* and *params* contain the updated values. The *visualization_destroy* method will be called when the Web Component is deleted or removed from the DOM.

The communication between the different microservices and the frontend is ensured with REST (Tilkov 2011) calls. This results in a loosely coupled architecture and a clean API between the two parts of the framework.

```
<dom-module id="my-chart">
    <template>
    </template>
    <script>
        Polymer({
            is: "my-chart",
            visualization_init: function(){ },
            visualization_update:
                function(data, params){ },
            visualization_destroy: function() { },
            ready: function () { }
        });
    </script>
</dom-module>
```

Fig. 5 Source code of visualization template

The framework has various advantages over a traditional approach:

- A visualization is highly customizable because every visualization instance can contain individual parameters. Additionally, a visualization template code can contain as many parameters as needed. Furthermore, a visualization template parameter has a default value which results in a very flexible parameter management. The user can apply the default values to immediately see a result but he or she can also modify every single parameter in order to obtain a highly customized solution.
- The programmer coding the HTML and JavaScript code behind a visualization template can use any Web technology that is embeddable in a Web Component, especially any JavaScript library (e.g. Highcharts, Google Charts, D3.js, Three.js, etc.). The only restriction is that the template code has to implement a valid polymer component with the three mandatory functions implemented (see above).
- A visualization instance can have multiple heterogeneous data sources which leads to a high support of different data sets.
- The integration of a new visualization is very easy because it can be achieved by adding a few lines of HTML/JavaScript code. The code runs in every browser that supports Web Components or that can be polyfilled. In addition, visualizations can be wrapped by other technologies, e.g. Web portlets, in order to reduce or eliminate even programming efforts to embed a visualization instance into a certain web page.
- Due to the support of any JavaScript library, the visualization can be implemented with high interactivity on desktop computers and mobile devices.
- The microservice based architecture allows the framework to be scalable. The different components can be deployed separately because they are only communicating using the REST interface.
- Once created, data sources, visualization templates and visualization instances can be reused.

3 The Framework in Use

In the following, usage of the framework will be explained with two scenarios. The workflow for applying the framework is as follows: firstly, a data manager creates a data source if a new data source is needed. In our example, the data source for a generic time series service based on OpenTSDB (2016) to store time series data e.g. about pollutants has already been created and it can be used out of the box. This time series service is developed in a second project at the WebIS group and allows storing arbitrary measurement data. It can serve time series data depending on a few parameters that have to be submitted with the request URL. Those parameters determine the start and end value of the time series and the resolution of the data (e.g. daily values, monthly values, etc.). The response is a JSON formatted HTTP response that contains the data.

Secondly, two different visualization templates will be used for our scenario by a web author wanting to embed visualizations of pollutants into web pages: a line chart that can display multiple values of different pollutants at the same location at a time (Figs. 6 and 7) and a bar chart (Fig. 8) which displays daily averages of one pollutant. By using the already available data source the web author can easily create these two new visualization instances with the configuration user interface of the

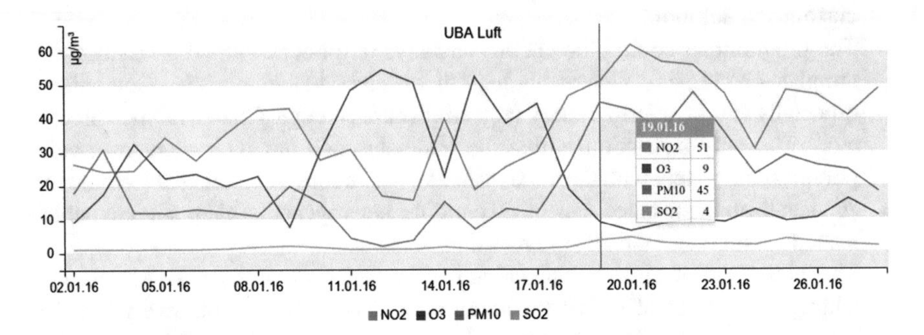

Fig. 6 Multiple pollutants in a line chart

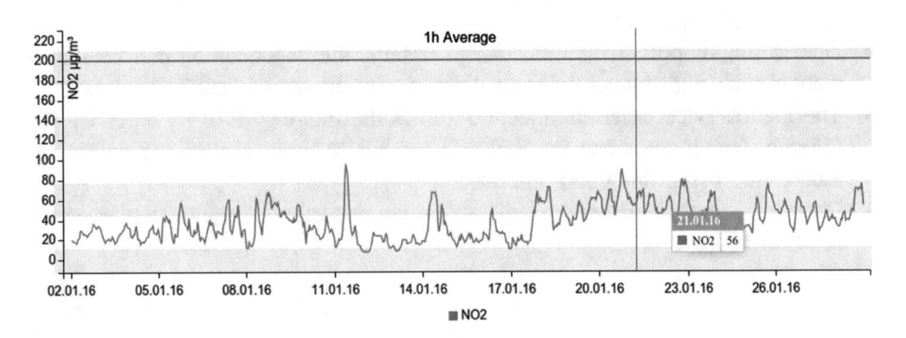

Fig. 7 Detailed visualization of NO2

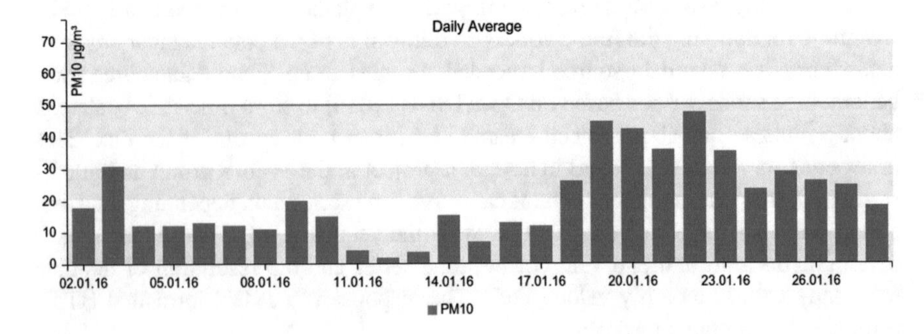

Fig. 8 Bar chart visualization of PM10

Visualization Instance Service. To create the first variant, the web author maps the request URLs of the different pollutant time series onto separate line diagrams of the visualization component. The time series data will also be filtered by location (in our case a measurement station at Ulm) and a certain time interval will be specified. Figure 6 shows the resulting visualization which shows the measurement data of the pollutants NO2, O3, PM10 and SO2 in the air at the measuring station Ulm.

The created visualization is dynamic and interactive. A reading web page user can hover over the time series lines to see a tooltip with more information (e.g. the values at the specific point). Additionally, the user can zoom in and out of the visualization and pan left and right to see other time ranges. These features allow the user to interact with the chart. Furthermore, the user can resize the browser window and the visualization is resized accordingly which is important for a modern responsive Web application.

Figure 6 shows an overview of the amount of pollutants in the air over a specific period of time. Additionally, the end user could be interested in more detail about a specific pollutant. Therefore, the second scenario provides two more visualizations which only show information about one pollutant. The data mapping for this example is partially shown in Fig. 3.

Figure 7 shows the hourly average value of NO2 in the air. The horizontal red line indicates the limit of 200 $\mu g/m^3$ which should not be exceeded at more than 18 days per year (Umweltbundesamt: Nitrogen dioxide 2013). This visualization instance can be created by using the already existing visualization template for single line chart with boundaries and mapping the time service data fields accordingly to the line diagram and its boundaries.

The last visualization (Fig. 8) should show the daily average of PM10 in the air. Similar to the previous example the red line indicates the limit of PM10 which is 50 $\mu g/m^3$. This value should not be exceeded more than 35 times a year (Umweltbundesamt: Particulate matter 2013). To create this visualization, the web author chooses the visualization template for a bar chart diagram with boundaries and configures the data source URL for accessing the pollutant data for PM10 to aggregate values to a daily average. The resulting aggregated data will then be mapped to the bar chart. Finally, the upper boundary of 50 $\mu g/m^3$ had to be configured (Fig. 9).

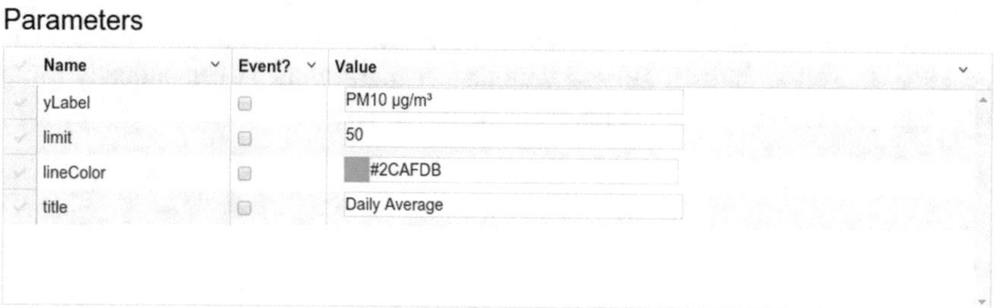

Parameters

Name	Event?	Value
yLabel		PM10 µg/m³
limit		50
lineColor		#2CAFDB
title		Daily Average

Fig. 9 Parameter configuration of the bar chart visualization

4 Summary and Outlook

The previous chapter shows that web users using the framework can create modern visualizations which have all the advantages over traditional approaches discussed in Chap. 2. In particular, the disadvantages of static images, which are used to visualize data, are all ruled out by the framework. The visualization can be resized accordingly, end users can interact with the visualization and a visualization instance will always provide actual values using a JavaScript API for fetching the data from a defined data source, if the data is regularly updated within the data source. In addition, the framework has many more advantages because a user is not required to program a visualization but he or she can use modern GUIs to configure it. This is only limited by the number and types of visualization templates already available and integrated into the framework. If a web author needs a completely new visualization with new underlying code, a programmer with HTML and JavaScript experience has to implement a new template.

However, this can be seen as another advantage because the framework can be completely customized to the point that one can create a new visualization from scratch which then can be integrated into the framework. And this is quite easy because arbitrary JavaScript frameworks already providing visualizations can just be used to create new codings for visualization templates. Additionally, visualizations can be made highly customizable at a higher level since different parameters for the visualization and different parameters for the data source can defined which can later be configured by the web user. Additional information about the framework can be found in the relating Master's Thesis (Braun 2015).

The framework described in this paper can be extended in many aspects. First of all, the framework not only manages visualizations but supports Web Components in general. Therefore, it can be updated to support any Web Component, whether it is a visualization or not.

Next, a framework to visualize data is only the first step towards a system that can handle data with ease. The next step is an improvement of the data management which has to be easy to use. Such an improvement leads to reusable data sources which are the key to handle a lot of different data sets without having too many heterogeneous data sources.

A third area which can be addressed is data semantics. The utilization of explicit concepts can help to identify, address, interconnect, and integrate related data from different sources. It is important that users can create rich visualizations that include metadata as well as links to related data and visualizations. A semantically aware data management can be the solution to such a problem.

References

Braun, E. (2015). A highly customizable and generic weg framework for data visualization (Master's Thesis, KIT 2015).

Liferay. (2016). Retrieved April 01, 2016 from https://www.liferay.com.

Newman, S. (2015). *Building Microservices*. Sebastopol: O'Reilly Media.

OpenTSDB. (2016). Retrieved June 06, 2016 from http://opentsdb.net/.

Polymer Project. (2016). Retrieved April 01, 2016 from https://www.polymer-project.org.

Spring Boot (2016). Retrieved April 01, 2016 from http://projects.spring.io/spring-boot/.

Tilkov, S. (2011). Rest and HTTP. dpunkt.verlag, Heidelberg.

Umweltbundesamt: Particulate matter. (2013). Retrieved April 01, 2016 from https://www.umweltbundesamt.de/en/topics/air/particulate-matter-pm10.

Umweltbundesamt: Nitrogen dioxide. (2013). Retrieved April 01, 2016 from https://www.umweltbundesamt.de/en/topics/air/nitrogen-dioxide.

W3: Web Components Current Status. (2016). Retrieved April 01, 2016 from https://www.w3.org/standards/techs/components#w3c_all.

Creating a Data Portal for Small Rivers in Rostock

Sebastian Hübner, Ferdinand Vettermann, Christian Seip and Ralf Bill

Abstract In the context of the project KOGGE an Open-Source Spatial data portal based on GeoNetwork is being developed. Firstly, this portal supports the project-internal data exchange in an OGC and INSPIRE conform way. Secondly, it will be used for civic participation. With the integration of thesauri there is a comfortable data search interface available. To animate users to contribute to the portal the effort of data-adding is reduced massively using metadata templates. Furthermore, these templates guarantee the compliance with ISO standards for spatial data. To avoid duplicate datasets from other spatial data portals the GeoNetwork harvesting technology is used. The portal also offers the possibility to view and edit spatial data by making use of web services (WMS, WFS and WPS). The portal is also used to involve the citizens of Rostock in the planning and management process of urban waters. The user will be able to view project results and comment actual hydrological developments. Also an integration of volunteered geographic information with social media is planned.

Keywords GeoNetwork · Spatial data infrastructure · Web Processing Service · Public participation · Urban water bodies

S. Hübner · F. Vettermann (✉) · C. Seip · R. Bill
Geodesy and Geoinformatics, University of Rostock, Rostock, Germany
e-mail: ferdinand.vettermann@uni-rostock.de

S. Hübner
e-mail: sebastian.huebner@student.HTW-Berlin.de

C. Seip
e-mail: christian.seip@uni-rostock.de

R. Bill
e-mail: ralf.bill@uni-rostock.de

V. Wohlgemuth et al. (eds.), *Advances and New Trends in Environmental Informatics*, Progress in IS, DOI 10.1007/978-3-319-44711-7_24

1 Introduction

Within the BMBF project (Federal Ministry of Education and Science) KOGGE ("Kommunale Gewässer Gemeinschaftlich Entwickeln") the development of a modern web-based spatial data infrastructure for the project partners and the civic participation presents one of the main objectives. Project partners are the University of Rostock (professorship of water management, professorship of hydrology, professorship of geodesy und geoinformatics), the "EURAWASSER Nord GmbH", the water and soil association "Untere Warnow-Küste" and "biota—Institut für ökologische Forschung und Planung GmbH". They cooperate with the city of Rostock, senate administration for construction and environment, the Warnow-water and wastewater association, the department of agriculture and environment of central Mecklenburg-Western Pomerania and the department for environment, nature conservation and geology Mecklenburg-Western Pomerania. With these partners, all stakeholders of water management in the area of Rostock participate in this project.

Concerning periodic floods in Rostock, an integrated drainage concept (INTEK) was developed (Mehl et al. 2015). To continue INTEK the project KOGGE started in 2015, especially to analyze the smaller water bodies in the urban area with regard to compliance with the water management (WFD) and flood management (Floods Directive) directives. The result will be a more sophisticated drainage concept for the city of Rostock (KOGGE 2015).

Because of the hydrological context there is a massive amount of spatial data for each stakeholder. Therefore it is important to realize a fluid interaction between all stakeholders and their data. A complete description of properties and creation history is mandatory for each dataset as well. An online addressable spatial data infrastructure will be the best solution to meet these requirements. The SDI will manage metadata, spatial data, data services, quality standards and access rights (GDI-DE 2015).

Further, the web-connection of the project is a good method to serve the function of civic participation, like other spatial data portals have shown in the past. An important aspect is to serve the interest of citizens in flooding areas and flood-risk-zones. Common for civic participation is also the use of social media, like Twitter (Bill 2016).

2 Creation of the SDI

First step is to plan and setup an SDI. The three main targets of the KOGGE SDI are data exchange, data editing and civic participation. For those, three service types have to be provided by the portal. The first type is a simple up- and download service. The second type is a Web Processing Service (WPS) to edit the provided spatial data. The third type are web services that cover the display of the provided features to the user via Web Map Service (WMS) and Web Feature Service

(WFS) (Korduan and Zehner 2008). For the last service type the integration of volunteered geographic information from social media is also useful. Firstly, the data can be displayed and secondly the inhabitants of Rostock participate in an indirect manner with their posts on social media platforms. Therefore the language of posts on different social media platforms has to be analyzed to get their correct location and their expression. But this related work will not be discussed in this paper, because the target of the social media analysis is civic participation and volunteered geographic data in general without focus on urban water bodies.

The first challenge is to integrate all these different modules and services into one platform. The second challenge is to make the spatial data easily accessible and editable. The main feature of the SDI will be providing and managing project relevant data and giving the possibility to view and edit these data sets for the project participants.

2.1 Requirements of the SDI

For easy administration and exchange of the data, the data portal has to fulfil the following requirements (compare GDI-DE 2015):

1. Format independent data access for everyone everywhere
2. User- and rights management
3. Overview about the data's metadata
4. Traceable changes
5. Finding datasets
6. Integration of new datasets
7. Visualization and editing

These requirements are best addressed by the software GeoNetwork 3, as Hübner (2016) pointed out. GeoNetwork is an Open-Source metainformation system, which was started in 2001 in conjunction with a Food and Agriculture Organization of the United Nations (FAO) project. Via the integration of GeoServer, it is able to provide WMS and WFS for displaying and WPS for processing data. The visualization is realized through OpenLayers 3 and the 3D extension Cesium.js. Further, GeoNetwork provides a harvesting function to integrate OGC and ISO conform data from other data portals. With this feature, redundant data provisioning is circumvented. The harvester supports different service types like WMS, WFS, Web Coverage Services (WCS), WPS, Catalogue Service Web (CSW) and Sensor Observation Services (SOS) (GeoNetwork 2015).

GeoNetwork offers also predefined template sets for the three metadata standards ISO 19139, ISO 19115 and Dublin Core. Each set has a form for different record types like text-based data or vector data. Further, every dataset could be checked for INSPIRE conformity (GeoNetwork 2015).

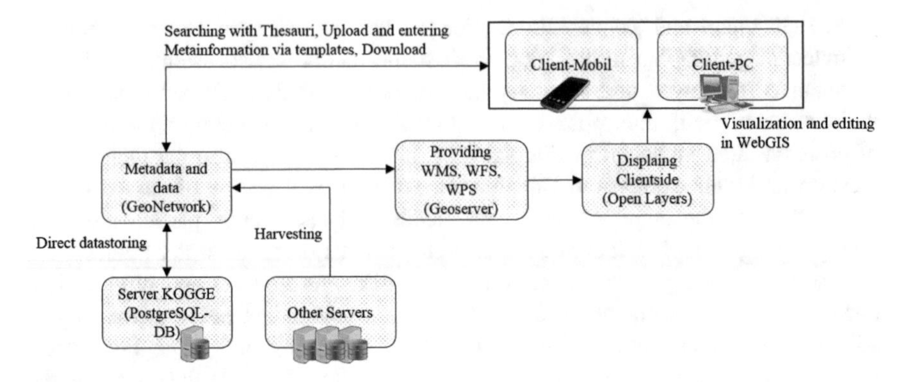

Fig. 1 Structure of the SDI

2.2 *Structure of the SDI*

To realize these requirements, a specific structure has been designed (Fig. 1). Data will be stored centralized and decentralized (GDI-DE 2015). Local data is stored in a PostgreSQL-Database on the projects' server and other datasets are integrated by harvesting other resources. The SDI combines both storages and makes it accessible.[1] Furthermore, up- and downloading of the datasets is integrated within this SDI.

To display and process spatial data sets, web services are provided by GeoServer. The client-side display of layers is realized with the JavaScript library OpenLayers 3. There are different access points available, the recommended stationary desktop PC or mobile clients like smartphones and tablets.

2.3 *Features of the SDI*

With the presented structure (Sect. 2.2) specific features are provided by the SDI. An important aspect is to ensure conformity with other data catalogs, which is provided by the harvesting function. Various data catalogues can be harvested and the harvested data can be integrated easily. The ISO and W3C standards for metadata ISO 19115/19139 (W3C-Standard for Spatial data) and Dublin Core (W3C-Standard for Documents, presentations, etc.) are used when data is integrated (GeoNetwork 2015).

Further, the thesauri, which are basically designed in the Simple Knowledge Organization System (SKOS) are used to organize the data structure of the portal (W3C 2009). For each topic a single thesaurus with specific keywords and a hierarchical order was created. Every provided dataset is categorized with the help

[1]http://webgis-kogge.auf.uni-rostock.de:8085/geonetwork/.

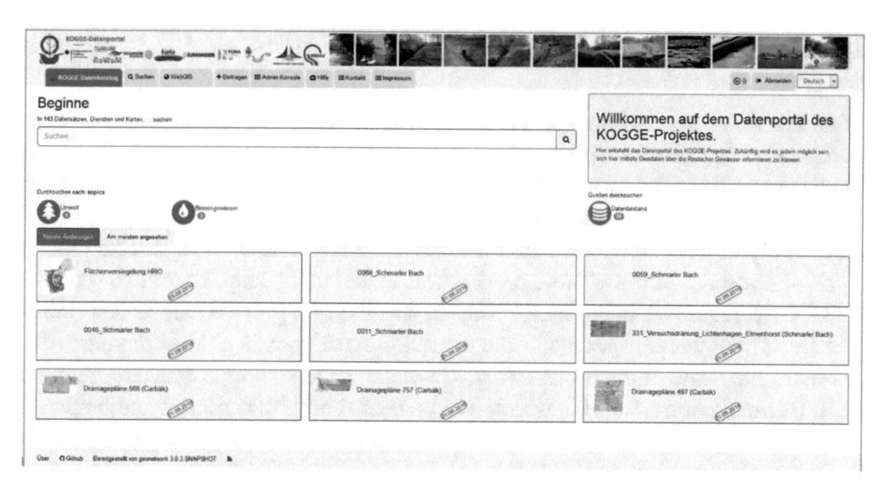

Fig. 2 Screenshot of the edited GeoNetwork user interface

of the assigned keywords and located by so-called facets in GeoNetwork (2015). GeoNetwork is improving the handling of keywords by limitation of available keywords for one datatype. Also, typing errors are ruled out. With this technique, a hierarchical multi-lingual order system, as is known for example from amazon, is provided by the SDI. This simplifies the data-search massively and make it more user-friendly (Moreno-Sanchez 2009).

To prevent errors, especially when uploading data, a template library for each topic is created. The goal is to reduce the effort of data publishing and the inhibition threshold of it. There is also the possibility to save a metadata entry as a template to reuse it for similar data. To improve the usability furthermore, a new user interface has been created (Fig. 2).

A core aspect of the portal is the integration of a WPS to give stakeholders the possibility to edit data online (Bill 2016). The next section describes the creation and integration of these services in GeoNetwork.

3 Integration of OGC Web Services

GeoServer provides the WPS through a separate, Java written plugin. As mentioned earlier (Sect. 2.2) GeoNetwork utilizes GeoServer to process its spatial data. Therefore the WPS plugin can be used to edit spatial data using the portal's user interface. The provided functionality is similar to a desktop GIS. A user can, for example, measure distances, manipulate and edit spatial data. The native Geo-Server plugin provides 186 processes that can be executed on spatial data, but most of them are redundant. After deduplication 93 processes were remaining. The integration of user programmed algorithms is possible as well.

3.1 Description of the Functionality

After installing the plugin the processes can be executed using a form of the GeoServer user interface (Open Source Geospatial Foundation 2015). The challenge is to provide the form in the map view of the portal and to ease the filling of the form with layer interaction for the user.

The default map view was edited so that these processes can be selected by the user. A button for each available process was added to the map. By clicking on it, a HTTP GET request is being made. The description of a process can be determined by a DescribeProcess[2] operation. The operation must contain at least the parameter identifier, the name of the process (e.g. JTS:buffer). GeoServer is answering with in XML (formally correct GML) which will be read through JavaScript and be parsed to JSON. Depending on the amount of information a form displayed in the GeoNetwork map view is being filled with this information by the script.

In order to execute a process the form has to be completed by selecting layers, since the spatial data is usually represented by layers. Thereby it is to differentiate if the layer is external (e.g. from GeoServer) or was created by the script itself (e.g. a result layer of an earlier executed process). If it is a self-created layer, the information can be extracted from the object used to add the layer to the map. Otherwise a GetFeatureInfo[3] request has to be send to the WMS. The returned information contains properties like the features, the coordinate referencing system or the bounding box. If a property is not readable, the parameter has to be entered manually by the user. After filling out all required fields of the form the process can be executed.

The form data is then parsed into XML. Finally a WPS execute request is being made. Requesting the WPS via HTTP Post with the XML as payload will result in a different response for each specific process executed. For example, when calculating a distance, the result is a string which contains the specific distance. By merging two layers, the response is a new geometry. The specific results are displayed to the user either as string or a newly added layer. In Fig. 3 the use case for online spatial data editing is shown as a UML sequence diagram.

The WPS offers the possibility to make simple hydrological analyses online. A common use case is buffering rivers or wetlands to set up a flooding zone for example.

In Fig. 4 the result of setting a buffer of 100 m around a wetland-shapefile is shown. Further, all available tools and their symbols are visible. But this use-case is not restricted to hydrological aspects. The WPS is usable to address issues for all kinds of spatial data.

---

[2]http://webgis-kogge.auf.uni-rostock.de:8085/geoserver/ows?service=WPS&version=1.0.0&request=DescribeProcess&identifier=JTS:buffer.

[3]http://webgis-kogge.auf.uni-rostock.de:8085/geoserver/wms?request=GetFeatureInfo&service=WMS&version=1.1.1&layers=example_layer.

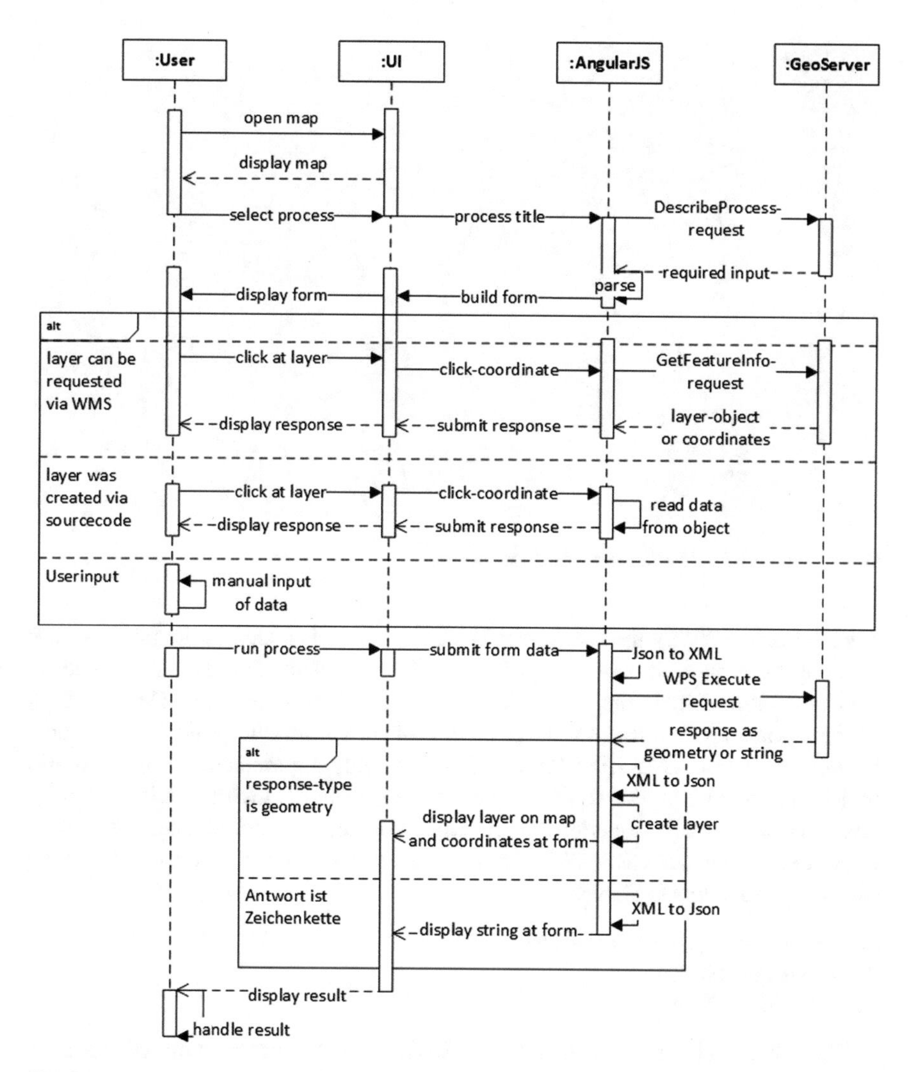

Fig. 3 Use case for online spatial data editing

3.2 *Implementation Challenges*

When working with shapefiles from different contributors it is most likely that different coordinate reference systems (CRS) are used. A common projection in web mapping is EPSG:3857 which is used inter alia by OpenLayers 3[4] (OL3) and OpenStreetMap (OSM). Contributed shapefiles often use other projections so coordinate transformation to EPSG:3857 is needed before displaying them. This is

[4]http://openlayers.org/.

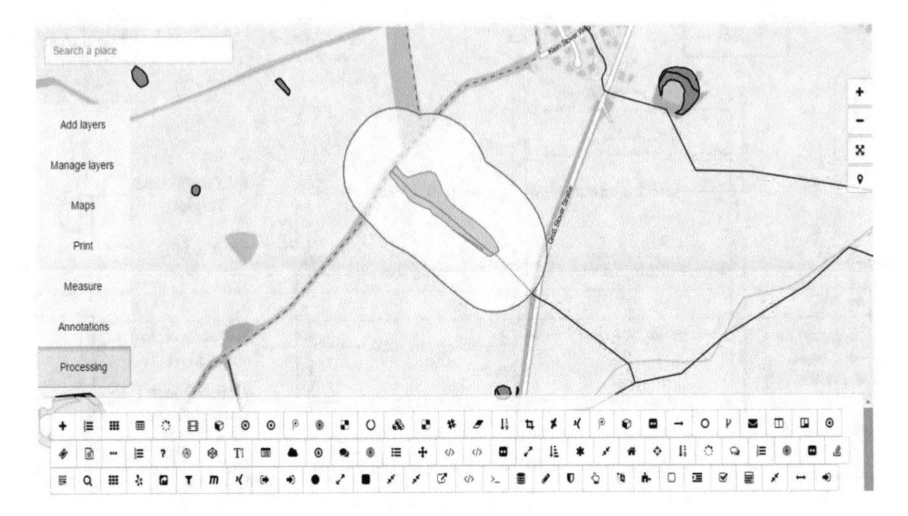

Fig. 4 Screenshot of buffering wetlands in Rostock

done automatically by the GeoServer WMS if a layer is added to the map, but not by using WPS processes or the spatial data fed into them. OpenLayers 3, used by GeoNetwork to display the map, offers a method called 'transform'. This function, as the name suggests, transforms given coordinates from one projection to a destination projection, EPSG:3857 in this case. The only requirement is that the origin projection must be known to OL3. The response from the GetFeatureInfo request, made when a layer is queried by the user, contains the coordinates as well as information about the CRS. With that in mind the transform method will be called like this code-snippets shows:

```
ol.proj.transform(coordinateArray
this.fromCode.srsCode
this.toCode.srsCode)
```

The other challenge is to handle the XML (GML) structure returned from the web service requests. As mentioned earlier, the code has to be parsed from and to JSON (GeoJSON) so that the web services and JavaScript code can handle it.

4 Summary and Outlook

The work at the KOGGE data portal will proceed continually over the whole project period (04.01.2015–03.31.2018). As of now, the thesauri, the template library and the WPS are integrated. This means that data can be down- and uploaded, documented and searched. With this, the core concepts of the data portal are realized. The focus is now being shifted towards the implementation of WebGIS functionality.

With the shop-like keyword-based search interface and the integration of templates, the usability of the SDI was improved. Thus, it was possible to lower the hurdle for the project participants to use the portal and to create an easy-to-use interface for uploading and editing metadata. About 10 GiB of project relevant data has been uploaded at the moment.

The SDI is not only for hydrological purposes but is developed in a hydrological context. The integration of WPS, thesauri and templates into the SDI can be adopted by any other kind of a SDI.

At this time it can be stated, that the combination of the four Open-Source-Software-packages GeoNetwork, GeoServer, Open Layers and PostgreSQL are able to establish a comfortable platform for data exchange and data documentation. Further, with the availability of WPS this WebGIS could be a serious alternative to desktop GIS distributions. With the presented way of integrating the GeoServer WPS in GeoNetwork, it becomes a lightweight and easily manageable WebGIS tool. Especially there are clearly less problems with different data-formats, data exchange and data documentation. With this SDI we provide a solution of project-intern data-management.

References

Bill, R. (2016). *Grundlagen der Geo-Informationssysteme*. Berlin-Offenbach: Herbert Wichmann Verlag.

GDI-DE: Geodatendienste im Internet (2015). Ein Leitfaden. Authors edition.

GeoNetwork Opensource: User Guide (2015). Retrieved January 25, 2016 from http://geonetwork-opensource.org/manu-als/trunk/eng/users/index.html.

Hübner, S. (2016). Web-Service-Orchestrierung beim Aufbau einer Geodateninfrastruktur zur Integration, Prozessierung und Dissemination verschiedenster Daten (Master Thesis, htw Berlin, University of Rostock, 2016).

KOGGE: Kogge-Website (2015). Kommunale Gewässer Gemeinschaftlich Entwickeln. Retrieved January 25, 2016 from http://kogge.auf.uni-rostock.de/ (2015).

Korduan, P., & Zehner, M. L. (2008). *Technologien zur Nutzung raumbezogener Informationen im WWW*. Heidelberg: Herbert Wichmann Verlag.

Mehl, D., Hoffmann, T. G., Schneider, M., Lange, A., Neupert, A., Badrow, U., et al. (2015). Gemeinschaftliches Handeln im kommunalen Hochwassermanagement: Das „Integrierte Entwässerungskonzept" (INTEK) der Hansestadt Rostock. *Hydrologie und Wasserbewirtschaftung, 11*(2015), 700–709. doi:10.3243/kwe2015.11.005.

Moreno-Sanchez, R. (2009). The geospatial semantic web. What are its implications for geospatial information users? In M. M Cruz-Cunha, F. F Oliveira, A. J. Tavares, & L. G. Ferreira (eds.) Handbook of research on social dimensions of semantic technologies and web services (pp. 588–609). Denver: University of Colorado.

Open Source Geospatial Foundation: WPS Request Builder (2015).Retrieved January 29, 2016 from http://docs.ge-oserver.org/stable/en/user/extensions/wps/requestbuilder.html (2015).

W3C: SKOS Simple Knowledge Organization System Reference (2009). Retrieved January 26, 2016 from https://www.w3.org/TR/2009/REC-skos-reference-20090818/.

Convergent Infrastructures for Municipalities as Connecting Platform for Climate Applications

Jens Heider and Jörg Lässig

Abstract Since topics such as climate protection come more and more in focus to municipalities, as they are in the role to implement related processes, technical support systems in this area are powerful tools. The most obvious requirement is the implementation of a monitoring infrastructure, which helps to measure the current state of energy consumption in buildings, properties and far beyond, focusing on various sectors such as industry, mobility, waste, etc. Apart, there are requirements and connected technologies and solutions in the fields of Ambient Assisted Living or Smart Home, which can be considered separately from the infrastructure perspective. But, in future, digital services will continuously grow together, simply by the need of various requirements for data exchange in many use cases. In the work at hand we show that in certain areas topics such as IoT in energy management, Ambient Assisted Living or Smart Home solutions should not be considered as separate platforms, but as convergent building blocks of a connected infrastructure landscape. In the context of municipal climate protection, this approach is illustrated in an example case study by presenting a central platform, which allows municipal energy management in conjunction with other services. The development is illustrated based on an innovative software environment that brings the idea in productive application.

Keywords Convergent infrastructure · IT infrastructure · Municipal climate protection · Energy efficiency · Energy management

1 Introduction

Municipalities have the role to implement national and international decisions taken by agreements and legislation decisions at the regional and local community level (Kern et al. 2005). In addition to an appropriate role model of the municipality,

---

J. Heider (✉) · J. Lässig
Department of Computer Science, University of Applied Sciences Zittau/Görlitz, Brückenstr. 1, 02826 Görlitz, Germany
e-mail: jheider@eadgroup.org

J. Lässig
e-mail: jlaessig@eadgroup.org

V. Wohlgemuth et al. (eds.), *Advances and New Trends in Environmental Informatics*, Progress in IS, DOI 10.1007/978-3-319-44711-7_25

which should have the effect that potential savings (i.e. renovation measures, service vehicles, user behavior) are truly realized, there may be impulses for local regulations, e.g. in the construction industry, which go beyond national legislation. Often monetary (and other) resources for actions are limited and preferably actions with a high potential in terms of the impact (costs and emissions) should be implemented. Accordingly, the consequences of the actions of decision makers in the local environment are often difficult to predict. Consequently, decision making should be done in an informed way and hence, services must be provided which assist this process. Various approaches and solutions are available to satisfy this need. Some tools support, e.g., greenhouse gas accounting in the local context, providing information about the current state in terms of greenhouse gas emissions. Other approaches support in terms of future planning and development. Scenarios regarding to potentials within the community are created to reduce emissions in future.

A significant factor in terms of resource consumption are municipal buildings and properties. Depending on their renovation status, this can vary considerably. Municipal properties vary in type of use, equipment and many other parameters. Hence, it is difficult to detect in which properties and processes resources are consumed. It stays unclear, where costs and emissions could possibly be saved, which are target mitigation measures. For an assessment of the actual state, energy management systems are applied. They monitor the current state of consumption on meter level and allow the normalized calculation of costs and emissions by integrating information of energy suppliers (e.g. prices and climate/emission factors). It is clear that there are many approaches, which can contribute as software-based planning and decision making tools in the context of local climate protection, but holistic approaches can be more effective and efficient.

The rest of the paper is structured as follows: Sect. 2 describes the role of municipalities in the context of climate change and discusses the municipal IT infrastructure for advanced digital services. In Sect. 3 the need of infrastructure convergence is motivated and established, looking at different current developments and requirements. We make this concrete in Sect. 4, showing it on the example of energy management and climate change mitigation. A case study that presents a platform for municipal and regional climate protection applications follows in Sect. 5. Reflecting the needs in research and development and future research work, we conclude the study in Sect. 6.

2 Focus on the Local IT Infrastructure for Advanced Digital Services

The requirements for technical support systems for municipalities are diverse. Current issues such as energy policy, connected to various demands from global agreements and policies at EU and national level, conservation objectives, targets for renewable energy production and improved energy efficiency requirements in the context of creating the smart grid and demand side integration (Felden 2013). At municipal level, appropriate targets need to be achieved by climate actions. Here, objectives are in different sectors such as

(S1) industry,
(S2) traffic,
(S3) waste management,
(S4) buildings and properties, etc.

The municipality has the role to shape this process and to act responsible as a local player. Municipalities are drivers of the change and have to actively promote this as a role model. The aspirations particularly in the energy and emissions domain are associated with other strands of mainly IT-driven development, as progress in the economy and society is nowadays often driven by progress in the IT sector (Rydge et al. 2015). So also concepts such as the Internet of Things (IoT) are becoming a municipal focus. While the industrial sector does focus on current issues such as Industry 4.0—IoT for industry (Shrouf et al. 2014), the focus in the field of building and property are rather issues such as Ambient Assisted Living and Smart Home (Wichert et al. 2012). Nevertheless there is a need for benchmarking of the energy efficiency (Lassig et al. 2014).

Current IT developments also bring new technology advances in the transport sector. Currently this is mainly determined by new transport concepts, but also the possibilities and opportunities through automated / autonomous driving or networked driving (Moradi-Pari et al. 2014) create new demands on the municipal infrastructure as the basis for these technologies. This can also be understood as IoT technology for the transport sector—i.e. Smart City as overall infrastructure construct for the requirements in different sectors and in particular in the transport sector (Neirotti et al. 2014).

The developments and possibilities mentioned above are heavily connected with the general progress in digitization, further increasing computing performance, introducing new methods and algorithms, progress in the field of (Big) Data processing and analytics, etc. Furthermore, there are increasing demands to the underlying infrastructure, network expansion, connectivity and the presence of interfaces between different systems as building blocks of a connected infrastructure landscape, Fig. 1.

We try to understand how the mentioned subjects

1. energy management in buildings and properties,
2. smart Home / Ambient Assisted Living,
3. smart Grid and Demand Side Integration,
4. smart City and changes in the sector traffic and
5. digitization and service orientation generally

stimulate each other, but create new demands on the connectivity and interface functionalities of the underlying infrastructure and architecture at the same time (Blaschke et al. 2013).

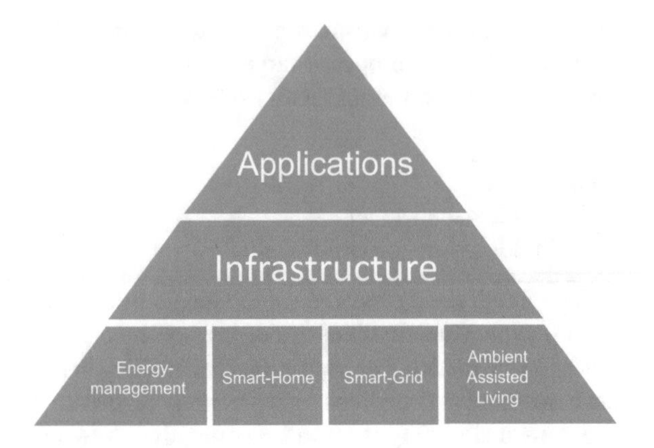

Fig. 1 Demand for connecting infrastructure components for different application fields

The factors IT and data security play an increasingly important role in this context (Eckert and Krauß 2012; Farhangi 2016). In the public perception and in media and politics, the subject of IT security is of constantly increasing importance. The benefits of the new technologies are very limited in their generally positive effects or even counterproductive, unless privacy and data security can be guaranteed. Issues such as security by design, cryptography, end to end encryption, private cloud, etc. occur frequently in the focus of development. With the increasing level of digitization thereby the demands on safety, trustworthiness and reliability of digital infrastructures and services increases constantly. The stated goal is to promote the development of trusted IT solutions and infrastructures.

The present work puts the developments as described above into one context, to reveal the new requirements of a modern municipal IT infrastructure as a basis for converged digital services. General goals are

- an improvement in the overall quality of life
- addressing special needs, e.g. demographics
- energy and climate protection
- improved and more efficient mobility.

The observation is carried out based on the requirements, needs and the development of services for the municipal sector and further described for the example of local climate and emissions mitigation.

3 Infrastructure Convergence

While new developments in different domains can initially be considered separately, typical use cases require a smooth interaction between corresponding infrastructure components.

Using a municipal energy and resource management that in particular considers consumption in buildings, connected with reduction goals, it is necessary to determine specific consumption values in the property, such as electricity, water and heat. For a machine evaluation and continuous improvement of the consumption profile, it is necessary to use data loggers to connect the properties to the internet and deploy transmission facilities for the collected data for central analysis. This should be seen as a simple expression of IoT, which relates primarily to sensor data. To ensure such functionalities, powerful infrastructure components are necessary at least in the area of each municipality, if not even in a broader sense.

If there is an appropriate infrastructure available, it is clear that a purely one-way interpretation of corresponding components is not enough. Thus, the topics Smart Grid and demand side integration should probably not be seen independently of topics such as monitoring and energy management in buildings. The accounting for electricity prices and network conditions as well as the desire to actively control consumption depending on the current energy supply, as realized e.g. for charging heat storage devices in different time frames—when energy is available, motivate the bidirectional exchange of relevant information. This may be extended by strategies that the energy management takes partly place in the cloud and therefore allows the application of new (IT) technologies and information processing in a big data context (in the cloud) or the implementation of other innovations with limited effort.

At the same time this yields to the seamless integration of additional components resulting in a convergence of the Smart Grid and Smart Home, for an extension of data logger technologies in the energy context in buildings and properties towards a bi-directional exchange and introduction of actuators on the one hand. Thus, the system may consider energy only in a decentralized way and therefore operationalizes the smart home concept. On the other hand, the exchange of parameters of each property is also necessary in order to allow the implementation of new ways of energy management at municipal or local level.

The issue of mobility is integrated here in a natural way. Autonomous Driving and electromobility combines mobility with the issue of energy management and smart grid, and even further the idea of a Smart City moves in the focus, because only the implementation of the network with traffic control devices, other vehicles and other external parameters may here enable the implementation of certain functionalities in the corresponding systems.

In the municipal sector these issues are brought together even more closely because of the administrative unit of the municipality as well as its role as a local contact and service point. Infrastructure convergence for municipalities, cities and counties means to provide various technical services in the mentioned areas (1)–(5). Due to the interconnected nature of modern IT systems, these are primarily web applications, smartphone apps, APIs for communication, data and control services, etc.

Because the services are particularly data and IT driven, these are to be built upon a flexible layer of infrastructure, which brings together and manages the information from partly different domains. In the following we want to limit the focus of this need again to keep the example comprehensible. We provide relevant applications and infrastructure components specifically for municipal climate protection and describe applications running on a standardized platform. Furthermore, the need for networking and data exchange is motivated by the example. On the basis of requirements which exceed the current implementation, we motivate on a specific use case the networking and necessary infrastructure convergence as described in the above section for the areas (1)–(5).

4 Convergence in Terms of Energy and Climate Protection

Local climate protection is simply a consequence of the objectives

(Z1) reduction of emissions, particularly through
(Z2) reduction of energy consumption and improved energy efficiency,
(Z3) the use of low-emission/-free power generation but also
(Z4) resource conservation in general through innovative processes.

Usually the objectives are expanded—sustainability of processes, recycling and environmental services, etc.

Systematic links and screws are the mentioned areas (S1)–(S4). Only measured devices and systems can be managed, which is why an emissions measurement and greenhouse gas accounting and the calculation of additional KPI in the municipal sector play a key role in municipal environmental management.

Measurability can take place bottom-up based on data or top-down based on existing statistics, based on emissions generated by consumption or within processes (polluter balance sheet) but also by the energy production (source balance). Various approaches are possible. What they all have in common is that the quality of the balance sheet is based on the quality of available data. Usually, more data is connected with more effort and an automated data collection can help to efficiently, accurately and continuously collect information on current emissions and other KPI in the municipal context. Thus, a central application is one, which supports greenhouse gas accounting for municipalities and counties (Heider et al. 2013; Will et al. 2014).

As described, the application can not be considered isolated. Firstly, there is of course the goal to take measurements to achieve the objectives (Z1)–(Z4). Therefore, an assistance system, e.g. to reduce emission figures, has to be based on information obtained in the accounting process, i.e. in close feedback with the KPI of the respective study area, associated with possible potentials, which can be strictly data-driven.

Secondly, it is useful to collect information on the sectors (S1)–(S4) on the basis of automated recovered data streams, i.e. from monitoring systems that collect information. Especially in the energy field and resource consumption, also a link with local energy management systems may be useful here, e.g. energy management systems for buildings and properties. This in turn is possible much easier, as these systems are networked and based on a single infrastructure.

Thus a central accounting and analysis component is combined with systems for data collection, processing and compression as well as with assistance systems for decision making and possibly also for decision implementation. All these systems are in a continuous exchange of data and therefore ideally based on a uniform IT infrastructure.

5 Municipal Climate Strategy: A Case-Study for Infrastructure Convergence in Local Context

To illustrate the infrastructure convergence described above, we introduce a platform for local climate protection which supports the mentioned targets (Z1)–(Z4) of Sect. 4 by the following functionalities.

(F1) consumption data acquisition manually or telemetrically,
(F2) centralized data storage, processing and analysis,
(F3) domain-specific integrated applications for local climate protection:

a. energy management in municipal buildings and properties,
b. creation of greenhouse gas inventories,
c. scenario generation.

As illustrated in Fig. 2, there is a data-driven platform to manage and collect sensor data. From IoT perspective a locally installed gateway (based on the Raspberry Pi hardware platform) which connects to the internet is required in each munici-

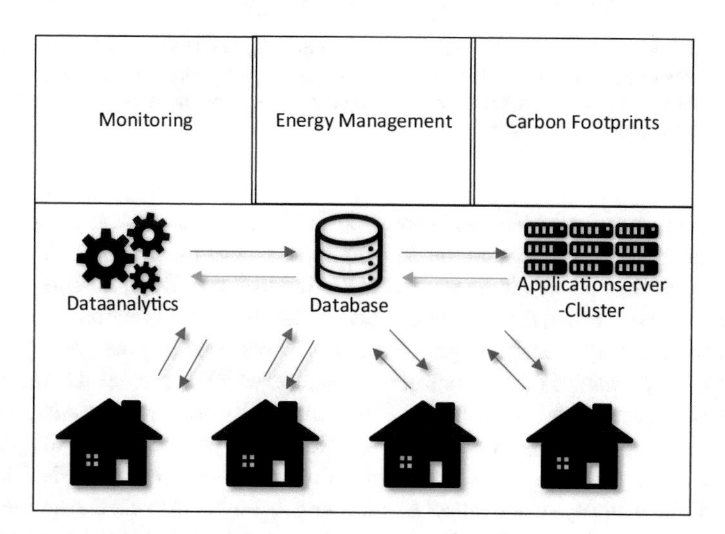

Fig. 2 The basis for the municipal climate strategy platform is a monitoring infrastructure, which connects locally installed gateways to the central cloud based data management platform. These gateways send sensor data continuously to the server where it is aggregated. This central infrastructure provides APIs for applications, such as a municipal energy management system and a system for greenhouse gas accounting

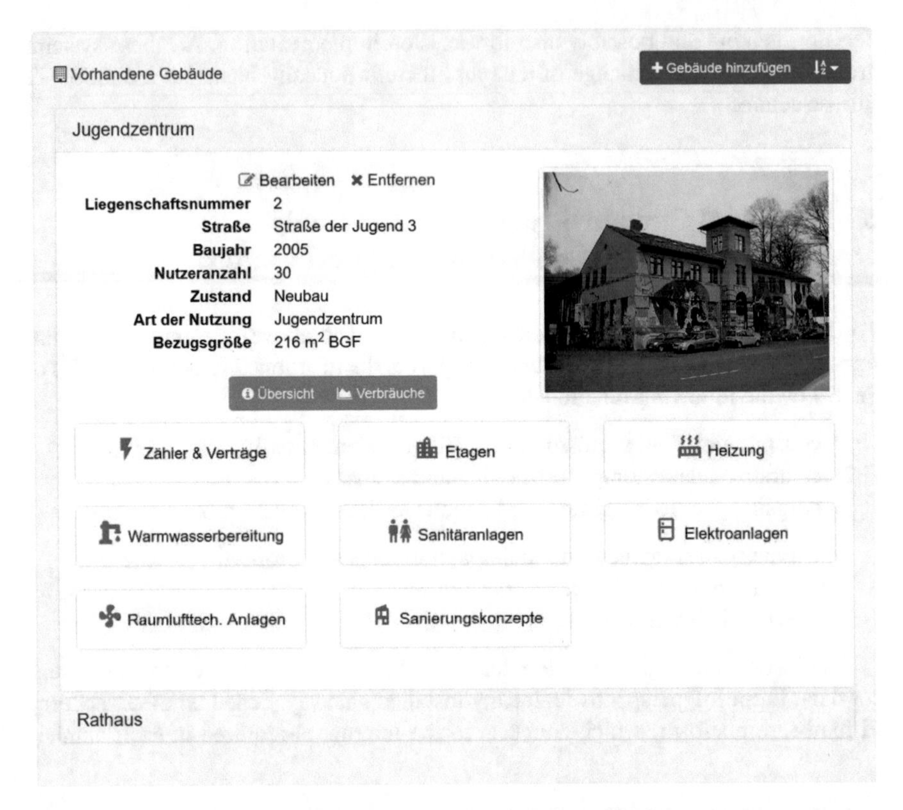

Fig. 3 The web-based energy management system for buildings and properties in municipalities is intended to manage all energy relevant data, e.g. meters and supplier contracts, consumption data for electricity, heat and water as well as technical equipment. Based on this data various management and reporting functions are provided

pal building or property. The gateway itself is connected to several sensors which record and send their consumption data continuously to the gateway. Thereby, the sensor itself relies on standards such as MBUS, wireless MBUS, S0-Bus. Once the infrastructure is present, the devices can also be used for a bi-directional connection. Hence, the gateway is also integrated e.g. the Z-Wave or ZigBee communication standards. This enables for the control of actuators, which combines the areas IoT in energy management and smart homes. The central platform also considers security related issues, since this plays an essential role in data exchange. E. g. all gateways in the buildings and the central data aggregation server are within a VPN. This prevents that the gateways can connect to other targets on the internet except the actual destination server. Required services, e.g. a time server, is provided by the central platform. In addition, the data transmission itself is encrypted.

However, the described monitoring platform serves exclusively as the base for specific applications. These are cloud-based and use the data that previously has been collected. On this side, first, a system for energy management is introduced

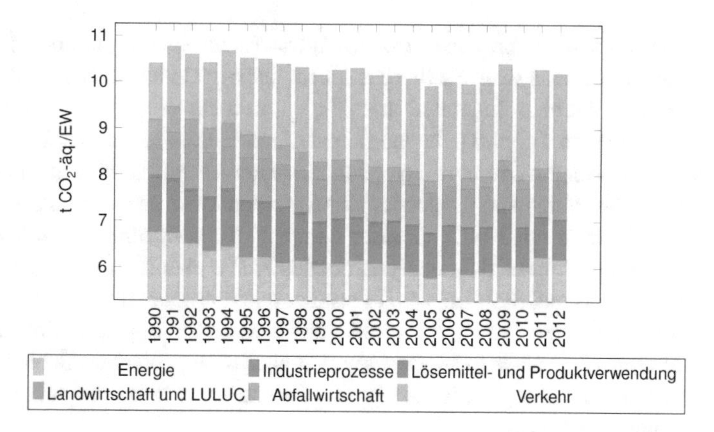

Fig. 4 For greenhouse gas accounting the measurement takes place bottom-up based on data for the sectors energy, transport, industry, agriculture, waste management, aggregated sources and atmospheric deposition. If this data is not available due to certain reasons, statistical data is integrated (top-down). Based on these information it is possible to generate greenhouse gas balances

(see Fig. 3). This allows for the mapping of municipal buildings and properties with their data. Furthermore, it analyzes the consumption data, e.g. for electricity, heating, gas and water, in real-time. In addition, supplier contracts can be integrated with their key data such as tariffs and emissions. This includes various types of resources. It is not only possible to capture these values via telemetrics, but also manually. For this purpose, the present meter structure of the buildings can be mapped in detail in the software. Furthermore various management and reporting functions are provided to take appropriate measures to decrease the consumption.

Since present buildings and properties are only a single part of the issues for climate protection in the municipal context, another application is presented, showing the concept extended to the sectors energy, transport, industry, agriculture, waste management, aggregated sources and atmospheric deposition. Thus, the second application is intended for greenhouse gas accounting based on this data. It is used for a user-centered collecting of data (see Fig. 4) and integrates several statistics to close any data gaps. Certain users such as experts can be integrated within the workflow and are able to generate corresponding greenhouse gas reports from the available information. The system supports and automates this in many ways and provides specific reports for various stakeholders such as e.g. mayors or facility managers.

6 Conclusion and Future Work

We discussed the need of supporting municipalities with certain digital services in climate protection. As we have seen, the services should not be considered separately while implemented and need to be on top of a connected infrastructure landscape. The term convergent infrastructures has been coined for this requirement, which is driven by the use cases in practice. If an IoT infrastructure is present once, it can be used by other services as well. So there is a need for certain interfaces to access the appropriate data. We also illustrated this concept in a case study for infrastructure convergence in the local context, the municipal climate strategy with applications for consumption monitoring of electricity, heat and water as well as systems for energy management in buildings and properties and for greenhouse gas accounting.

Current and future work will consider concepts for controlling of actuators as well as more integration of data analysis concepts. We plan to analyze the data of building renovations and their impacts to predict this for other similar buildings and make automated suggestions. The overall goal is to create a generic and uniform platform for several services to support municipal climate protection, which is built on top of current technologies, is easy to use for end-users in municipalities and is affordable in terms of costs.

References

Blaschke, R., Suhrer, S., Engel, D. (2013). Serviceorientierte Architekturen für Smart Grids. In *IT für Smart Grids*. dpunkt.verlag.

Eckert, C., Krauß, C. (2012) Sicherheit im smart grid. Sicherheitsarchitekturen für die Domänen Privatkunde und Verteilnetz unter Berücksichtigung der Elektromobilität. Alcatel-Lucent Stiftung, Stiftungsreihe, *96*.

Farhangi, H. (2016) Cyber-security vulnerabilities: An impediment against further development of smart grid. In *Smart grids from a global perspective* (pp. 77–93). Springer.

Felden, C. (2013). Smart grids-die intelligenz hat es manchmal schwer. *HMD Praxis der Wirtschaftsinformatik*, *50*(3), 6–15.

Heider, J., Tasche, D., Lässig, J., Will, M. (2013). A greenhouse gas accounting tool for regional and municipal climate change management. In *Sustainable Internet and ICT for Sustainability (SustainIT)* (pp. 1–3). IEEE.

Kern, K., Niederhafner, S., Rechlin, S., & Wagner, J. (2005). *Kommunaler klimaschutz in deutschland: Handlungsoptionen, entwicklung und perspektiven*. WZB Discussion Paper: Tech. rep.

Lassig, J., Will, M., Heider, J., Tasche, D., Riesner, W. (2014) Energy efficiency benchmarking system for industrial enterprises. In *Energy Conference (ENERGYCON), 2014 IEEE International* (pp. 1069–1075). IEEE.

Moradi-Pari, E., Tahmasbi-Sarvestani, A., Fallah, Y. P. (2014) A hybrid systems approach to modeling real-time situation-awareness component of networked crash avoidance systems.

Neirotti, P., De Marco, A., Cagliano, A. C., Mangano, G., & Scorrano, F. (2014). Current trends in smart city initiatives: Some stylised facts. *Cities*, *38*, 25–36.

Rydge, J., Jacobs, M., Granoff, I. (2015) Ensuring new infrastructure is climate-smart.

Shrouf, F., Ordieres, J., Miragliotta, G. (2014) Smart factories in industry 4.0: a review of the concept and of energy management approached in production based on the internet of things paradigm. In*2014 IEEE International Conference on Industrial Engineering and Engineering Management (IEEM)* (pp. 697–701). IEEE.

Wichert, R., Eberhardt, B. (2012). *Ambient assisted living*. Springer.

Will, M., Lässig, J., Tasche, D., Heider, J. (2014). Regional carbon footprinting for municipalities and cities. In *28th International conference on informatics for environmental protection (EnviroInfo 2014)* (pp. 653–660).

Part IX
Others

ICT Support of Environmental Compliance—Approaches and Future Perspectives

Heiko Thimm

Abstract The obligation to conform to environmental regulations requires from companies a set of diverse tasks. A classification of these tasks into central tasks, collaborative tasks, and departmental tasks is proposed and corresponding examples are described. Then, an overview is given of state-of-the-art ICT support approaches for central tasks that are usually performed by cross-departmental compliance managers. The overview includes traditional ICT-support approaches but also advanced approaches that attempt to offer smart and active assistance to compliance managers. Furthermore, the future perspective of ICT support approaches for compliance management are described considering recent ICT advancements.

Keywords Environmental compliance management · Environmental compliance information systems · Environment health and safety information systems

1 Introduction

Over the last decades corporate environmental compliance management for many companies has evolved to a work area with a high complexity. The overall goal of the environmental compliance efforts is threefold and can roughly be described as follows. First, the organization needs to ensure that at all time it knows all the relevant legal regulations for environmental protection. Second, for all relevant regulations the organization has to determine and implement measures that are required in order to fulfill given requirements. Third, the organization needs to document all considerations, measures, and actions targeted on environmental protection (IMPEL 2012, Thimm 2015). This documentation task has to be completed in a way that enables the organization at any time to proof the fulfillment of the previously described two obligations.

H. Thimm (✉)
School of Engineering, Pforzheim University, Pforzheim, Germany
e-mail: heiko.thimm@hs-pforzheim.de

V. Wohlgemuth et al. (eds.), *Advances and New Trends in Environmental Informatics*, Progress in IS, DOI 10.1007/978-3-319-44711-7_26

The growing complexity of environmental compliance management tasks is mainly a fact that arises from today's globalized business environment. In this environment, companies act in increasingly regulated markets, engage with many other business partners in global supply chains, and develop and produce products of high complexity. These conditions require companies to constantly evaluate and improve their organizational design in order to be more agile and competitive for example through new forms of collaboration such as out-tasking and body leasing (Gunningham 2011). A further reason for the increasing complexity is the fact that industry associations, non-governmental organizations, and regulatory agencies at various levels (community level, state level, country level, region level, and global level) are permanently generating new and more stringent regulations and are also frequently revising existing regulations based on performance and stakeholder expectations. The regulations are directed at the typical subjects of environmental regulations such as water, land, waste, radiation, emission, but also fire, and occupational safety. Many companies attempt to cope with these challenges by the use of Information and Communication Technology (ICT) based solutions. It is expected that performance improvements of business processes that are directed at the enforcement of environmental regulations can be achieved through the use of proper ICT solutions. Diverse ICT solutions for environmental compliance management are available covering individual software tailored to a company's specific needs, standard solutions of software vendors, and enterprise EH&S (Environment, Health and Safety) and sustainability management software, that include functions for environmental compliance management (NAEM 2011, Teuteberg and Straßenburg 2009, Thimm, 2015). Furthermore, other solutions that are not primarily developed for compliance management tasks offer compliance enforcement functionalities, too. For example, Computer Aided Design (CAD) software often supports functionality to perform material substitution analyses by product developers in order to reduce hazardous product components.

The remainder of this article is organized as follows. The next section describes compliance management from an organizational perspective and givens an overview of typical tasks performed by cross-departmental compliance management units and other functional units. In Sect. 3 different classes of ICT-support approaches for corporate compliance management tasks are identified. Section 4 contains new opportunities for ICT-based environmental compliance management that are enabled by recent ICT advancements and Sect. 5 concludes the article.

2 Environmental Compliance Management as a Company Obligation

The obligation to conform to environmental regulations requires from companies to complete a diverse set of tasks that are directed at the enforcement of the company's relevant set of regulations. Many organizations attempt to fulfill this obligation by establishing an environmental management system (EMS) according to the ISO Standard 14000 (2009). Welch's handbook with the title "Moving Beyond Environmental Compliance" (1997) among others gives a comprehensive overview of environmental compliance management tasks. Central environmental compliance responsibilities and practical aspects have also been described by Nicolson (2016). Insights into the compliance practice of the manufacturing industry are given in a study of the Aberdeen Group (2011).

In large companies, but also increasingly in smaller companies, often a special organizational unit—called for example "compliance management department"—has been established in order to assure that all environmental compliance management tasks are performed. Usually, the members of this unit who are referred in the following by "compliance managers" are experts with certified permits. The compliance managers initiate, coordinate, control, and manage compliance enforcement tasks. Also, they delegate tasks to and assist other units with respect to environmental compliance duties.

In companies, environmental compliance departments are often defined to be cross-departmental organizational units. It is expected that this approach especially promotes that top priority attention is given to the company's compliance obligations. Furthermore, it is expected that proper resources are allocated to the compliance tasks (CM Tasks) which can be divided into three types. In the organizational model of Fig. 1 these three types of compliance management tasks are illustrated.

Concrete examples of these task types are given in Table 1 starting with several central tasks. The full responsibility for these central tasks is given to the central compliance management unit. The task fulfillment only in rare cases involves other departments. The second task type corresponds to collaborative tasks that require a

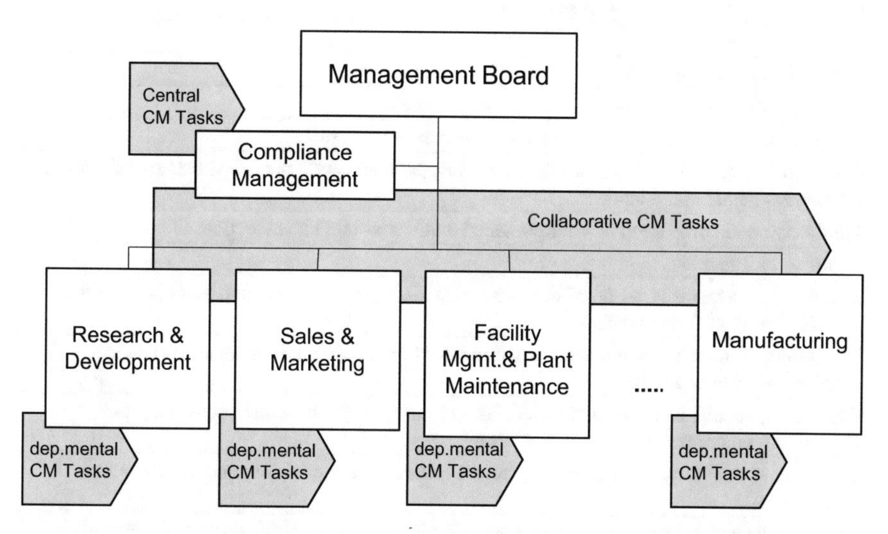

Fig. 1 A differentiation approach for environmental compliance management tasks

close collaboration between the central compliance unit and other units. The third task type corresponds to departmental tasks that are allocated as core tasks to other departments. In the corresponding section of Table 1, in parentheses, examples of departments that perform departmental compliance tasks are given.

The central compliance management tasks can be divided into two complementary task sets. One of the two task sets is directed at the "translation" of generally defined environmental regulations into concrete company-specific obligations and corresponding enforcement measures. This requires that for any regulation it is checked how the regulation fits to the scope of the company considering several general aspects such as size, structure, product portfolio, production processes, logistics and transportation processes. The possible results of the scoping check are: 1. that the regulation is out of the current company scope and does not need further attention, 2. that the regulation is within the current company scope

Table 1 Examples of typical environmental compliance management tasks

Central task
Completion of an initial environmental compliance review in order to establish the company's baseline and a central regulation register
Monitoring of announcements by rule setters at all different levels
Checking the relevance of new regulations and revised regulations
Maintaining the company's central regulation register
Monitoring of environmental compliance-specific indicators and performing escalation management if indicators violate defined thresholds
Reporting information about the company's compliance status to management and to external parties such as customers and regulation agencies
Preparation of mandatory safety data sheets, e.g., for transportation providers
Completing specific inspections such as tank inspections
Collaborative task
Determination of compliance enforcement measures
Implementation of compliance enforcement measures
Checking the effectiveness of compliance enforcement measures
Continuous acquisition of monitoring data about the company's operation that are of relevance for environmental compliance
Preparation and completion of environmental audits and regular inspections
Departmental task
Material optimization in terms of hazardous material being used through material substitution analyses (product development)
Monitoring compounds and machines with respect to certain environmental parameters (production/process control)
Completing special inspections directed, e.g., at emergency plans, fire protection, and fire-fighting equipment (facility management)
Managing special storage facilities for certain hazardous material (inventory/warehouse management)
Managing hazardous waste by use of waste disposal boxes (material/waste management)

which calls for a relevance assessment in order to judge if the regulation is relevant at all. Moreover, one needs to judge whether simple information measures (e.g. instructions, trainings, preparation of safety data sheets) or more substantial measures are required in order to enforce the regulation. Examples of substantial measures include the analysis and revision of processes, modifications of production equipment and transportation vehicles, modifications of compounds and buildings, and even modifications of products and production processes.

The second set of central compliance management tasks consists of tasks to ensure that the enforcement measures are completed according to given temporal constraints and that they lead to specified results. Obviously, the enforcement measures can suffer from all sorts of failures. The typical failures sources are human task executants who possess an inadequate qualification profile or who suffer from human factors, problems inherent to group work, material defects, and malfunctioning of infrastructure components.

The central task "completion of an initial environmental compliance review" targets to establish the company's current position or baseline with respect to environmental regulations (Nicolson 2016). In a first step, the full set of all regulations that are within the scope of the company are obtained. Then, in a second step the relevance of these regulations is determined. The initial compliance review also targets to establish an initial version of a company specific central information repository that is often referred by "regulation register". This register typically contains the regulations that are within the scope of the company, the revisions of the regulations, the results of relevance checks, the enforcement measures, and special company allowances.

The departmental task "monitoring compounds and machines with respect to certain environmental parameters" (e.g. noise, energy consumption, cooling water temperature, waste water and waste air pollution) can involve many different monitoring parameters. The parameter values are obtained either manually by human inspections or automatically through sensors or/and metering devices.

3 ICT Support Approaches for Central Tasks of Environmental Compliance Managers

In the following an overview of selected ICT-based approaches to support central tasks of environmental compliance managers is given. The overview contains traditional approaches for which corresponding commercial software solutions are already available, more recently developed individual solutions, research prototypes, and visionary approaches that so far exist only at a conceptual level.

Workflow Automation for Collaborative Compliance Management Tasks. As described above, the compliance managers' set of tasks include routine tasks which involve not only activities of the central compliance department but also activities of employees from other company areas. For a successful completion of these collaborative asynchronous tasks by possibly dislocated actors, information such as background information, instructions, and results needs to be shared. Also a system instance is required to schedule, monitor, and moderate the activities as well as to enforce given deadlines.

An example of a collaborative task is the determination of measures that are necessary in order to assure that the company conforms to a new regulation. Especially for costly measures, the use of a group decision process seems to be an adequate approach in order to find a well-fitting compliance enforcement measure. Another example is the collaborative completion of measure effectiveness checks. From the departmental specialists, it is expected to check if implemented measures indeed lead to defined results and to prepare a test report. The further administration of the report is usually a responsibility of the compliance managers.

It is possible to automate the routine collaborative tasks of compliance managers such as described above through the use of Workflow Management Technology (Van der Aalst and Van Hee 2004). This requires to specify corresponding workflow models. On the basis of these models individual tasks, i.e. workflow instances are performed with automation support from a workflow engine. The workflow automation approach for routine compliance management tasks is a relatively popular approach. Many compliance management information systems offer integrated workflow support (NAEM 2015) that includes predefined workflow models for typical compliance tasks. The predefined models can be adapted and also they can be complemented by new models in order to meet company specific needs. Workflow support can also be found in more narrow generic software tools such as tracking systems. Some offered software solutions for environmental compliance such as ECOCION's ACTS system (Ecocion 2016) include a tracking system with predefined process models for typical tracking tasks.

Information Provisioning for Environmental Reporting and Informed Decision Making. Information about the company's conformance to environmental regulations can be required for many reasons. Typically these information are needed for reports that are demanded by governmental authorities, business agencies, business partners, shareholders, and the public in general. The information are also needed for internal and external auditors, for reoccurring decisions, and also for internal monitoring tasks. Especially, this includes the monitoring of typical company risks and the risk of non-compliance (Thimm 2015). Naturally, it is the responsibility of the central compliance management department to frequently prepare these information. The traditional approach is directed at typical management reports which should promote a fast comprehension of the entire status and also of abnormal situations by corresponding indicators, tables, and diagrams.

In general, the acquisition of these information requires data selection operations from corresponding data sources, data transformation operations, and further operations such as data aggregation and generation of diagrams. For reoccurring information demands, an automation of these steps through an information provisioning approach can lead to substantial benefits. The automation of the data processing operations described above can be achieved in various ways.

One can assume that at some companies a general full-fletched corporate Data Warehouse infrastructure is available. Through corresponding extensions of this infrastructure, up-to-date compliance content can be automatically delivered to the users as interactive reports and dashboards. This approach, for example, has been studied by Freundlieb and Teuteberg (2009). However, in companies often there is no such central data analytics infrastructure available. One can establish compliance information provisioning with less user convenience and also fewer functional features as compared to the former approach by the use of a general reporting tool and/or a dashboard creation tool.

The compliance managers' demand for automated information provisioning has already been recognized by software suppliers such as Locus Technologies (Locus Technologies 2016). Many solutions are available that offer standardized reports and dashboards for typical compliance management content (NAEM 2015). Some solutions also offer integrated reporting tools to analyze special compliance issues.

Guided Information Access for Complex Judgement and Selection Tasks. As described above, compliance managers need to complete sophisticated tasks in which a complex analysis problem or decision problem has to be addressed. In order to solve these problems properly, compliance managers are required to select the best fitting alternative out of a set of possible alternatives. The best fit is typically achieved through expert analyses, expert judgements, and rational decision making. For example, the determination of the relevance of new regulations and of revisions of existing regulations requires a careful judgement and situation analysis. The determination of measures to enforce environmental compliance presents another example of a non-trivial selection task that compliance managers need to deal with. Limited guidance and experience for the task might be available from measures implemented in the past. However, companies can be confronted with regulations for which it is not sufficient to obtain measures only based on historic information. Sometimes more insights, considerations, analytics, and additional constraints need to be considered and pondered. For example, often customer constraints, cost restrictions, and company standards, which may even conflict with each other, need to be considered in the search for appropriate compliance enforcement measures.

ICT-based assistance can help compliance managers to deal with these sophisticated tasks by offering guided access to relevant information stored in a comprehensive compliance management register and other information sources. The guided information access enables a direct and fast access to exactly the background information that is needed. Examples are functionality for smart information filtering and functionality for navigating through the compliance repository along semantic relationships between the information objects. A typical exploration path through the repository starts with the regulations from which it is navigated to associated enforcement measures and then it is moved further to the results of measure effectiveness checks. Also demanded are functions for intelligent information retrieval, queries to retrieve possible options/ "recommendations", and ad hoc queries. Furthermore, guided information access can also be given in the form of context-specific advice for information that should be considered by the compliance managers in specific compliance management situations.

Alerts and Recommendations for Counteracting Actions to Abnormal and Critical Situations. Compliance managers need to monitor and document compliance indicators that are for example obtained by sensors and metering devices. This includes temperature and pollution parameters of waste air, waste water, and fill levels of special waste disposal containers. Furthermore, some companies' compliance managers also monitor the perimeter of the company through mobile sensing and metering devices in order to detect discharges and to counter act at an early stage. Frequently, monitoring activities are also directed at workforce incidents and reported dangerous spots in the company.

The monitoring of ongoing individual compliance management tasks also belongs to the duties of compliance managers. This duty is required because of the fact that many of the compliance tasks are high-priority and long running tasks that need to satisfy specified temporal constraints and also tasks constraints (Thimm 2015). Missing compliance management tasks and overdue tasks can lead to non-compliance similar to compliance violation through breaches against environmental regulations. Therefore, it is inevitable that the set of scheduled and ongoing central compliance management tasks are frequently monitored.

The monitoring data as described above is made available for the analyses and detection of specific situations. When a critical or abnormal situation is detected special attention and proper counteracting actions are required. Obviously, an automated data checking and alerting of relevant situations can most likely lead to benefits such as a lower risk for unrecognized abnormal/critical situations, a fast resolution of critical situations, and time savings. In general, the principle of data checking and alerting when critical situations are detected can be found in several software solutions such as workflow management systems, project management tools, control centers for order flow, and shop floor control stands. Also in several compliance management systems, monitoring and alerting is supported. For example, Dakota Software's ProActivity Suite performs automated email alerts from compliance calendars and information about action items directed at incidents, audits, and compliance profiles (Dakota Software 2016).

4 Future Perspectives of ICT-Support for Environmental Compliance Management

For many years innovations in the field of ICT core technologies and ICT applications have been proposed with a remarkably high speed. These new innovations are creating new options and advancements for ICT-support of environmental compliance management tasks. In the following the current status of several of these options are described. This is followed by a discussion of major environmental compliance issues that in the light of the Industry 4.0 vision (Wang et al. 2016) need further attention by both researchers and practitioners.

Vendors of dedicated compliance management software solutions have already started to offer cloud-based solutions that support typical Web Clients but also Mobile Clients for tablets and smart phones. Several compliance management tasks can substantially benefit from the possibility to access compliance data from mobile devices. For example, it is possible to fulfill inspection tasks more easily when the needed data can be accessed from anywhere and at any time.

Support of geographical information similar to Geographical Information Systems (GIS) can also be found in various software solutions for environmental compliance management. For example, GIS approaches are used to precisely indicate on a digital map the position of objects relevant for environmental compliance. This includes hazardous materials, danger zones of industrial compounds, exhaust pipes, waste water paths, and special sensing and monitoring devices. In some solutions the digital map is augmented with monitoring information such as temperature, fill levels, air pollution parameters. GIS information are helpful for inspection scenarios, maintenance and repair scenarios, and also for security enforcement in general.

It is possible to advance today's environmental compliance management systems towards "Knowledge-Based Systems" (Zaki & Daud 2001; Dinesh et al. 2008) that provide context-specific recommendations and supervision. For example, the recommendations may help compliance managers to better understand critical situations and to guide their choice for enforcement measures.

The workflow management and process management capabilities of today's compliance management support software, among others, builds on logging data about executing workflows. The use of these data in order to check compliance with business rules and regulations through process mining technologies has been investigated by several research teams (Caron et al. 2013; Weidlich et al. 2011). It seems to be a promising approach to use the corresponding research results in order to develop a new generation of environmental compliance management software that offers capabilities for automated realtime compliance checking.

Sensor-network based monitoring and surveillance of large compounds such as chemical plants, industrial parks, and refineries only recently started to obtain attention in the practice. Through corresponding sensor devices especially environmental data can be permanently monitored including places that are not easy to reach by human inspectors. The monitoring data is delivered as continuous data stream for realtime analysis, realtime presentation, and storage. This can even include mobile sensing data from sensor equipped drones. A typical use case for these approaches is to perform comprehensive data analyses that are directed at discharges. When a discharge is detected, alerts are generated in order to trigger a fast reaction according to corresponding emergency plans. Furthermore, the stored monitoring data can be analyzed in order to identify points of weakness in the infrastructure itself, the security system, and the organization in general.

Big Data Technologies and machine learning approaches can be used in order to support complex judgement tasks of compliance officers. For example, relevance checks for new regulations and for revisions of existing regulations could be prepared by a corresponding "intelligent system component" that computes relevance proposal. The proposal generation could involve not only the announcement of the rule setter and the content of the regulation register. Following the principle idea of "Collective Intelligence" or "Wisdom of the Crowd" the intelligent system component can also use judgements and opinions of others such as comments and recommendations from agencies, consulting companies, and other companies.

A lot of attention is paid to the use of cyber physical system approaches (Guan, et al. 2016) in industry which is often described as the fourth industrial revolution or in short Industry 4.0 (Wang, et al. 2016). The Industry 4.0 vision draws on the concept of the fully digitized factory in which the flow of work, material flow, resource demand and consumption, and environmental drains are fully digitally managed. For the envisioned new cross company Industry 4.0 scenarios, new issues directed at environmental compliance are to be addressed. This includes for example the extension of Industry 4.0 collaboration protocols to deal with discharge events and to comply with corresponding notification obligations.

5 Concluding Remarks

The fully digitized global business world builds on highly intelligent systems that massively interact with each other at all levels of business entities. As of today, to a large degree this business world exists only as a vision. However, with more and more initiatives to promote this vision the current research prototypes of corresponding systems will advance and will eventually reach the maturity required for roll outs at a large scale. It seems that so far environmental compliance management in this future digitized business world has not gained a lot of attention in the research community. It can also be assumed that only a minority of the software vendors of environmental compliance management solutions are with a high pace re-aligning their traditional products or are developing new products for this future. This is a surprising fact since it appears that through proper ICT-support companies can effectively cope with current and forthcoming compliance management challenges.

References

Aberdeen Group. (2011). *Compliance Management in Environment, Health and Safety, White Paper 6991*. Boston, MA: AberdeenGroup.

Caron, F., Vanthienen, J., & Baesens, N. (2013). Comprehensive rule-based compliance checking and risk management with process mining. *Decision Support Systems*, 1357–1369.

Dakota Software, 2016. *ProActivity Suite*. http://www.dakotasoft.com/what-we-do/proactivity-suite/metrics. Accessed 28 February 2016.

Dinesh, N., Lee, I., & Sokolsky, O. (2008) Reasoning about conditions and expectations to laws in regulatory conformance checking. In *9th International Conference on Deontic Logic in Computer Science (DEON)* (pp. 110–124). Springer.

Ecocion. (2016). *Ecocion—Environmental Solutions.* Retrieved January 19, http://ecocion.com/main.html.

Freundlieb, M., & Teuteberg, F. (2009) Towards a reference model of an environmental management information system for compliance management. In Wohlgemuth et al. (Ed) *EnviroInfo 2009 (Berlin) Environmental Informatics and Industrial Environmental Protection: Concepts, Methods and Tools* (pp. 139-214). Shaker Verlag.

Guan, X., Yang, B., Chen, C., & Dai, W. (2016). A comprehensive overview of cyber-physical systems: from perspective of feedback system. *IEEE/CAA Journal of Automatica Sinica*, 1–14.

Gunningham, Neil. (2011). Enforcing environmental regulation. *Journal of Environmental Law, 23*(2), 169–201.

IMPEL. (2012). *Compliance assurance through company compliance management systems 2011/04.* European Union Network for the Implementation and Enforcement of Environmental Law (IMPEL).

ISO. (2009). *Environmental Management: The ISO 14000 Family of International Standards.* Geneva, Switzerland: ISO Central Secretariat 1.

Locus Technologies. (2016). *Executive Environmental Dashboard.* Retrieved on February 28th, 2016 from http://www.locustec.com/environmental_solutions/solutions_dashboard.html (accessed February 28th, 2016).

NAEM. (2011). *2011 EHS MIS Survey—Report, Benchmarking Corporate EHS Management Information Systems*, Washington, DC: National Association for Environmental Management (NAEM).

NAEM. (2015). *2015 EHS and Sustainability Software Buyers Guide, March 2015.* Washington, DC: National Association for Environmental Management (NAEM).

Nicolson, I. T. (2016). Environmental audit in environmentl management. In D. Sarkar, R. Datta, A. Mukherjee & R. Hannigan (Eds.) *An Integrated Approach to Environmental Management* (pp. 465–520). Hoboken, New Jersey: John Wiley & Sons.

Teuteberg, F., & Straßenburg, J. (2009). State of the art and future research in Environmental Management Information Systems—a systematic literature review. In *Information Technologies in Environmental Engineering, Proc. 4th Int. ICSC Symposium Thessaloniki, Greece* pp. 64–77. Berlin: Springer.

Thimm, H. (2015). A continuous risk estimation approach for corporate environmental compliance Management. In *Proc. IEEE 15th International Conference on Environmental and Electrical Engineering, Rome, Italy* (pp. 83–88).

Van der Aalst, W. M. P., Van Hee, K. M. (2004). *Workflow Management: Models, Methods, and Systems.* Cambridge, MA: MIT press.

Wang, S., Jiafu W., Li, D., & Zhang, C. (2016). Implementing smart factory of industrie 4.0: An outlook. *International Journal of Distributed Sensor Networks*, 10.

Weidlich, M., Ziekow, H., Mendling, J., Guenther, O., Weske, M., & Desai, N. (2011). Event-based monitoring of process execution violations. In *Proceedings of Business Process Management (BPM 2011), France* (pp. 182–198).

Welch, T. E. (1997). *Moving Beyond Environmental Compliance—A Handbook for Integrating Pollution Prevention with ISO 14000.* Boca Raton: Lewis Publishers, CRC Press.

Zaki, N. M., & M. Daud. (2001). Development of a computer-aided system for environmental compliance Auditing. *Journal of Theoretics.*

Communicating Environmental Issues of Software: Outline of an Acceptance Model

Eva Kern

Abstract During the last years, the research activities regarding software and its environmental impacts could find their way into the field of "Green IT". Thus, researchers became aware of the fact that software is one of the drivers of the energy consumption by ICT. However, the awareness for these aspects could be much higher—especially in the non-scientific area. On the one side, software developers should be aware of green strategies of software engineering. On the other side, those using the products need to be responded to the effects of using ICT products onto the environment. One idea to transfer the environmental effects of software into a social issue is to create an eco-label for software products. Next to defining criteria, such a label could be laid on, and methods to evaluate software products, it seems to be helpful to identify aspects influencing the acceptance of a certification for green software products. In this context, acceptance means taking the eco-label into account while searching for new software. Hence, the following paper aims at identifying those aspects by applying the Technology Acceptance Model (TAM 2) to the specific case of labelling green software products. We will present a first version of an acceptance model for a label for green software products. It is still work in-progress and needs to be evaluated as a next step. Generally, the aim is to create a tool that can be used to develop an eco-label for software products that will be strongly accepted.

Keywords Green software · Eco-label · Communication · Acceptance model

E. Kern (✉)
Leuphana University Lueneburg, Scharnhorststrasse 1, 21335 Lueneburg,
Environmental Campus Birkenfeld, P.O. Box 1380, 55761 Birkenfeld, Germany
e-mail: mail@nachhaltige-medien.de

V. Wohlgemuth et al. (eds.), *Advances and New Trends in Environmental Informatics*, Progress in IS, DOI 10.1007/978-3-319-44711-7_27

1 Introduction

According to Stobbe et al. (2015), the energy demand of ICT reduced about 15 % in Germany from 2010 to 2015 and might continue to decrease in the next years. This trend is mainly caused by two aspects: (1) the technical improvement of the devices used in households and on workplaces and (2) the regulation of product characteristics based on the European Ecodesign Directive and the European Ecolabel.

Regarding the relation between ICT and the environment, ICT can be generally understood as "part of the problem" and as "part of the solution". (Hilty and Aebischer 2015) That means, on the one hand, using ICT does have (negative) effects onto the environment, e.g. energy and resource consumption. These effects should be reduced ("Green IT"). On the other side, ICT can be used to support activities for the environment ("Green by IT"). Examples for both aspects can be found e.g. in (Hilty and Aebischer 2015; Naumann et al. 2011).

Current implemented activities mainly focus on the hardware side of ICT. Indeed, software issues do also have an environmental impact. (Naumann et al. 2011) To address environmental issues of software, the terms "Green Software" and "Sustainable Software" became established in research contexts. According to (Dick and Naumann 2010), "Sustainable Software is software, whose impacts on economy, society, human beings, and environment that result from development, deployment, and usage of the software are minimal and/or which have a positive effect on sustainable development." Next to this definition, one can find a few more definitions in the current literature with slightly different foci. However, a standardized characterization for green software products and corresponding criteria are still missing (Kern et al. 2015).

By including the software side, the above mentioned effects of the technical improvements and ecofriendly product designs of ICT should be enhance. To do so, the principle of benchmarking and informing about environmental impacts to encourage more energy efficient hardware products will be transferred to software products. In that way, the positive effects should be increased in total.

This paper focusses on the end users of software. The question is how to draw attention to the environmental issues of software on the user side. The idea is to create a label for green software products and, thus, inform about environmental issues of software. As far as we know, there is no eco-label for software products so far. We assume that the information process leads to a more environmental-friendly usage of ICT and especially software products. This could further reduce the negative environmental impacts in this area.

Generally, the effort of developing a label is more justifiable if the target group's interest in the label is high and it is accepted by them. In this context, acceptance stands for taking the label into account while searching for new software products. In order to find out (1) which aspects influence the acceptance of a label for green software products and (2) how these factors are related to each other, we outline an acceptance model for a label for green software products (Labelling Green Software Products - LGSP). This model will be introduced in the following paper. It is still work in-progress. That means, we will introduce a first version of the model. In order to assess it, the next step will be a survey evaluating the proposed factors. Overall, the aim is to create a tool that can be used during the development process of an eco-label for software products and leading to a label that will be strongly accepted.

The paper is structured as follows: Sect. 2 presents current research activities in the context of environmental impacts and labelling of software as well as information about the acceptance of eco-labels. Based on that, the idea of the label and potential stakeholders in this context will be pointed out in Sect. 3. The description of the acceptance model, including the influencing aspects and related hypotheses, represents the main part of the paper (Sect. 4). Following to that, we will present a conclusion and an outlook how to go on in researching the acceptance and the development of a label for green software products in Sect. 5.

2 Related Work

Additionally to researching environmental impacts of hardware products current research activities are paying more and more attention to the software side. Thus, the number of publications dealing with criteria and metrics for sustainable or rather green software products is growing (e.g. (Albertao 2004; Taina 2011; Bozzelli et al. 2013; Kern et al. 2013; Calero et al. 2015)). This is just one aspect of sustainable software and its engineering. Further topics are, among others, procedure models, energy issues and measurements, quality aspects, sustainability issues in requirements engineering and general impacts of ICT on the environment. The idea of labelling green software products is, in comparison with these topics, relatively new (Kern et al. 2015). Accordingly, many research questions arge to be answered. One of these questions is, next to the development of the label itself, the analysis of the potential acceptance and the effectiveness.

It is not yet possible to make a clear statement about the general acceptance or rather the effectiveness of eco-labels. This is mainly due to missing data. (Rotherham 2005) Above, there can be neither a common proposition about the effects of eco-labels nor about the intention to buy labelled products. (Rahbar and Abdul Wahid 2011) Indeed, Loureiro et al. (2001) conclude, referring to different researchers, that “a change in labeling or information can change consumers’ perceptions and behavior”. The label seems to go towards influencing the preferences of the purchaser more than the choice of products. (Loureiro et al. 2001) Overall, the eco-labels can be seen as communication and information tools for consumers. In this context, Rotherham talks about “a tool that communicates expectations and requirements to whoever is interested”, calls it a “catalyst for change” and points out that it “can stimulate a process of environmental awareness raising in companies and the general public” (Rotherham 2005). To do so, working as a complement to other initiatives seems to be sensible.

Apart from the lack of data, a general statement is challenging caused by the multiple characteristics of eco-labels and the varying contexts the labels are used in

(Loureiro et al. 2001; Rotherham 2005; Horne 2009; Rahbar and Abdul Wahid 2011). The effectiveness also depends on the kind of product (Loureiro et al. 2001). Thus, giving generic recommendations for creating and using eco-labels is not possible. However, there are various types of studies addressing product labels to spread environmental issues. The foci of these studies differ: they deal with selected product categories (e.g. food products (Loureiro et al. 2001; Grunert et al. 2014), vehicles (Teisl et al. 2008)), specific labels (e.g. ENERGY STAR (EPA Office of Air and Radiation, Climate Protection Partnerships Division 2015), Blue Angel (Dirksen 1996), EU Energy Label (Heinzle and Wüstenhagen 2012), influence of and consumer trust in eco-labels and productlabels in general (Thøgersen 2000; Borin et al. 2011; Atkinson and Rosenthal 2014, London Economics and Ipsos 2014) or green market and their effects in specific areas (e.g. Malaysia (Rahbar and Abdul Wahid 2011) or Europe (Allison and Carter 2000)).

Since there is no eco-label for software products so far, these research results refer to other products, contexts, and local regions. Nevertheless, they could be taken as inspiration while developing a green software product label. Besides, developing an eco-label is accompanied by coming up with new methods and generating new data. This new-generated knowledge might also contribute to an environmental awareness and positive effects regarding environmental issues.

3 Green Software Products: Labelling and Stakeholders

In the following, we will shortly describe the idea of the label and its potential stakeholders. This accounts for the statement by Rotherham (2005): “It is time to shift the discussion away from ecolabels generally and towards their specific characteristics”.

1. The label in mind should award environmental friendly products, i.e. the underlying criteria evaluate environmental impacts of the product but do not include social or economic aspects.
2. It should award the product itself, i.e. without the development and distribution process, the surroundings, and the organization developing the product. The label refers to the whole life cycle of the product but does not include anything that is not part of the product. That means it values e.g. the algorithm and the modules of an application software but not the packing of it.
3. It should award “green in software” aspects. That means the product itself should be as environmental friendly as possible. Products that support environmental friendly processes or rather make them more environmental friendly (“green by software”) but are not green itself are not included.

In a general sense, we see developers, purchasers, administrators, and users as stakeholders regarding green software (Naumann et al. 2011). Regarding an interest in environmental products D’Souza et al. (2006) describe four types of consumers: environmentally green consumers, emerging green consumers, price sensitive green consumers and conventional consumers. Supplementary, Rotherham (2005) indicates that, next to the final consumer, retails and governments are also important purchasers. As a first step, our acceptance model (Sect. 4) generally takes every possible person interested in software products into account. If the proposed constructs, their connections to each other or the relevance of them differ depending on the stakeholder, their attitude towards green issues and/or the context (private or professional purchaser) will be analyzed in a next step by implementing an appropriate survey.

4 Acceptance Model "Labelling Green Software Products"

The proposed acceptance model regarding a certification of green software products (LGSP) is oriented towards the Technology Acceptance Model 2 (TAM 2) by Venkatesh et al. (2000). Whereas the original TAM states that *Perceived Usefulness* and *Perceived Ease of Use* influences the *Attitude Toward Using* and thus the *Usage of a Technology*, TAM 2 is expanded by factors divided into *Social Influence* and *Cognitive Instrumental Process*. We choose this model since the field of application is environmental computer science: the idea of our model is to find a construct to analyze factors influencing the acceptance of an eco-label for software products and thus the acceptance of green software products in contrast to non-green software products. We assume that both, the attitude toward using those products and social factors, play a significant role in the decision process of buying green or non-green software products. The model can be seen as a tool being useful in the development of green software products and their certification. As a first step, we check TAM 2 for applicability regarding labelling green software products.

Figure 1 shows the proposed model. The theoretical constructs and the causal relationships of the model will be presented in the following sections. Additionally, we propose hypotheses about the relation of the constructs. These hypotheses need to be validated in a next step. They refer to the stakeholders described in Sect. 3 and are related to a specific label for green software products rather than general eco-labels. Finally, the proposed model will be compared to TAM 2.

4.1 Constructs Related to the Technology Acceptance Model

The core of the LGSP Acceptance Model comprises the well-known Technology Acceptance Model (TAM) by Davis (1986). Here, we use the construct names presented in (Venkatesh and Davis 2000) (e.g. "Intention to Use" instead of "Attitude towards Using" and "Usage Behavior" instead of "Actual System Use"). In this context, the construct *Usage Behavior* is understood as the potential use of a

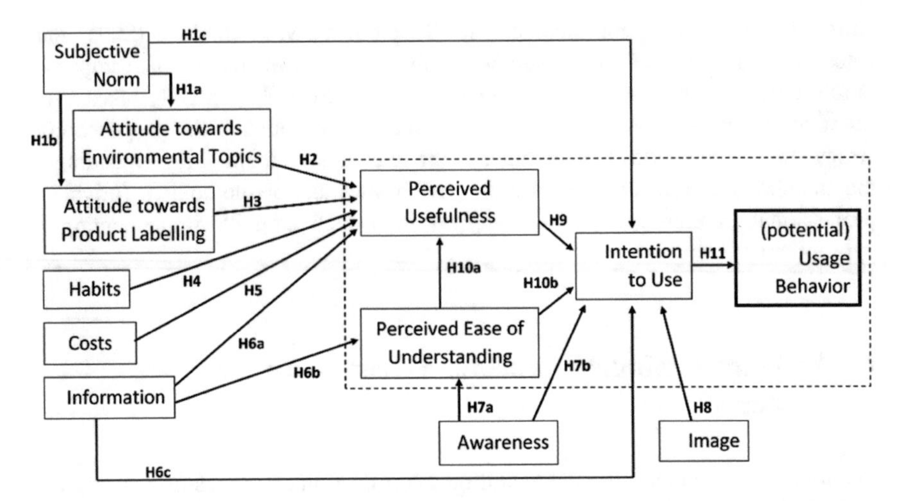

Fig. 1 Outline of an acceptance model for a label for green software products (LGSP Acceptance Model)

label for green software products. We call it "potential" since there is no label for these kinds of software products so far. Using a label means that the label is taken into account while searching for new software products. Precedent to that, there needs to be an *Intention to Use*. Regarding a label for green software products this construct stands for the general notion that an attention to a certification of environmental impacts caused by software makes sense. The *Intention to Use* is caused by the *Perceived Usefulness* and the *Perceived Ease of Understanding*. Here, the proposed model differs from TAM in calling the construct *Perceived Ease of Understanding* instead of talking about "Use". We see the importance in understanding the label including its statements or rather the idea behind the label. It seems to be decisive of paying attention to green issues of software.

H9 The higher the *Perceived Usefulness* that is produced by the certification, the higher the *Intention to Use*.
H10a The higher the *Perceived Ease of Understanding* of the certification, the higher the *Perceived Usefulness*.
H10b The higher the *Perceived Ease of Understanding* of the certification, the higher the *Intention to Use*.
H11 The higher the *Intention to Use*, the higher the potential *Usage Behavior*.

4.2 Constructs Related to the Person of Interest

Next to the four constructs of the TAM, the proposed model contains aspects related to the person who should take the green software product label into account.

According to (D'Souza et al. 2006), having knowledge about environmental issues has the potential to increase the awareness and consumer's positive attitude towards green products. Thus, generating knowledge seems to be the first step to create an interest into environmental topics. It can result in a high involvement concerning environmental issues and favorable influence a person's eco-behavior. (D'Souza et al. 2006, Teisl et al. 2008, Rahbar and Abdul Wahid 2011) Hence, we assume that it has a positive effect onto the acceptance of the label if the person has a positive attitude towards environmental topics as well as towards product labelling in general. The construct *Attitude towards Environmental Topics* refers to the personal interest in environmental topics and the relevance of environmental questions for the personal life. It can be positive or negative. Similarly, *Attitude towards Product Labelling* considers the personal view on product certifications. The last one is to be seen in conjunction with trust that can be understood as one premise of environmentally friendly purchase decisions (Teisl et al. 2008). Additionally to the individual believes, a person is guided by her own habits and trades off new against already-known aspects. D'Souza et al. (2006) point out that "Consumers appear to be somewhat less inclined (31.6 %) to consider known brands as being environmentally safe and seem to rely more on their own experience (66.5 %) in selecting environmentally safe products". This conduct is represented by *Habits*. Not only personal interests and practices may have an influence on the acceptance of a product label. Above, the social circumstances may play a role. Venkatesh et al. (2000) talk about *Subjective Norm* and define it, according to (Fishbein and Ajzen 1975), as a "person's perception that most people who are important to him think he should or should not perform the behavior in question". The social environment of a person brings together personal experiences and leading options of the ones living with the person. Further explanations to the "faith in the co-behavior of others" (Teisl et al. 2008) can be found e.g. in (Berger and Corbin 1992; Gould and Golob 1998).

H1a The more positive the *Subjective Norm* is about environmental topics, the more positive is the personal *Attitude towards Environmental Topics*.
H1b The more positive the *Subjective Norm* is about product labelling, the more positive is the personal *Attitude towards Product Labelling*.
H1c The more the *Subjective Norm* tends to pay attention towards certifications, the higher the *Intention to Use*.
H2 The more positive the personal *Attitude towards Environmental Topics*, the higher the *Perceived Usefulness*.
H3 The more positive the *Attitude towards Product Labelling*, the higher the *Perceived Usefulness*.
H4 The more important individual *Habits* for the person, the lower the *Perceived Usefulness* of the certification.

4.3 Constructs Related to the Product Itself

Additionally to the personal interests and preferences, the label itself matters in order to find out if a label for green software products is taken into account while searching for new products or not. First of all, we think that the awareness of the product label plays a role. The construct *Awareness* covers if the label itself – in other contexts or related to other products – is generally known. Whereas *Image* is related to the person of interest in the TAM 2 by Venkatesh et al. (2000) the construct concerns the reputation of the label in the proposed LGSP model. Is this kind of label generally accepted in the society and well-reputed? Next to the standing, financial issues might have an influence onto the acceptance of a green software label. Certified products might be more expensive than non-certified ones. According to (Loureiro et al. 2001; Rotherham 2005), it is unclear if there is a higher willingness to pay for a product with an eco-label. The construct *Costs* brings this aspect into the proposed model. Furthermore, information material about the certification, criteria the rating is laid on, etc. might be relevant for the acceptance of the label. Above, the construct *Information* covers the relevance of the information content of the product certification itself. The aspect is related to the fact that, according to (Horne 2009), consumers are looking for simple eco-labels providing clear information for decision making. However, one should be aware of oversimplifying the label in order to keep the label's credibility and effects (Teisl et al. 2008; Horne 2009).

H5 The higher the *Costs* for the certified software product, the higher the *Perceived Usefulness*.
H6a The more *Information* about the certification are available, the higher the *Perceived Usefulness*.
H6b The more *Information* about the certification are available, the higher the *Perceived Ease of Understanding*.
H6c The more *Information* about the certification are available, the higher *the Intention to Use*.
H7a The higher the certification *Awareness* in general, the higher the *Perceived Ease of Understanding*.
H7b The higher the certification *Awareness* in general, the higher the *Intention to Use*.
H8 The better the *Image* of the certification, the higher the *Intention to Use*.

4.4 Comparison to the Technology Acceptance Model 2

As already mentioned, the proposed model is based on the TAM 2, introduced by Venkatesh et al. (2000). In comparison to TAM 2, the constructs *Experience*, *Job Relevance*, *Output Quality*, and *Voluntariness* were deleted. Concerning a product label instead of a technical system, we assume that those who are looking for (new) software products, are used to gather information, so that experiences in using different kinds of information sources are taken for granted. Similarly to that, we consider additional information to be generally sensible for the job to find new products and thus do not mention the *Job Relevance*. Taken a product label into account while choosing a product, does not lead to an output comparable to a technical system as addressed by TAM. Thus, we neither include the *Output Quality* nor *Result Demonstrability* in our model. Since it will not be possible to label every software product, the option to choose a green-labelled product is voluntary. Hence, the *Voluntariness* might only have an influence on the acceptance of a green software product label if every product would be certified (as green or non-green).

Although both models—TAM 2 and LGSP—contain the construct *Image*, the definition of it differs. Whereas Venkatesh et al. take the image of a person into account, our model refers the image to the product, here the label for green soft-ware products, itself. Nevertheless, there might be a connection between both views since it seems to be possible that the personal "status in one's social system" (Venkatesh and Davis 2000) benefits from using products labelled having a good image. Indeed, we will not analyze this connection here.

Venkatesh et al. (2000) differentiate between social influence processes and cognitive instrument processes. According to that, it is possible to similarly class *Subjective Norm* in our model as social influenced process. The other constructs that are related to the person (*Attitude towards Environmental Topics* and *Product Labelling* as well as *Habits*) could be understood as more individual, but might also be influenced by social forces. Those constructs of LGSP that are related to the label itself might be seen as equivalences to the cognitive instrument determinants in TAM 2. However, we think, the differentiation between person- and product-related fits better in case of analyzing the acceptance of a label for green software products.

5 Conclusion and Outlook

Summarizing, the paper presents an outline of an acceptance model regarding a certification of green software products. The model is based on TAM 2 by Venkatesh et al. (2000) and theoretically transferred to the new case of application. Whereas the model constructs related to the person should be especially taken into account while promoting the label for green software products, the constructs related to the product are important while developing the label. So far, there is no such kind of certification. Hence, the idea is to find out the aspects influencing the acceptance of the label and incorporate them into the labelling development process. For example, if the hypothesis "The higher the certification *Awareness* in general, the higher the *Intention to Use*" turns true, the development of a certification for green software should be based on well-known eco-labels. On the other hand, the promotion of the label could e.g. focus on those who are already interested in environmental topics if the assumption that the acceptance of those who care about such issues is confirmed.

The next step is to evaluate the proposed acceptance model by conducting a user survey. Doing so, the hypotheses will be validated. Based on the results of the evaluation, the model will be adjusted and completed, if reasonable. Overall, the idea is to construe practical recommendations for labelling green software product from the resulting model (similarly to the given example above). In that way, the consumer should be integrated into the development process of an eco-label for software products. The aim is to design an information strategy including software users instead of simply presenting the results to them (Horne 2009). Simultaneously to the label development, researching the environmental impacts of producing, using and deactivating software should go on since having accurate criterion for an eco-label is very important (D'Souza et al. 2006).

References

Albertao, F. (2010). *Sustainable Software Engineering*. Retrieved November 30, 2010, from http://www.scribd.com/doc/5507536/Sustainable-Software-Engineering#about.

Allison, C., & Carter, A. (2000). *Study on different types of Environmental Labelling (ISO Type II and IN Labels): Proposal for an Environmental Labelling Strategy*. Retrieved January 16, 2016, from http://www.food-mac.com/Doc/200411/study_ecolabel_en.pdf.

Atkinson, L., & Rosenthal, S. (2014). Signaling the green sell: the influence of eco-label source, argument specificity, and product involvement on consumer trust. *Journal of Advertising, 43*(1), 33–45.

Berger, I.E., & Corbin, R.M. (1992). Perceived consumer effectiveness and faith in others as moderators of environmentally responsible behaviors *Journal of Public Policy & Marketing*, 79–89 (1992).

Borin, N., Cerf, D. C., & Krishnan, R. (2011). Consumer effects of environmental impact in product labeling. *Journal of Consumer Marketing, 28*(1), 76–86.

Bozzelli, P., Gu, Q., & Lago, P. (2013). *A systematic literature review on green software metrics*. VU University Amsterdam: Technical Report.

Calero, C., Moraga, M.Á., Bertoa, M.F., & Duboc, L. (2015) Green Software and Software Quality. In Calero, C., Piattini, M. (Eds) *Green in Software Engineering*, (pp. 231–260). Springer.

D'Souza, C., Taghian, M., & Lamb, P. (2006). *An empirical study on the influence of environmental labels on consumers Corporate Communications: An International Journal, 11*(2), 162–173.

Davis Jr, F.D. (1986). A technology acceptance model for empirically testing new end-user information systems: Theory and results. Ph.D. thesis, Massachusetts Institute of Technology.

Dick, M., & Naumann, S. (2010). Enhancing Software Engineering Processes towards Sustainable Software Product Design. In K. Greve & A. B. Cremers (Eds.), *EnviroInfo 2010: Integration of Environmental Information in Europe: Proceedings of the 24th International Conference on Informatics for Environmental Protection, October 6–8, 2010, Cologne/Bonn, Germany* (pp. 706–715). Aachen: Shaker.

Dirksen, T. (1996) HEWLETT-PACKARD's experience with the German BLUE ANGEL Eco-label. In *Proceedings of the 1996 IEEE International Symposium on Electronics and the Environment, 1996. ISEE-1996*, (pp. 302–304).

EPA Office of Air and Radiation, Climate Protection Partnerships Division: National Awareness of ENERGY STAR for 2014: Analysis of CEE Household Survey (2015).

Fishbein, M., & Ajzen, I. (1975) *Belief, attitude, intention, and behavior: An introduction to theory and research.*

Gould, J., & Golob, T. F. (1998). Clean air forever? A longitudinal analysis of opinions about air pollution and electric vehicles. *Transportation Research Part D: Transport and Environment, 3*(3), 157–169.

Grunert, K.G., Hieke, S., & Wills, J. (2014). Sustainability labels on food products: Consumer motivation, understanding and use. *Food Policy 44*, 177–189.

Heinzle, S. L., & Wüstenhagen, R. (2012). Dynamic Adjustment of Eco-labeling Schemes and Consumer Choice-the Revision of the EU Energy Label as a Missed Opportunity? *Business Strategy and the Environment, 21*(1), 60–70.

Hilty, L.M., & Aebischer, B. (2015). ICT for Sustainability: An Emerging Research Field. In Hilty, L.M., & Aebischer, B. (Eds) *ICT Innovations for Sustainability: Advances in Intelligent Systems and Computing*, (vol. 310, pp. 3–36). Springer.

Horne, R. E. (2009). Limits to labels: The role of eco-labels in the assessment of product sustainability and routes to sustainable consumption. *International Journal of Consumer Studies, 33*(2), 175–182.

Kern, E., Dick, M., Naumann, S., & Filler, A. (2015). Labelling Sustainable Software Products and Websites: Ideas, Approaches, and Challenges. In V. K. Johannsen, S. Jensen, V. Wohlgemuth, C. Preist, & E. Eriksson (Eds.), *Proceedings of EnviroInfo and ICT for Sustainability 2015: Copenhagen, September 7–9, 2015* (Vol. 22, pp. 82–91)., Advances in Computer Science Research Amsterdam: Atlantis Press.

Kern, E., Dick, M., Naumann, S., Guldner, A., & Johann, T. (2013). Green Software and Green Software Engineering—Definitions, Measurements, and Quality Aspects. In L. M. Hilty, B. Aebischer, G. Andersson, & W. Lohmann (Eds.), *ICT4S ICT for sustainability: proceedings of the first international conference on information and communication technologies for sustainability, ETH Zurich, February 14–16, 2013* (pp. 87–94). ETH Zurich: University of Zurich and Empa, Swiss Federal Laboratories for Materials Science and Technology, Zürich.

London Economics: Study on the impact of the energy label-and potential changes to it-on consumer understanding and on purchase decisions, London (2014).

Loureiro, M.L., McCluskey, J.J., & Mittelhammer, R.C. (2001). Assessing consumer preferences for organic, eco-labeled, and regular apples. *Journal of agricultural and resource economics*, 404–416 (2001).

Naumann, S., Dick, M., Kern, E., & Johann, T. (2011). *The GREENSOFT Model: A Reference Model for Green and Sustainable Software and its Engineering SUSCOM, 1*(4), 294–304.

Rahbar, E., & Abdul Wahid, N. (2011). Investigation of green marketing tools' effect on consumers' purchase behavior *Business Strategy Series 12*(2), 73–83 (2011).

Rotherham, T. (2005). The trade and environmental effects of ecolabels: Assessment and response United Nations Environment Programme.

Stobbe, L., Proske, M., Zedel, H., Hintemann, R., Clausen, J., & Beucker, S. (2015). Entwicklung des IKT-bedingten Strombedarfs in Deutschland: Studie im Auftrag des Bundesministeriums für Wirtschaft und Energie Projekt-Nr. 29/14.

Taina, J. (2011). Good, bad, and beautiful software. *Search of Green Software Quality Factors CEPIS UPGRADE XII*(4), 22–27 (2011).

Teisl, M. F., Rubin, J., & Noblet, C. L. (2008). Non-dirty dancing? Interactions between eco-labels and consumers. *Journal of Economic Psychology, 29*(2), 140–159.

Thøgersen, J. (2000). Psychological determinants of paying attention to eco-labels in purchase decisions: Model development and multinational validation. *Journal of Consumer Policy, 23* (3), 285–313.

Venkatesh, V., & Davis, F. D. (2000). *A Theoretical Extension of the Technology Acceptance model: Four Longitudinal Field Studies Management Science, 46*(2), 186–204.

Partial Optimization of Water Distribution System Accounting for Multiobjective System Safety

Marcin Stachura

Abstract The paper presents an application of an evolutionary multi-objective optimization method in the estimation a Pareto optimal set of solutions of pipe replacement schedule. An optimization task is a problem with total investment cost, system entropy, hydraulic reliability and water losses as the objective functions. The paper shows that it is possible to optimize a part of a water distribution system without a deterioration of the crucial parameters in the rest of it, which also results in reduction of a computational complexity. Next, the paper presents how to select the pipes that have to be replaced by application of specific coding of decision variables. The results, based on real life example shows that the method is able to identify the pay-off surface characteristic between the proposed objectives.

Keywords Optimization · Reliability · Water distribution system

1 Introduction

Rehabilitation of water supply system has been a research subject for a long time i.e. investigated in (Arulraj and Rao 1995; Engelhardt et al. 2000; Daniela et al. 2008; Alegre et al. 2012; Stachura et al. 2012; Fajdek et al. 2014). Most of the water distribution networks (WDN) were developed to operate in near to optimal conditions. However, during the operation of the water distribution system a number of failures occurs, which are mainly caused by deterioration of pipes and other hydraulic components. Thus, it is very important to frequently improve operation of the water distribution system. Improvements in a system performance can be achieved through replacing, rehabilitating or repairing selected pipes or other components of the system, which is a complex non-linear, multi-objective problem.

M. Stachura (✉)
Institute of Automatic Control and Robotics, Warsaw University of Technology,
Andrzeja Boboli 8, 02-525 Warsaw, Poland
e-mail: m.stachura@mchtr.pw.edu.pl

V. Wohlgemuth et al. (eds.), *Advances and New Trends in Environmental Informatics*, Progress in IS, DOI 10.1007/978-3-319-44711-7_28

2 Problem Formulation

Properly designed water distribution network should supply water for each customer in the required quantity and with the required pressure. This means that under extreme operating conditions occurring in rush hours, the pressure at each point of the network should fit constraints of required economic pressure and the maximum permissible pressure. A placement of the monitoring devices that would detect pressure anomalies is the other problem to be solved (Stachura and Fajdek 2014; Stachura et al. 2015). On the other hand daily, weekly and annual pressure changes can be predicted (Stachura and Studzinski 2014; Studzinski et al. 2013), but during the maintenance of each water supply system hydraulic parameters of water pipes change, due to higher hydraulic resistance of the pipes. Hence, in order to supply customers water with the required pressure, it is necessary to increase pumps head in order to overcome the additional pressure loss and to carry incidental large peak flows required for firefighting (Filion 2008). The pump scheduling problem can be solved in different ways using evolutionary (Fajdek et al. 2015) or nonlinear least squares algorithms (Pytlak et al. 2013).

On the other hand, the deteriorated pipes can be replaced. The (sub)optimal diameters of new pipes have to be selected with regard to minimize investment costs and simultaneously maximize network reliability. Herein a hydraulic reliability is the extent to which a network is able to service demands under mechanical and/or hydraulic failures (Prasad and Park 2004; Mays 1996). In principle, a network layout itself could be subject to optimization as well. However, due to extreme complexity of this task which is additionally largely constrained by a location of local infrastructure, this is mostly left out of consideration in pipe network optimization (Prasad and Park 2004).

Difficulty in resolving a pipe replacement task is mainly with a large space of possible solutions (Coello et al. 2006), which computation complexity is $\sim K^{N_w}$ where N_w is a number of pipes that might be replaced and K is a number of available pipe types (in other words a type of new pipes that can be installed instead of an old deteriorated pipe). So, the problem of revitalization of water supply system is a *NP* hard problem (Yates et al. 1984), which solution with use of conventional methods may cause numerical problems. Another issue is an implementation of planned revitalization. Commonly urban WDN cannot afford to optimize whole network. Hence the paper shows that it is possible to optimize a part of WDN without a deterioration of the crucial parameters in the rest of it (which also results in reduction of a computational complexity). At last the paper presents how to select the pipes that have to be replaced by application of specific coding of decision variables.

3 Optimization Objectives

Walski (2000) pointed out three competing goals for water distribution system operation: (1) maximize reliability, which is achieved by keeping the maximum amount of water in storage; (2) minimize costs of energy used by pumps, which is achieved by operating pumps against as low a head as possible; and (3) meet water quality standards, which involves minimizing the time the water is in the distribution system. Herein the optimization is accounting for system safety, hence WDN reliability was defined as follows: it is necessary to deliver a water to all customers which can be described as maximizing: (a) the number of ways through which water is delivered to the consumers and (b) available surplus hydraulic power, that might be used in case of emergency. These objectives can be mapped to system entropy and resilience indices (see below). Next it is necessary to minimize: (c) the costs of possible leaks which was mapped into water losses index and (d) costs of investments necessary to rebuild the network. The water age was omitted, because its mainly driven by water consumption and proper control of pump stations. These optimization objectives are described below:

System entrophy refers to the idea that water demand at a given internal node of a network is ideally met using multiple different paths to that node. Required flow should be distributed over these routes as evenly as possible. Hence, if one pipe would be damaged, alternative routes would exist and demand could still be supplied. Let I_s be the index of entropy of the entire network:

$$I_s = \sum_{j=1}^{N} \frac{\widetilde{q}_j}{\widetilde{Q}} \left(I_{s_j} - \ln \frac{\widetilde{q}_j}{\widetilde{Q}} \right) \tag{1}$$

where: $\widetilde{q}_j = \sum_{i:j=dest(i)} |q_i|$ is water flow transported towards node j by the incoming pipes i connected to it (thus for reservoir nodes $q_j = 0$), $I_{s_j} = \sum_{i:j=dest(i)} \frac{-|q_i|}{\widetilde{q}_j} \ln \frac{|q_i|}{\widetilde{q}_j}$ is entropy index of node j, $\widetilde{Q} = \sum_{j=1}^{N} \widetilde{q}_j$ is the sum of all incoming flows in the network.

Resilience Index refers to a concept of resilience, which was introduced by Todini (2000) to account for the fact that WDN are designed as looped systems in order to increase the hydraulic reliability and the availability of water during pipe failure. In a case of change in demand or a pipe failure water flow will change and an original network is transformed into a new one with higher internal energy losses. Providing more power than required at each node could be one of the possible ways to avoid this problem. This surplus power has been used by Todini (2000) to characterize the resilience of the looped networks. The resilience index can be written as:

$$I_r = \frac{\sum_{i=1}^{N} q_{d,i}(h_i - h_{req,i})}{\left(\sum_{j=1}^{N_F} q_{r,j} h_{r,j} + \sum_{j=1}^{N_P} P_k/\gamma \right) - \sum_{i=1}^{N} q_i h_{req,i}} \tag{2}$$

where $q_{(r,j)}$ and $h_{(r,j)}$ are the discharge and the head, at each reservoir j, P_k is the power introduced into the network by the pump k, γ is the specific weight of water and N_P is the number of pumps.

Water losses might be replaced with a risk coefficient which takes into consideration impact of a leak together with a possibility that a failure might occur. The aggregated coefficient is calculated for each pipe as follows:

$$I_{WL,i} = T_{ef} q F \tag{3}$$

where: i index of a pipe, q is a flow of water losses $q = C\left(\frac{p_p+p_k}{2}\right)^{\gamma}$, p_p is a pressure in a node which is a beginning of a pipe, p_k is a pressure in a node which is the end of a pipe $\gamma = 0.5$, F is an annual frequency of failures of the selected ith pipe type and T_{ef} is a predicted exploitation time of a pipe: $T_{ef} = \sum_{t=1}^{T} \frac{1}{(1+D)^t}$, where: T is a current exploitation time in years, D is a discount rate (2–5 %). Hence total water losses in a network can be estimated as follows:

$$I_{WL} = \sum_{i}^{N_p} I_{WL,i} \tag{4}$$

where N_p is the number of pipes in a network.

Total cost of investment includes the capital costs of pipes together with the value of the energy consumed by pumps during a specified period. The present worth of energy costs is based on an interest rate of 12 % and an amortization period of 20yr.

$$I_{TCI} = \sum_{j=1}^{m} l_j k_{k,j} + E_p \tag{5}$$

where: I_{TCI} value of the optimized indicator, l_j is a length of the jth pipe, $k_{k,j}$ is unit cost of the k type link selected from the catalog of pipes that could be inserted in place of link k, E_p is an energy consumed by pumps.

Failure index can be used both to identify infeasibilities during the optimization process and to evaluate and compare the effect of pipe failures.

$$I_f = \frac{\sum_{i=1}^{N} I_{f,i}}{\sum_{i=1}^{N} q_{d,i} h_{req,i}}, \quad I_{f,i} = \begin{cases} 0 & \forall i : h_i \geq h_{req_i} \\ q_{d,i}\left(h_i - h_{req,i}\right) & \forall i : h_i < h_{req_i} \end{cases} \tag{6}$$

4 Multiobjective Evolutionary Optimization

If one wants to compare candidate solutions on quality, it is possible to map the objective functions to a single quality measure, and then let the optimization algorithm search for a single best solution. This *a priori* approach requires preferences to be formulated beforehand regarding the objectives involved (Beume et al. 2007). On the other hand, by adopting the *a posteriori* approach, the decision maker postpones the final decision on preferences between conflicting objectives until having been presented with a set of compromise solutions. Such sets can be obtained by Pareto optimization (Coello et al. 2006; Deb 2001). The aim herein is to approximate the Pareto-optimal set that consists of all solutions that cannot be improved in any one objective without degradation in another (Zitzler et al. 2000).

In the paper the following objective functions were assumed to be minimized:

$$\mathbf{y} = \left(-\exp I_S, -I_r, I_{TCI}, I_{WL}\right) \quad (7)$$

subject to:

$$g_1(\mathbf{x}) = I_{f,i} \leq 0. \quad (8)$$

The practical application of evolutionary algorithms requires the adoption of appropriate presentation method for the links that are selected for replacement. The most natural is to save them in a vector of integers. The essence of this form is to create a one-dimensional array with a number of columns equal to the number of selected links. Each of the array element can take values from 1 to L_t (number of possible new pipe types). During each calculation of the objective function simulation is executed, in which each link from the vector is replaced by the link from the available new types. This type of coding is commonly applied. To fasten the computation process and replace only a part of pipes from the whole set, a novel coding was proposed. The genome can be presented in the following form:

$$\mathbf{x} = [N_R, \xi, \mathbf{k}] \quad (9)$$

where N_R is a number of pipes selected to be replaced and $\xi = [\xi_1, \xi_2, \ldots, \xi_{N_p}]$, $\mathbf{k} = [k_1, k_2, \ldots, k_{N_p}]$, N_P is a number of pipes that might be replaced, ξ is a float vector of each pipe "weight factor" in planned revitalization, $\xi_i \in (0, 1)$, $\mathbf{k}$ integer vector of the pipe types (the type of pipe that will be put instead of the replaced one), $k_i \in (0, L_t - 1)$, where L_t is the number of possible pipe types.

The algorithm of decoding the above presented genome is as follows:

1. Sort decently the pipes according to their "weight factors" ξ,
2. Select first N_R pipes,
3. Divide vector $\mathbf{k}$ into N_p/N_R parts,
4. Every part of the vector $\mathbf{k}$ have to be summed up and divided modulo by L_t. The result is an index of a new pipe type that the respective pipe have to be replaced with.

For example the above algorithm was presented in Fig. 1, where the number of pipes N_p is 6 and number of new pipe types L_t is 9. In the example the number of pipes that are selected to be replaced N_R is 3. Taking into consideration the values of ξ vector pipe with indexes $i = 1, 4, 5$ will be replaced. Pipe $i = 1$ will be replaced by a pipe type 2, pipe $i = 4$ will be replaced by a pipe type 3 and Pipe $i = 5$ will be replaced by a pipe type 2.

In the context of multiobjective optimization various evolutionary algorithms are applied in literature, the most popular being `NSGA-II` by Deb et al. (2002), `SPEA2` by Zitzler et al. (2001), and `MOGA` by Fonseca and Fleming (1993). However, in (Tiwari et al. 2011) it was shown that the `AMGA2` by Tiwari et al. (2011) procedure has better performance on the ZDT, DTLZ, and practical engineering problems. `AMGA2` procedure operates on a small population while maintaining a large archive

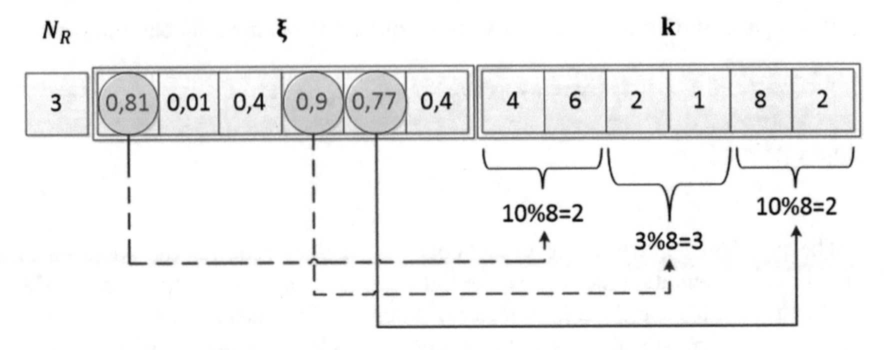

Fig. 1 The example of a proposed genome decoding

of already achieved good solutions. Such an approach to the problem has several significant advantages (Tiwari et al. 2011) so this algorithm was used in the presented research.

5 Results

The tests of the presented methodology were performed on the example of water distribution system in Glubczyce, Fig. 2. It is a town in the Opole province, Poland, in the district of Glubczyce situated on the river Psina. The town is inhabited with 23 778 people. Water production in 2011 was estimated at 2.782 m^3/day. The WDN in Gubczyce was selected mainly due to a relatively low complexity of a network, which helped to conduct the simulations and to analyze the results. The necessary simulations were performed in EPANET2 toolkit. Next a specialized software was implemented so as to calculate the above presented quality indices and to couple the simulation software with an optimization algorithm.

Fig. 2 A group of pipes that were selected for revitalisation

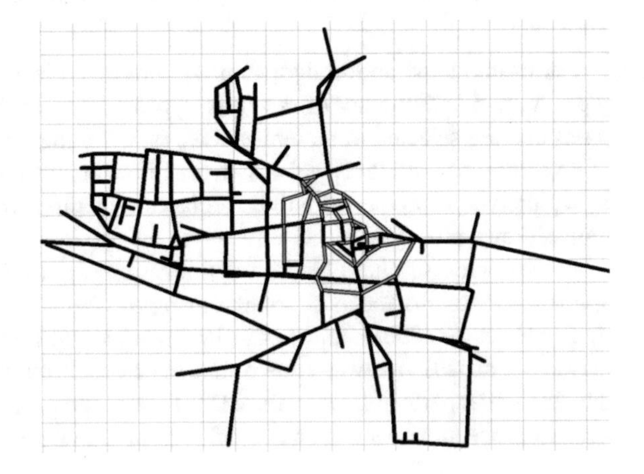

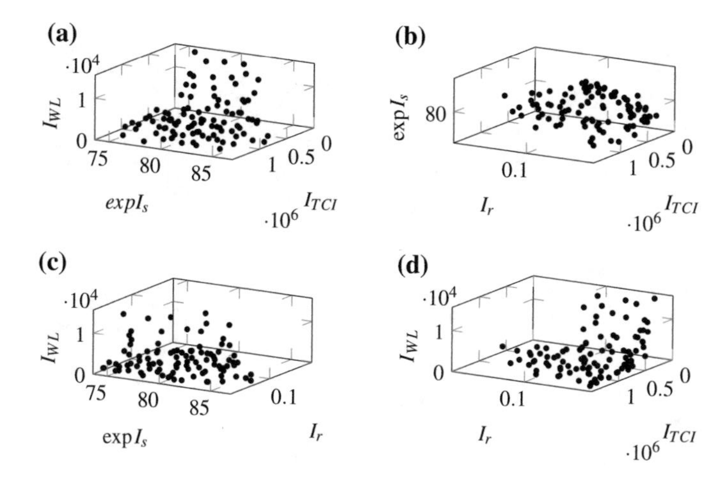

Fig. 3 Pareto optimal solutions of Glubczyce case study

The pipes that were taken into consideration are marked with red color (Fig. 2). Results of the total investment cost, resilience index, system entropy and water losses impact are presented in Fig. 3. The solution space and the Pareto optimal front for the Glubczyce case study is a surface in the four dimensional objective space. This indicates that the tradeoffs among the objectives are not dominated by one objective. The main differences in the solution costs are from the pipe costs, whereas a pumping cost was not leading factor that differentiate the solutions. The water losses can be decreased together with increasing total cost of investments, Fig. 3a, b. Next, in order to provide higher reliability in the network, its capacity has been increased to respond to possible leaks. However, increase in resilience index had a negative effect on water losses index, Fig. 3b. At last system entropy was the factor that differentiate the computed solutions in terms of Pareto optimality.

The objective functions cannot be simply mapped into boosting up or decreasing the values of pressure at each node, Fig. 4a, b. The results of simulations conducted with extreme solutions of each of the objective functions shows that in some nodes achieving the lowest water losses, in whole network, is related with boosting the pressure up in an independent node, Fig. 4a, and in others the same effect takes place when the pressure is lowered in the independent node, Fig. 4b. The corresponding effect was observed with reliability index. At last the conducted computations show that it is possible to optimize a part of a network without deterioration of the parameters in the rest network nodes. Figure 4a, b shows that pressures in an area where pipes were replaced have significantly changed, whereas the pressure in a node which lies outside this region, Fig. 4c, remain almost the same for all tested scenarios.

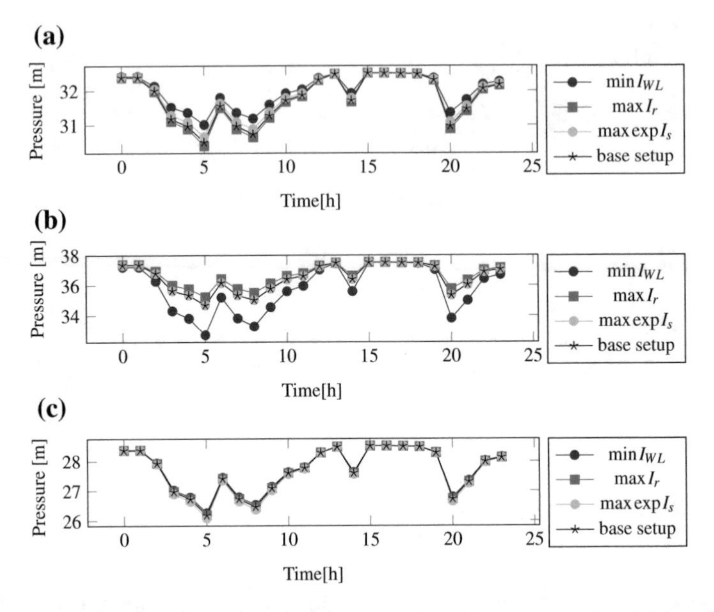

Fig. 4 Pressures in selected nodes within the area where pipes were replaced (**a**) and (**b**) and outside this area (**c**)

6 Conclusions

The paper investigated the application of an evolutionary multiobjective optimization method in the estimation a Pareto optimal set of solutions of network rehabilitation. The optimization task was a problem with total investment cost, network reliability and water losses as the objective functions. The design variables were parameters describing which pipe have to be replaced with which new pipe type. The results, based on Glubczyce case study shows that the method is able to identify the pay-off surface characteristic between the proposed objectives. Presented simulation results show that the task of optimization the network can be distributed into smaller tasks in which only a part of the network is being modified. The presented approach is going to be incorporated in a decision support system implemented in WDN in Gubczyce. Next, the methodology is going to be implemented in similar systems in other WDNs in Poland, for example in Upper Silesian region.

Acknowledgments The research was co-supported by the Polish National Center for Research and Development (NCBiR) project no. PBS3/B3/31/2015 entitled "ICT system for urban water supply system safety using the tools of computer science and systems analysis".

References

Alegre, H., Covas, D., Coelho, S. T., do Almeida, M. C., & Cardoso, M. A. (2012). An integrated approach for infrastructure asset management of urban water systems. *Water Asset Management International*, *8*, 10–14.

Arulraj, G. P., & Rao, H. S. (1995). Concept of significance index for maintenance and design of pipe networks. *Journal of Hydraulic Engineering*, *121*(11), 833–837.

Beume, N., Naujoks, B., & Emmerich, M. (2007). Sms-emoa: Multiobjective selection based on dominated hypervolume. *European Journal of Operational Research*, *181*(3), 1653–1669.

Coello, C., Lamont, G., & Veldhuizen, D. V. (2006). *Evolutionary Algorithms for Solving Multi-Objective Problems (Genetic and Evolutionary Computation)*. Secaucus, NJ: Springer-Verlag New York Inc.

Daniela, F. H., Gerald, G., Birgit, K., Jorg, K., Johannes, H., & Harald, K. (2008). Pirem-pipe rehabilitation management developing a decision support system for rehabilitation planning of water mains. *Water Practice & Technology*, *3*(1)

Deb, K. (2001). *Multi-Objective optimization using evolutionary algorithms*. New York, NY: Wiley.

Deb, K., Pratap, A., Agarwal, S., & Meyarivan, T. (2002). A fast and elitist multiobjective genetic algorithm: Nsga-ii. *IEEE Transactions on Evolutionary Computation*, *6*(2), 182–197.

Engelhardt, M. O., Skipworth, P. J., Savic, D. A., Saul, A. J., & Walters, G. A. (2000). Rehabilitation strategies for water distribution networks: a literature review with a uk perspective. *Urban Water*, *2*(2), 153–170.

Fajdek, B., Stachura, M., & Studziski, J. (2014). Optimization of water supply network rehabilitation using genetic algorithms. In *Industrial Simulation Conference.*

Fajdek, B., Stachura, M., & Studziski, J. (2015). Water distribution network optimization using genetic algorithm. In *13th Annual Industrial Simulation Conference*

Filion, Y. (2008). Single-objective deterministic versus multi-objective stochastic water network design: Practical considerations for the water industry. In *Proceedings of IMechE, vol Part O: Journal of Risk and Reliability*, (pp. 655–665)

Fonseca, C., & Fleming, P. (1993). Genetic algorithms for multiobjective optimization: Formulation, discussion and generalization. In *Proceedings of the 5th International Conference on Genetic Algorithms, Morgan Kaufmann*, (pp. 416–423)

Mays, L. (1996). Review of reliability analysis of water distribution systems. In Tickle K (Ed.) *Stochatic Hydraulics 96: Proceedings of the Seventh IAHR International Symposium*

Prasad, T., & Park, N. S. (2004). Multiobjective genetic algorithms for design of water distribution networks. *Journal of Water Resources Planning and Management*, *130*(1), 73–82.

Pytlak, R., Tarnawski, T., Fajdek, B., & Stachura, M. (2013). Interactive Dynamic Optimization Server (IDOS) connecting one modeling language with many solvers. *Optimization Methods and Software (May)*, 1–21. http://www.tandfonline.com/doi/abs/10.1080/10556788.2013.799159

Stachura, M., & Fajdek, B. (2014). Planning of a water distribution network sensors location for a leakage isolation. In: *Proceedings of the 28th EnviroInfo 2014 Conference*, (pp. 715–722)

Stachura, M., & Studzinski, J. (2014). Forecasting hydraulic load of urban water supply system using tsk fuzzy models. *Ochrona Srodowiska*, *36*(1), 57–60.

Stachura, M., Fajdek, B., & Studzinski, J. (2012). Model based decision support system for communal water networks. In *Proceedings of Industrial Simulation Conference ISC, Brno*

Stachura, M., Studzinski, J., & Fajdek, B. (2015). Model based leakage isolation in water distribution system: A neural classifier approach. In *Proceedings of the 29th EnviroInfo and 3rd ICT4S Conference 2015*, (pp. 142–147)

Studzinski, J., Bartkiewicz, L., & Stachura, M. (2013). Development of mathematical models for forecasting hydraulic loads of water and wastewater networks. *EnviroInfo 2013* (pp. 736–748). Environmental informatics and renewable energies. Proceedings. Pt. II: Shaker Verlag.

Tiwari, S., Fadel, G., & Deb, K. (2011). Amga2: Improving the performance of the archive-based micro-genetic algorithm for multi-objective optimization. *Engineering Optimization*, *43*, 377–401.

Todini, E. (2000). Looped water distribution networks design using a resilience index based heuristic approach. *Urban Water*, *2*, 115–122.

Walski, T. (2000). *Hydraulic design of water distribution storage tanks* (p. 10). New York, chap In Water Distribution Systems Handbook: McGraw-Hill.

Yates, D., Templeman, A., & Boffey, T. (1984). The computational complexity of the problem of determining least capital cost designs for water supply networks. *Engineering Optimization*, *7*(2), 143–155.

Zitzler, E., Deb, K., & Thiele, L. (2000). Comparison of multiobjective evolutionary algorithms: Empirical results. *Evolutionary Computation*, *8*(2), 173–195.

Zitzler, E., Laumanns, M., & Thiele, L. (2001). Spea2: Improving the strength pareto evolutionary algorithm. Technical report 103. Technical report, Computer Engineering and Networks Laboratory (TIK) Swiss Federal Institute of Technology (ETH), Zurich, Switzerland

Towards Environmental Analytics: DPSIR as a System of Systems

Corrado Iannucci, Michele Munafò and Valter Sambucini

Abstract DPSIR (Driving forces, Pressures, States, Impacts, Responses) is a framework addressing the needs of environmental data reporting and assessment, extensively exploited by the European Union, its Member States and associated countries. The DPSIR framework does not model the environment, however it implies a systems model. Making such model explicit allows users to better understand the semantics of the five DPSIR components, therefore improving the harmonization and quality of the reported data, in particular in order to process them as Big Data. To such purpose, the tools provided by the systems theory are useful, specifically the system-of-systems (SoS) point of view. On such basis, a conceptual model of the relationships between the constituent systems of the socio-ecological system (SES) is proposed. As a result of such analysis some possible evolutions of DPSIR are envisaged, in terms of expanded and refined environmental data flows. Such evolutions aim to build a better integrated environmental knowledge base, as a prerequisite to a wider exploitation of analytics for the relevant decision processes and management activities.

Keywords DPSIR · Socio-ecological system · System-of-systems · Environmental reporting · Big Data

C. Iannucci (✉)
IPTSAT Srl, Via Sallustiana 23, 00187 Rome, Italy
e-mail: corrado.iannucci@gmail.com

M. Munafò · V. Sambucini
ISPRA Istituto Superiore per la Protezione e la Ricerca Ambientale (Italian National Institute for Environmental Protection and Research), Via Vitaliano Brancati 48, 00144 Rome, Italy
e-mail: michele.munafo@isprambiente.it

V. Sambucini
e-mail: valter.sambucini@isprambiente.it

V. Wohlgemuth et al. (eds.), *Advances and New Trends in Environmental Informatics*, Progress in IS, DOI 10.1007/978-3-319-44711-7_29

1 Introduction

Since the mid-Nineties, increasing amounts of both structured and unstructured environmental data are being collected on a routinely basis in Europe. Such collected data increasingly show eminent features of Volume (the amount of data and the throughput of data flows), Variety (the semantic heterogeneity of data), Velocity (the frequency of creation, modification and use of data) that are three characteristics associated to the Big Data approach (Pietsch 2013; Chen et al. 2014). The Big Data approach relies upon a large set of logical, mathematical and statistical tools (usually referred to as analytics), potentially able to deal with complex social and physical problems, meeting the evolving needs of the various stakeholders (citizens, industries, administrations) and therefore eliciting a fourth characteristic of Big Data, Value (the usefulness of data processing).

Kaisler et al. (2014) list sixteen classes of such tools (explicitly including those pertaining to the systems science) and advocate their joint exploitation as "advanced analytics" in order to get actionable descriptive, predictive and prescriptive results and therefore to support decisions. Sound decisions rely upon good data; achieving and safeguarding the data quality is a task to be carried out during the whole life cycle of the data, from their generation by the producers to the processing by the final users. The effort to be associated to such task is amplified when dealing with large and ever increasing data: in such context, the Veracity (in terms of data consistency and trustworthiness) is mentioned as the fifth characteristic of Big Data (Demchenko et al. 2013). Specifically, the management of the environment requires timely and reliable data. Taking into account that the earliest phases of the data life cycle prove to be the most critical for the quality, it appears to be of interest to review the conceptual framework that streamlines the data acquisition.

The European Environment Agency (EEA) is in charge of managing network-centric infrastructures, from EIONET (connecting the data providers) to SEIS (the EU-wide environmental information system) and to others, that allow the collection and sharing of quantitative data delivered by the 28 Member States of the European Union, as well as other countries with whom it holds protocols of collaboration (EEA 2015). The data flows have the Reporting Obligations (stored in a dedicated ROD database at http://rod.eionet.europa.eu/index.html) as legal basis; their exploitation requires the adoption of a framework of indicators that permit the comprehension and sharing of information (EEA 2014).

EEA has developed its own framework, relying on the previous work of the Organization for Economic Co-operation and Development (OECD 1994) and of the United Nations (UN 1997). Such framework includes five components: Driving forces (or Drivers), Pressures, States, Impacts, Responses, whose initials form its acronym DPSIR. At times, the practical application of DPSIR comes up against perplexities, apparently deriving from the operational definitions currently adopted for the components of the framework. In other cases, taking concrete situations into account, it has been deemed necessary to extend and/or modify the framework. That does not, of course, affect the overall validity of the DPSIR framework; although it highlights the need to proceed with the definition of the conceptual basis upon which DPSIR is built.

The DPSIR framework has been developed with explicit reference to systems theory (Smeets and Weterings 1999). Therefore, it seems natural to test if the methodological instruments proper to this theory may contribute to the elucidation of the conceptual basis whereon the framework itself stands. In other words, it is deemed useful to explicate the DPSIR framework's underlying system model.

2 The Model Implied by DPSIR

DPSIR streamlines the data flows, does not model the environment; however, DPSIR implies a model i.e. a conceptual description of its object. The analysis of such model elicits relevant features of the framework that have an impact upon the characteristics of the data flows (specifically, upon the Veracity as defined in the above). As a starting point, Fig. 1 shows a graphic schema of the dynamic interrelations between the DPSIR components as represented by Smeets and Weterings (1999). The schema represents DPSIR as a set of components (indicated in the ovals) connected by arrows, symbolizing the dynamical actions that originate from a given component and produce effects on another.

The relevant literature shows that the definitions of the DPSIR components are in some cases adapted to suit specific needs; this can jeopardize the semantic interoperability of the collected data. It should be noted that the EEA Environmental Terminology and Discovery Service (http://glossary.eea.europa.eu/) provides explicit definitions for the components D, P, S and R only (mainly in union to the term "indicator"); Smeets and Weterings (1999) define also the component I, however neglecting the role of a reference baseline in expressing it. On the basis of such sources, the following definitions of D, P, S, I and R are here assumed:

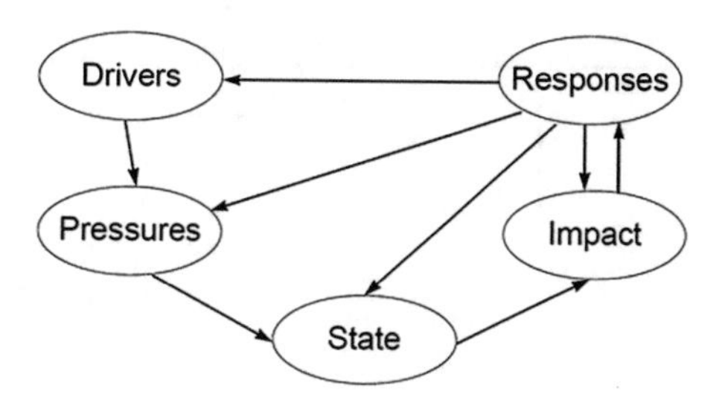

Fig. 1 Dynamical relationships between DPSIR components (Reproduced from Smeets and Weterings 1999)

- Drivers (**D**): social, demographic and economic developments in societies and the corresponding changes in life styles, overall levels of consumption and production patterns;
- Pressures (**P**): expression of human activities and/or natural pressures that could change the states of the environment in space and time;
- States (**S**): parameters describing the different environmental compartments;
- Impacts (**I**): effects of human activities and/or natural pressures on the ecosystems, measured with reference to a baseline;
- Responses (**R**): actions by groups (and individuals) in society, as well as government attempts to prevent, compensate, ameliorate or adapt to changes in the state of the environment.

In the following, it is also assumed that each DPSIR component has an associated multidimensional vector of quantitative measures.

The schema in Fig. 1 represents a dynamical system (i.e. a system evolving in time and space) in which there are loops of feedback (e.g. the actual states of the system are compared with some reference baseline in order to define suitable controlling responses). In particular, Lovelock and Margulis (1974) put the existence of feedback loops and the related concept of homeostasis among the initial postulates of their overall description of the environment. The interactions between the natural environment and human activity, too, have been modeled on the basis of feedback loops (Meadows et al. 1972). The relevant systems archetype for the causal feedback loop is shown in Fig. 2 in terms of a block diagram.

Upon those premises, the model implied by DPSIR represents an evolving system that tends to be stable, possibly with more than one positions of equilibrium. It has been shown (Iannucci et al. 2011) how, relying upon the concepts of ecosystem services (Daily 1997) and of Socio-Ecological System (SES) (Gallopin 1991), it is possible to elicit a model as in Fig. 3 (where the generic links between the blocks are labeled by a single lowercase letter; the letter is uppercase if a link conforms to a DPSIR component).

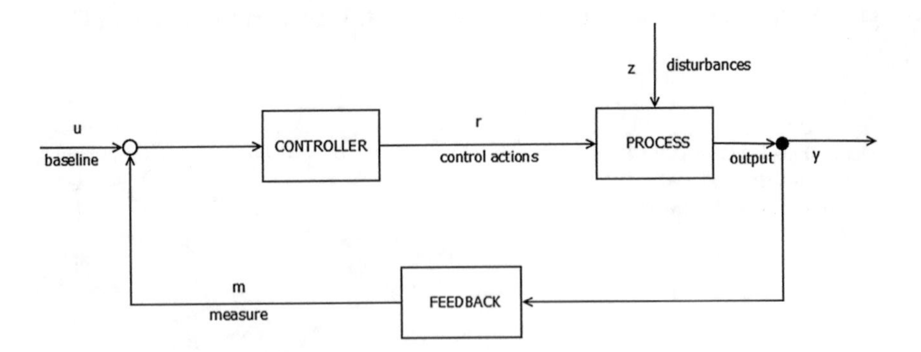

Fig. 2 Archetype of the causal feedback loop

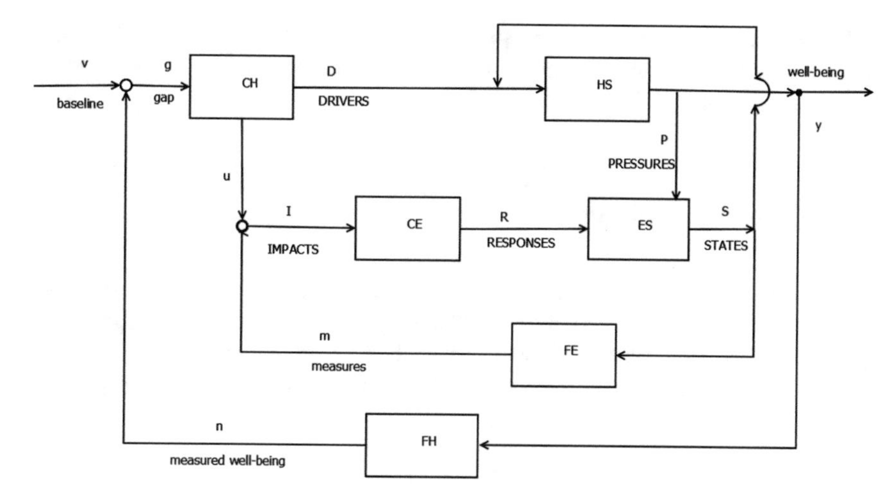

Fig. 3 Integrated model of DPSIR components and of ecosystem services

As DPSIR is being mostly applied to prevent or to manage the crisis of inter-relations between man and nature, such model assumes that:

- the nature has mainly lost its resilience, therefore the responses can be only provided by the man (as societal responses);
- the natural pressures are negligible, in comparison to the pressures generated by the human activities.

The socioeconomic system **HS** and the ecosystem **ES** are intertwined (even if still loosely coupled). The human society defines only the baseline **v** as an exogenous one, with reference to the aimed well-being (whose measures are still a research theme, as in Berger-Schmitt and Noll 2000; however, useful results are appearing, as in EEA 2013). Upon the feedback of the actually attained well-being, a second baseline **u** (i.e. the level of protection of the ecosystem needed to provide the ecosystem services requested by the socioeconomic system) is defined as an endogenous one by the control block **CH**.

The block diagram in Fig. 3, however, is not homologous to the graph in Fig. 1, that has given rise to this analysis. Figure 3 shows only five DPSIR actions (**I** on **R**, **R** on **S**, **D** on **P**, **P** on **S**, **S** on **I**) while other three actions (**R** on **I**; **R** on **P**; **R** on **D**) are missing.

With reference to these three actions the following considerations may be made. From one hand, it is clear, for obvious reasons of causality, that the response **R** cannot act on the value of the impact **I** (specifically, **R** is produced from **I**, which is the difference between **u** and **m**). Therefore, the existence of this action appears to be questionable. From the other hand, it is not immediate to show the other two actions (**R** on **P** and **R** on **D**) in Fig. 3. Both such actions appear to be structurally different from the action **R** on **S**, so that they cannot be captured by the causal feedback archetype alone. Therefore, it is necessary to extend the approach adopted for the creation of the SES model.

3 SoS as a Paradigm

Having shown in the above that DPSIR implies a dynamical model (whose components can exceed those currently included in the DPSIR data flows) the following question arises: what type of system is described by the model in Fig. 3?

The starting point is here given by the definition provided by Bertalanffy (1968) for a system:

> A system is a set of elements in interaction.

For ISO15288 (2015) a system is seen as:

> A combination of interacting elements organized to achieve one or more stated purposes.

Both definitions join the feature of a complex set of components (a.k.a. elements, things, parts…) with the feature of relationships (a.k.a. interactions, connections, links…), while ISO15288 (2015) pinpoints the discriminating feature of the purpose (a.k.a. mission, aim, function…): a complex set of interacting parts is a system if it has a mission to fulfill or a purpose to serve.

Both Bertalanffy (1968) and ISO15288 (2015) allow for hierarchies of systems, i.e. each of the elements composing the set can be a system in itself and have its own purpose. In such case, the purpose of the global system may be (and usually is) quite different from the purposes of the component systems; the issue of coordination arises. When the coordination is high, the complexity level is usually very low: the linked systems are bound to provide exactly the functionalities necessary to the global system; therefore, their individual behavior is strongly limited or even annulled. On the contrary, when more systems cooperate for a common mission without losing their identity, the resulting complexity is high; the global system shows properties that cannot be reduced to the sum of the properties of the cooperating systems and, moreover, are mostly new or even unintended.

The latter case includes global systems whose properties and behavior are related to the amount and types of relationships connecting the cooperating systems. In such context, the paradigm of system-of-systems (SoS) arises. For sake of avoiding confusion with the DPSIR "components" previously referred, in the following the term "constituent" (as suggested by Boardman and Sauser 2006) will be used to indicate each of the systems cooperating and grouped as a SoS. INCOSE (2012) provides the following definition:

> System-of-systems applies to a system […] whose *constituents* are themselves systems; typically these entail large scale inter-disciplinary problems with multiple, heterogeneous, distributed systems.

A more operationally suitable definition is provided by Maier (1998):

> A system would be termed a system-of-systems [...] when: (1) its *constituents* fulfilled valid purposes in their own right and continued to operate to fulfill those purposes if disassembled from the overall system, and (2) the *constituent* systems are managed (at least in part) for their own purposes rather than the purposes of the whole.

Recalling the concept of hierarchies of systems stated in the above, a SoS constituent can be a SoS in itself. Moreover, it should be noted that the specification of the system boundaries is a key issue: drawing boundaries that include or exclude some elements may result in observing a quite different system. Such definitions apply directly to the above described SES system. It is possible to pinpoint that both the single blocks in Fig. 3 and some of their unions represent individual systems according to Bertalanffy (1968) and ISO15288 (2015).

Evidently, the SES system (which, as previously stated, is the model of the relationships between man and the environment in the context implied by DPSIR) is a set of systems; some of them (namely, **ES** and **HS**) are autonomous, i.e. they have the additional properties of operational independence and of managerial independence (implying their evolutionary independence) requested by Maier (1998). According to Fisher (2006), the existence of autonomous constituent systems is both necessary and sufficient to define the global system SES as a system-of-systems. The relationships between such constituent systems allow SES to pursue new goals, not reachable by the separate constituents (in other words, SES exhibits an emergent behavior). Therefore, SES is a system-of-systems.

The above can be confirmed relying upon the taxonomy of increasingly complex systems provided by Baldwin et al. (2011). Such taxonomy shows that complex adaptive systems (CAS) appear to be SoS (while the reverse is not necessarily true). It has been shown that SES exhibits the features associated to a CAS (Iannucci and Munafò 2012); accordingly, it is a SoS.

In a SoS the constituent systems are autonomous and loosely coupled, by definition. Fisher (2006) underlines that:

> A system-of-systems depends on distributed control, cooperation, influence, cascade effects, orchestration, and other emergent behaviors as primary compositional mechanisms to achieve its purpose.

Therefore, the causal links postulated inside a feedback control loop (whose archetype is in Fig. 2 above) are only a subset of the mutual interactions between the constituent systems. Among the different interaction types in a SoS, here of specific interest is the influence, about which Fisher (2006) states the following:

> Where an entity is autonomous, it can only be influenced, not controlled, by outside forces. Influence is any mechanism by which one entity interacts with another in a way that changes the physical, informational [...] state of the other. Influence can be negative as well as positive. Influence can be cooperative, adversarial, or neutral.

From one hand, the statement above can be exploited in order to understand the nature of the actions of **R** on **P** and on **D**; from the other hand, such actions indicate what links have to be added to the block diagram of Fig. 3 aiming to complete the model making it as homologous as possible to the graph in Fig. 1.

In terms of systems theory, the influence described by Fisher (2006) can be expressed by the archetype of the feedforward control, i.e. a control that directly translates the disturbances into an action, relying upon an a priori knowledge of the process to be steered and not upon the gap between desired result and actual one (Seborg et al. 2011). Sometimes, this sort of control is termed "ballistic" to express its independence from the actual effects. DPSIR takes into account only one response **R**, while its implied model shows the need to separately tackle with different (causal and influential) responses, as it will be shown in the following.

Let the DPSIR Response be the societal control **R1** (as when a law extends the amount of protected areas, aiming to better preserve the availability of ecosystem services) that causally forces the ecosystem **ES** to modify the levels of its states **S** (e.g. increasing the abundance of species). **R2** and **R3** are the influences that the drivers **D** and the pressures **P** respectively receive (e.g., for biodiversity-related ecosystem services: **R2** could be associated with educational efforts addressing the importance of limiting the urbanization; **R3** could be an increase of penalties and other costs for those not complying with the safeguard of the protected areas). **R2** and **R3** appear to be different as far as **D** (control actions on **HS**, i.e. on the lifestyles) and **P** (disturbances, i.e. byproducts of the human activities) are different; however, according to the feedforward control archetype, both of them originate from a block **CJ** having the **P** as input.

The drivers **D** are part of the feedback loop that compare the aimed well-being level (the exogenous baseline **v**) to the actually achieved one; therefore, they depend on such baseline: a lower baseline reduces the extent of the control action (if, as usual, the achieved well-being tends towards the baseline but is always lesser). **R2** is represented by a link from the block **CJ** to the point where the exogenous baseline **v** is compared to the feedback link **n** from **FH**: this can be interpreted as the means of lowering the expectations of an increased well-being in the context of safeguarding the ecosystem services; therefore, **R2** is a Fisher's cooperative influence. This is the case of commuting by public transport vs. by private cars: convincing the commuters to refrain from using private cars will improve the common air quality, at the cost of renouncing part of the individual well-being.

For **R3**, the approach is somehow different. **P** are the (usually unwanted) by-products of the societal processes that, in **HS**, produce the well-being; as far as they are unwanted, they are out of the range of the control actions (i.e. there are no values of **D** able to modify them without modifying the well-being at the same time; **P** are practically insensitive to **R2**). Therefore, **R3** can act only as a disturbance *on* **HS**, assuming that the effects of **R3** *in* **HS** will somehow counter the **P** generated *by* **HS**: **R3** is a Fisher's adversarial influence.

It should be assumed that, in parallel to **P**, **CJ** can receives input also from **CE** (on the basis of **I**), in the form of another influence **R4** that **CJ** merges into **R3** and transfers to **HS** as a sort of the cascade effect (Fisher 2006) mentioned in the above. Both **R3** and **R4** can be easily seen in the case of the protection of the cultural ecosystem service associated with the bathing waters. If in a touristic resort the

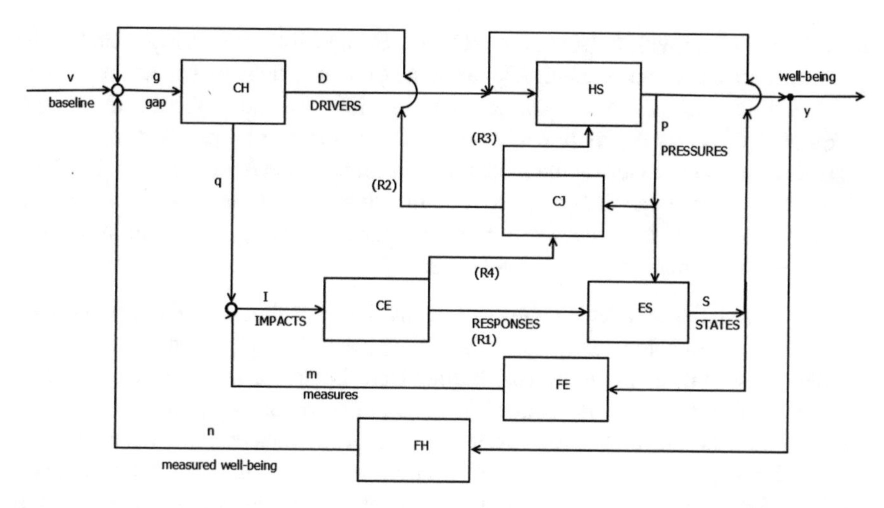

Fig. 4 Model implied by DPSIR, within a SoS approach

quality parameters of the bathing waters do not meet the prescribed level (this, usually detected as **I**, is caused by e.g. an inadequate or poorly maintained wastewater treatment plant, measureable as **P**), the recreational use of the sea (e.g. for swimming) is legally forbidden. Such prohibition, of course, does not directly modify the states **S** of the bathing waters; on the contrary, it sends an influence **R3** on **HS** (possibly triggered by **R4** on **CJ**), reducing the attractiveness of the sea resort and therefore lowering the income (as part of the well-being) of the local community that hopefully will be forced to properly restore the wastewater treatment.

Accordingly, the block diagram of Fig. 3 evolves as shown in Fig. 4. Such block diagram has been derived by analysis of the components and of their mutual relationships of the causal chain in Fig. 1. As a result, such block diagram appears to be congruent with the causal networks approach suggested by Niemeijer and de Groot (2008) for the indicators.

4 Conclusions

The adoption of a SoS viewpoint allows to draw a conceptual systems model that conforms to the DPSIR framework. While properly taking into account the specific interactions among the DPSIR components, this model calls for an extension of the framework, in order to:

- activate new data flows of measured parameters, with reference to the links that in Fig. 4 do not currently correspond to any of the DPSIR components; this implies the inclusion of a new set of components (specifically, it should be noted that the reference baselines do not explicitly appear in the current DPSIR);

- separately deal with responses (as **R1**, **R2**, **R3** and **R4**) that, as suggested in the above, appear to be semantically not homogeneous; their dynamical properties should be assessed, in comparison to the stability and the resilience of the SoS; the role of **R1** as the main causal control action has to be pinpointed;
- review the role as well as the associated meaning of the drivers **D**, moving from the current semi-qualitative view to a full quantitative view of **D** as the controlling action of the processes **HS** that the human society carries out in order to ensure the wanted levels of well-being.

With such extension (enhancing its quantitative aspects) the DPSIR framework will be freed from some inconsistencies, mainly related to the semantics. Therefore, the data flows will be more natively harmonized. In such a way, DPSIR would be able to better streamline the collection and storage of massive amounts of data. Taking into account the foreseen large increase of the data Volume (as a result of mainly the improved resolution of the environmental sensors), the relevant stakeholders (EEA, public administration, academia, business actors, citizens) will benefit of a description of the environment congruently expressed in a quantitative way and exploitable by advanced analytics.

In a context of fully quantitative data, the transfer functions pertaining to each block can be identified. On such basis, robust modeling methods could be exploited in modeling the environment. In their seminal paper, Rittel and Webber (1973) point out the inherent difficulties of applying formal methods (as the system approach) to "wicked problems" of the societal planning, compared and contrasted to "tame problems" usually dealt with by the technical professionals. This is specifically true with reference to the definition of policies and goals for the well-being of the social system. Such *caveat* should be treasured; however, it should not suggest to refrain from carefully exploiting the systems methods, at least to provide support in the formulation and in the validation of the proposals arising from the various articulations of the social system.

Moreover, the quantitative DPSIR framework could support the activities related to quality assurance and quality control that influence the data Value. The transfer functions could be exploited to estimate the quality of actual data flows both in terms of logical congruence and of descriptive capacity, also in comparison with results that may be derived from the numerous models of the sector available in literature (EEA 2008). E.g., if the formal analysis of the system showed that its dynamic behavior tends to be unstable when, in reality, it appears to be sufficiently stable, one would need to carefully check the data collected in DPSIR from many different sources.

Such kind of checks on the quality and congruence of data may not be derived in any other way, apart from the analysis of the dynamical evolution of the system. This confirms the potential usefulness of the approach presented here. However, the above would imply to make adequate efforts to properly understand the dynamical features of the data flows and even to rethink (in terms of semantics, autocorrelation, temporal and spatial resolution) the classes of the data themselves.

Finally, the analysis carried out in the above (whose results are graphically shown in Fig. 4) pinpoints the dependence of the environment protection initiatives from the aimed level of well-being. Surely, this aspect ought to be seen from different perspectives; further research efforts would be of interest on such issue.

References

Baldwin, W. C., Felder, W. N., & Sauser, B. J. (2011). Taxonomy of increasingly complex systems. *International Journal of Industrial and Systems Engineering, 9*(3), 298–316.

Berger-Schmitt, R., & Noll, H. H. (2000). *Conceptual Framework and Structure of a European System of Social Indicators*. Centre for Survey Research and Methodology (ZUMA), Mannheim.

Boardman, J., & Sauser, B. (2006). System of systems—the meaning of *of*. In *Proceedings of the 2006 IEEE/SMC International Conference on System of Systems Engineering, Los Angeles CA* (pp. 118–123).

Chen, M., Mao, S., & Liu, Y. (2014). Big data: A survey. *Mobile Networks and Applications, 19* (2), 171–209.

Daily, G. C. (1997). Introduction: what are ecosystem services? In G. C. Daily (Ed.), *Nature's services: Societal dependence on natural ecosystems* (pp. 1–10). Washington DC: Island Press.

Demchenko, Y., Grosso, P., De Laat, C., & Membrey, P. (2013). Addressing big data issues in scientific data infrastructure. In *Proceedings of the 2013 IEEE/CTS International Conference on Collaboration Technologies and Systems* (pp. 48–55).

EEA. (2008). *Catalogue of forward-looking indicators from selected sources*. Copenhagen: European Environment Agency. (Technical report No 8/2008).

EEA. (2013). *Environmental Pressures from European consumption and production; A study in integrated environmental and economic analysis*. Copenhagen: European Environment Agency. (Technical report No 2/2013).

EEA. (2014). *Digest of EEA indicators*. Copenhagen: European Environment Agency. (Technical report No 8/2014).

EEA. (2015). *EIONET priority data flows, May 2014–April 2015*. Copenhagen: European Environment Agency. (Corporate document No 2/2015).

Fisher, D. (2006). *An emergent perspective on interoperation in systems of systems*. Pittsburgh, PA: Software Engineering Institute, Carnegie Mellon University. (CMU/SEI-2006- TR-003).

Gallopin, G. C. (1991). Human dimensions of global change: Linking the global and the local processes. *International Social Science Journal, 130*, 707–718.

Iannucci, C., Munafò, M., & Sambucini, V. (2011). A system approach to the integration of ecosystem services with DPSIR components. In W. Pillmann, S. Schade, & P. Smits (Eds.), *Innovations in sharing environmental observations and information* (pp. 122–129). Aachen: Shaker Verlag.

Iannucci, C., & Munafò, M. (2012). The Response component in DPSIR and the SES dynamical stability. In H. K. Arndt, G. Knetsch, & W. Pillmann (Eds.), *Man environment Bauhaus: Light up the ideas of environmental informatics* (pp. 311–317). Aachen: Shaker Verlag.

INCOSE. (2012). *Systems engineering handbook: A guide for system life cycle processes and activities, v. 3.2.2*. San Diego, CA: International Council on Systems Engineering.

ISO15288. (2015). Systems and Software Engineering—System Life Cycle Processes. ISO/IEC/IEEE 15288:2015(E).

Kaisler, S. H., Espinosa, J. A., Armour, F., & Money, W. H. (2014). Advanced analytics—issues and challenges in a global environment. In *Proceedings of the 47th IEEE/HICSS Hawaii International Conference on System Sciences* (pp. 729–738).

Lovelock, J. E., & Margulis, L. (1974). Atmospheric homeostasis by and for the biosphere; the Gaia hypothesis. *Tellus, 26*(1), 2–10.
Maier, M. W. (1998). Architecting principles for system-of-systems. *Systems Engineering, 1*, 267–284.
Meadows, D. H., Meadows, D. L., Randers, J., & Behrens, W. W. (1972). *The limits to growth, a report for the Club of Rome's project on the predicament for mankind*. New York NY: Universe Books.
Niemeijer, D., & de Groot, R. S. (2008). Framing environmental indicators: Moving from causal chains to causal networks. *Environment, Development and Sustainability, 10*(1), 89–106.
OECD. (1994). *Environmental Indicators—OECD Core Set*. Paris: Organisation for Economic Co-operation and Development.
Pietsch, W. (2013). Big data—the new science of complexity. In *Proceedings of the 6th Munich-Sydney-Tilburg Conference on Models and Decisions, Munich*, 10–12 April 2013.
Rittel, H. W., & Webber, M. M. (1973). Dilemmas in a general theory of planning. *Policy Sciences, 4*(2), 155–169.
Seborg, D. E., Edgar, F. E., Mellichamp, D. A., & Doyle, F. J, I. I. I. (2011). *Process dynamics and control*. Hoboken, NJ: Wiley.
Smeets, E., Weterings, R. (1999). *Environmental indicators: typology and overview*. Copenhagen: European Environment Agency. Technical report No 25.
UN. (1997). *From theory to practice: Indicators for sustainable development*. New York, NY: United Nations (Division for Sustainable Development).
von Bertalanffy, L. (1968). *General system theory: Foundations, development, applications* (Rev ed.). New York NY: Braziller.

Printed in the United States
By Bookmasters